ESD

ESD Series
By Steven H. Voldman

ESD: Design and Synthesis
ISBN: 9780470685716
March 2011

ESD: Failure Mechanisms and Models
ISBN: 9780470511374
July 2009

Latchup
ISBN: 9780470016428
December 2007

ESD: RF Technology and Circuits
ISBN: 9780470847558
September 2006

ESD: Circuits and Devices
ISBN: 9780470847541
November 2005

ESD: Physics and Devices
ISBN: 9780470847534
September 2004

Upcoming titles:

ESD Basics: From Semiconductor Manufacturing to Use

ESD: Test and Characterization

ESD
Design and Synthesis

Steven H. Voldman
IEEE Fellow, Vermont, USA

A John Wiley and Sons, Ltd, Publication

This edition first published 2011

Registered office
John Wiley & Sons Ltd, The Atrium, Southern Gate, Chichester, West Sussex, PO19 8SQ, United Kingdom

For details of our global editorial offices, for customer services and for information about how to apply for permission to reuse the copyright material in this book please see our website at www.wiley.com.

Library of Congress Cataloging-in-Publication Data

Voldman, Steven H.
ESD : Design and Synthesis / Steven H. Voldman.
p. cm.
Includes bibliographical references and index.
ISBN 978-0-470-68571-6 (hardback)
1. Semiconductors–Protection. 2. Integrated circuits–Protection. 3. Electrostatics.
4. Analog electronic systems–Design and construction. I. Title.
TK7871.85.V6525 2011
621.3815'2–dc22

2010048032

A catalogue record for this book is available from the British Library.

Print ISBN: 9780470685716
E-PDF ISBN: 9781119991144
O-book ISBN: 9781119991137
E-Pub ISBN: 9781119992653

Set in 10/12pt, Times by Thomson Digital, Noida, India
Printed and bound in the UK by CPI Antony Rowe, Chippenham, Wiltshire

To My Daughter
Rachel Pesha Voldman

Contents

About the Author

Dr Steven H. Voldman is the first IEEE Fellow in the field of electrostatic discharge (ESD) for "Contributions in ESD Protection in CMOS, Silicon On Insulator and Silicon Germanium Technology." He received his B.S. in Engineering Science from the University of Buffalo (1979); a first M.S. EE (1981) from Massachusetts Institute of Technology (MIT); a second EE Degree (Engineer Degree) from MIT; an M.S. Engineering Physics (1986) and a Ph.D. in electrical engineering (EE) (1991) from the University of Vermont under IBM's Resident Study Fellow program.

He was a member of the IBM development team for 25 years, working on semiconductor device physics, device design, and reliability (e.g., soft error rate (SER), hot electrons, leakage mechanisms, latchup, and ESD). Steve Voldman has been involved in latchup technology development for 27 years. He worked on both technology and with-product development in bipolar SRAM, CMOS DRAM, CMOS logic, silicon on insulator (SOI), BiCMOS, silicon germanium (SiGe), RF CMOS, RF SOI, smart power, and image processing technologies. In 2008 he was a member of the Qimonda DRAM development team, working on 70, 58, and 48 nm CMOS technology. In 2008 he initiated a limited liability corporation (LLC), and worked at headquarters in Hsinchu, Taiwan for Taiwan Semiconductor Manufacturing Corportion (TSMC) as part of the 45 nm ESD and latchup development team. He is presently a Senior Principal Engineer working for the Intersil Corporation on ESD and latchup development.

Dr Voldman was Chairman of the SEMATECH ESD Working Group from 1995 to 2000. In his SEMATECH Working Group, attention focused on ESD technology benchmarking, the first transmission line pulse (TLP) standard development team, strategic planning, and JEDEC–ESD Association standards harmonization of the human body model (HBM) Standard. From 2000 to 2010, as Chairman of the ESD Association Work Group on TLP and very-fast TLP (VF-TLP), his team was responsible for initiating the first standard practice and standards for TLP and VF-TLP. Steve Voldman has been a member of the ESD Association Board of Directors and Education Committee. He initiated the "ESD on Campus" program, which was established to bring ESD lectures and interaction to university faculty and students internationally; the ESD on Campus program has reached over 32 universities in the United States, Singapore, Taiwan, Malaysia, the Philippines, Thailand, India, and China.

Dr Voldman teaches short courses and tutorials on ESD, latchup, and invention in the United States, China, Singapore, Malaysia, and Israel. He is a recipient of over 210 issued US patents, in the area of ESD and CMOS latchup. He has served as an expert witness in patent litigation cases associated with ESD and latchup.

Dr Voldman has also written articles for *Scientific American* and is an author of the first book series on ESD and latchup: *ESD: Physics and Devices*, *ESD: Circuits and Devices*, *ESD: RF Technology and Circuits*, a fourth text, *Latchup*, and a fifth text, *ESD: Failure Mechanisms and Models*. He is also a contributor to the book *Silicon Germanium: Technology, Modeling and Design*. There are international Chinese editions of the book *ESD: Circuits and Devices* and the text *ESD: RF Technology and Circuits.* He is also a chapter contributor to the text *Nanoelectronics: Nanowires, Molecular Electronics, and Nano-devices*.

Preface

The text *ESD: Design and Synthesis* is targeted at the semiconductor chip "architect", team lead floorplan engineer, circuit designer, design layout support, ESD engineer, and computer aided design (CAD) integration team. In this text, a balance is established between design synthesis, design integration, layout engineering, and design checking and verification.

The first goal of the text *ESD: Design and Synthesis* is to teach the "art" of ESD chip design for a semiconductor chip.

The second goal is to demonstrate a step-by-step process to provide ESD protection to a semiconductor chip. The flow of the text addresses floorplanning, architecture, power rails, ESD networks for power rails, ESD signal pin solutions, guard rings, and examples of implementations. This flow is significantly different from the approach taken in most texts, but is the actual flow of how a design team proceeds through the ESD implementation.

The third goal is to expose the reader to the growing number of architectures and concepts being discussed today. Examples of DRAM, SRAM, image processing chips, microprocessors, mixed-voltage to mixed-signal applications, and floorplans will be shown.

The fourth goal is to address topics that are not discussed in other ESD textbooks. These topics include power bus architecture, guard rings, and floorplanning. For many ESD engineers and circuit designers, this is common knowledge; for others, it is not. A significant part of the ESD design and synthesis is spent on placement, floorplans, and integration.

This text, *ESD: Design and Synthesis*, contains the following:

- Chapter 1 introduces the reader to an overview of the language and fundamentals associated with ESD design. In this chapter, ESD concepts are introduced from layout, circuits, to design rule checking. A "sampler" of concepts is laid out to the reader, to begin viewing the ESD design synthesis from a broader perspective. ESD design synthesis extends from the smallest contact, to full-chip integration. With this awareness, it is possible to realize the extent of the ESD design discipline in semiconductor design.

  For the next chapters, the text is structured as primarily a "top-down" ESD approach. This starts with floorplanning, bus architecture, ESD power clamps, ESD input circuits, and guard rings. The text will close with more examples of floorplanning and design integration. Most previous ESD texts focus on a "bottom-up" approach to ESD design integration; in "real-life" semiconductor integration, it typically starts from the "top down".

- Chapter 2 discusses chip architectures. In this chapter, the discussion focuses on ESD architecture and floorplan concepts. The chapter focuses on "peripheral I/O" and "array I/O" architectures, and how they influence the placement of the various elements for the whole-chip design integration. The chapter addresses native-voltage, mixed-voltage, and mixed-signal chip integration.
- Chapter 3 focuses on power grid design. In this chapter, the discussion continues to address issues associated with full-chip ESD design synthesis. The chapter focuses on the interconnects, power grid layout, and design itself. It addresses interconnect robustness, interconnect failure, and key metrics in the whole-chip ESD design synthesis. The chapter addresses the issue of integration with the ESD power clamps. This naturally flows into the next chapter.
- Chapter 4 addresses ESD power clamps on power domains and power pads. In this chapter, ESD power clamp circuits are discussed. ESD power clamp classification, key parameters, issues, and specific designs are discussed. How the ESD power clamps are integrated with the semiconductor chip will become more apparent.
- Chapter 5 focuses on ESD signal pad networks. In this chapter, ESD signal pin device layout and integration with bond pads are discussed. ESD signal pin classification, key parameters, issues, and specific designs are covered. The chapter focuses on ESD integration with the bond pad, from structures next to pads, adjacent to bond pad, partially-under, and under bond pads. All types of arrangements and orientation tradeoffs will be discussed. The chapter focuses on device layout and integration.
- Chapter 6 focuses on guard rings and guard ring integration. In this chapter, a "top-down" design synthesis approach for guard rings is shown for a semiconductor chip, starting with the seal ring, to domains, standard cell-to-standard cell, within-standard cell, and down to the individual devices. A "bottom-up" approach starts with the individual devices and works its way up to the full-chip implementation. Special structures and cases are shown as examples of how to further isolate both domains and devices. A small taste is given to show what is possible with the guard ring design synthesis and integration with both devices to full-chip implementations.
- Chapter 7 provides examples of different chip floorplans and architectures. In this chapter, the focus is on examples of design synthesis in full-chip implementations. Examples of DRAM, SRAM, microprocessors, mixed-voltage, mixed-signal, and RF applications will be shown. As part of the ESD design synthesis, the layout is key to a successful design implementation for both ESD and CMOS latchup. These examples will provide some understanding of the challenges in the ESD full-chip integration issues. By combining the knowledge of Chapters 1 through 6 with this chapter, the whole-chip design strategy should be better understood for any semiconductor chip architecture perspective.

This text is part of an ESD book series on electrostatic discharge protection. To establish a strong knowledge of ESD protection, it is advisable to read the other texts on ESD and latchup as well. For this text, *ESD: Design and Synthesis*, hopefully we have covered the trends and directions of ESD design synthesis.

Enjoy the text, and enjoy the subject of ESD design synthesis.

Baruch HaShem (B"H)

Dr Steven H. Voldman
IEEE Fellow

Acknowledgments

In the area of ESD and latchup design, I would like to acknowledge the years of support from the SEMATECH, the ESD Association, the IEEE, and the JEDEC organizations. I would like to thank the IBM Corporation, Qimonda, Taiwan Semiconductor Manufacturing Corporation (TSMC), and the Intersil Corporation. This text comes from 30 years of working with bipolar memory, DRAM memory, SRAM, NVRAMs, microprocessors, ASICs, mixed-voltage, mixed-signal, RF, and power applications. I was fortunate to work in a wide number of technology teams, and with a wide breadth of customers. I was very fortunate to work in bipolar memory, CMOS DRAM, CMOS logic, ASICs, silicon on insulator (SOI), and silicon germanium (SiGe) from 1 μm to 45 nm technologies. I was very fortunate to be a member of talented technology and design teams that were innovative, intelligent, and inventive. This provided the opportunity to explore experimental concepts, and try new ideas in ESD design in applications and products.

I would like to thank the institutions that allowed me to teach and lecture at conferences, symposiums, industry, and universities; this gave me the motivation to develop the texts. I would like to thank faculty at the following universities: MIT, Stanford University, University of Central Florida (UCF), University Illinois Urbana-Champaign (UIUC), University of California Riverside (UCR), University of Buffalo, National Chiao Tung University (NCTU), Tsin Hua University, National Technical University of Science and Technology (NTUST), National University of Singapore (NUS), Nanyang Technical University (NTU), Beijing University, Fudan University, Shanghai Jiao Tung University, Zheijang University, Universiti Sains Malaysia, Chulalongkorn University, Mahanakorn University, Kasetsart University, Thammasat University, and Mapua Institute of Technology.

I would like to thank – for the years of support and the opportunity to provide lectures, invited talks, and tutorials – the Electrical Overstress/Electrostatic Discharge (EOS/ESD) Symposium, the International Reliability Physics Symposium (IRPS), the Taiwan Electrostatic Discharge Conference (T-ESDC), the International Electron Device Meeting (IEDM), the International Conference on Solid-State and Integrated Circuit Technology (ICSICT), and the International Physical and Failure Analysis (IPFA) in Singapore.

I would like to thank my many friends for 20 years in the ESD profession – Professor Ming Dou Ker, Professor J.J. Liou, Professor Albert Wang, Professor Elyse Rosenbaum, Timothy J. Maloney, Charvaka Duvvury, Eugene Worley, Robert Ashton, Yehuda Smooha, Vladislav Vashchenko, Ann Concannon, Albert Wallash, Vessilin Vassilev, Warren Anderson, Marie

Denison, Alan Righter, Andrew Olney, Bruce Atwood, Jon Barth, Evan Grund, David Bennett, Tom Meuse, Michael Hopkins, Yoon Huh, Keichi Hasegawa, Nathan Peachey, Kathy Muhonen, Augusto Tazzoli, Gaudenzio Menneghesso, Marise BaFleur, Jeremy Smith, Nisha Ram, Swee K. Lau, Tom Diep, Lifang Lou, Stephen Beebe, Michael Chaine, Pee Ya Tan, Theo Smedes, Markus Mergens, Christian Russ, Harold Gossner, Wolfgang Stadler, Ming Hsiang Song, J.C. Tseng, J.H. Lee, Michael Wu, Erin Liao, Jim Vinson, Jean-Michel Tschann, David Swenson, Donn Bellmore, Ed Chase, Doug Smith, W. Greason, Stephen Halperin, Tom Albano, Ted Dangelmayer, Terry Welsher, John Kinnear, and Ron Gibson. I would also like to thank graduate students who are engaged in the study of ESD protection: Tze Wee Chen, Shu Qing Cao, Slavica Malobabic, David Ellis, Blerina Aliaj, and Lin Lin.

I would like to thank the ESD Association office for support in the area of publications, standards developments, and conference activities. I would also like to thank the publisher and staff of John Wiley and Sons, for including the text *ESD: Design and Synthesis* as part of the ESD book series.

To my children, Aaron Samuel Voldman and Rachel Pesha Voldman, good luck to both of you in the future.

To my wife, Annie Brown Voldman, thank you for the support of years of work.

And to my parents, Carl and Blossom Voldman.

Baruch HaShem (B"H)

Dr Steven H. Voldman
IEEE Fellow

1 ESD Design Synthesis

1.1 ESD DESIGN SYNTHESIS AND ARCHITECTURE FLOW

In the ESD design synthesis process, there is a flow of steps and procedures to construct a semiconductor chip [1–13]. In many cases, the floorplanning process is a function of the type of semiconductor chip. The following design synthesis procedure is an example of an ESD design flow needed for semiconductor chip implementations:

- **I/O, Domains and Core Floorplan:** Define floorplan of regions of cores, domains, and peripheral I/O circuitry.
- **I/O Floorplan:** Define area and placement for I/O circuitry.
- **ESD Signal Pin Floorplan:** Define ESD area and placement.
- **ESD Power Clamp Network Floorplan:** Define ESD power clamp area and placement for a given domain.
- **ESD Domain-to-Domain Network Floorplan:** Define ESD networks between the different chip domains area and placement for a given domain.
- **ESD Signal Pin Network Definition:** Define ESD network for the I/O circuitry.
- **ESD Power Clamp Network Definition:** Define ESD power clamp network within a power domain.
- **Power Bus Definition and Placement:** Define placement, bus width, and resistance requirements for the power bus.
- **Ground Bus Definition and Placement:** Define placement, bus width, and resistance requirements for the ground bus.
- **I/O to ESD Guard Rings:** Define guard rings between I/O and ESD networks.

ESD: Design and Synthesis, First Edition. Steven H. Voldman.

- **I/O Internal Guard Rings:** Define guard rings within the I/O circuitry.
- **I/O External Guard Rings:** Define guard rings between I/O circuitry and adjacent external circuitry.

1.1.1 Top-Down ESD Design

In the ESD design synthesis, the implementation can be thought of as a "top-down ESD design" process. Figure 1.1 is an example of a "top-down ESD design flow." In the ESD design synthesis process, there is a flow of steps and procedures to construct a semiconductor chip. In my experience, in the planning stages of a semiconductor chip, the circuit team leader addresses the ESD design synthesis from a procedure as shown. With a "top-down ESD design synthesis" the integration, placement, sizing, and requirements are addressed. This process will be independent of whether the semiconductor chip is for digital logic [1–7,11], analog design [14,31–33], power electronics [26–30,35–38], or radio frequency applications [8,39–41].

1.1.2 Bottom-Up ESD Design

In the ESD design synthesis, the implementation can also be addressed as a "bottom-up ESD design" process. Figure 1.2 is an example of a "bottom-up ESD design flow." In a bottom-up ESD design synthesis process, the circuits are defined, and the corresponding ESD networks.

One of the difficulties of ESD and the latchup design synthesis process is that the ESD design synthesis requires some "top-down" procedures, some "bottom-up" thinking, and integration. This will become more apparent throughout this text.

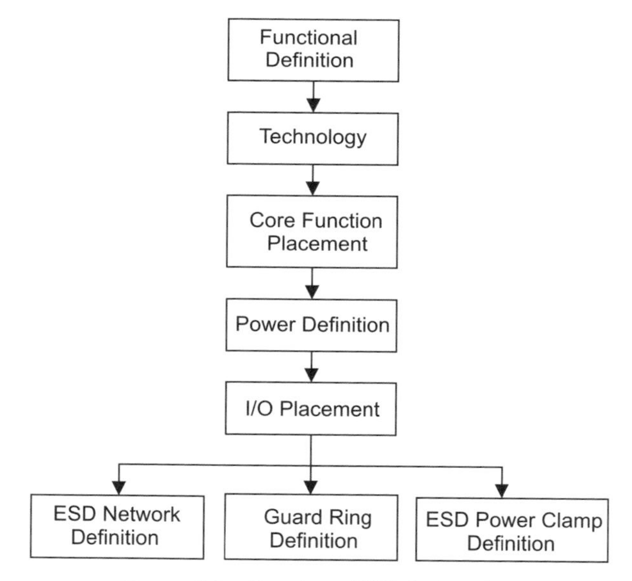

Figure 1.1 Top-down ESD design flow

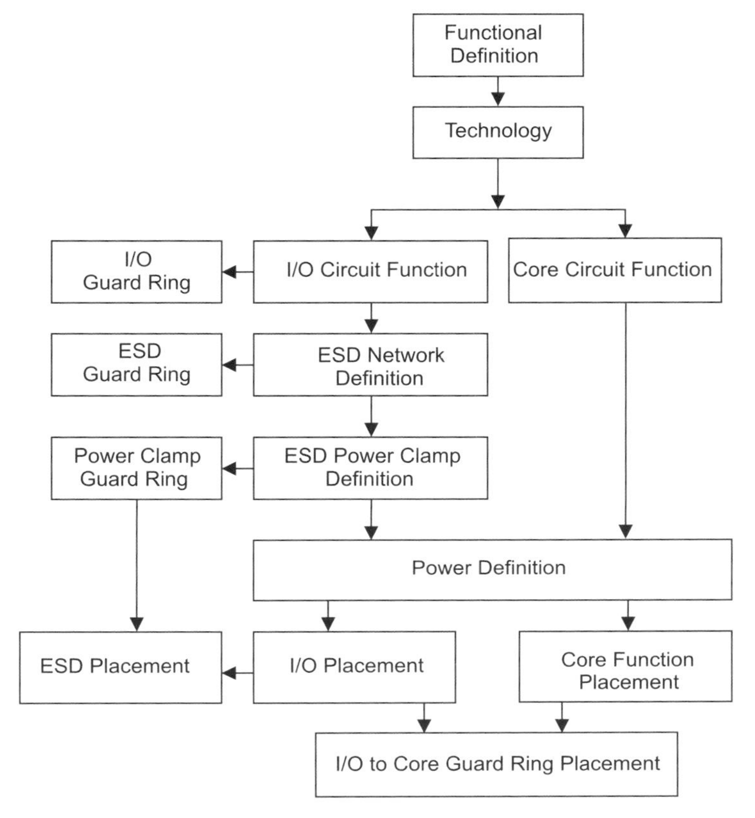

Figure 1.2 Bottom-up ESD design flow

1.1.3 Top-Down ESD Design – Memory Semiconductor Chips

In the ESD design synthesis of a memory chip, the thought process is a "top-down ESD design" process, with the floorplanning driven by the array region. These designs are "array-dominated" designs, with the focus on the array [7]. The I/O region is limited in physical area, and the architecture is driven by the number of output pins, how to integrate it with the packaging, and how to support the I/O and ESD in the least amount of space. Figure 1.3 is an example of a "top-down ESD design flow" for a memory chip.

1.1.4 Top-Down ESD Design – ASIC Design System

In the ESD design synthesis of an applications-specific IC (ASIC) architecture, the procedure for the ESD design integration is significantly different. In the ASIC environment, the chip size, the number of I/O, and its ESD integration is dependent on the chip size. In this "top-down" methodology, the number of I/O, supported bus locations, placement of the I/O cells, integration of the ESD elements, and power are all synthesized in a different flow. Figure 1.4 is an example of a "top-down ESD design flow" for an ASIC methodology.

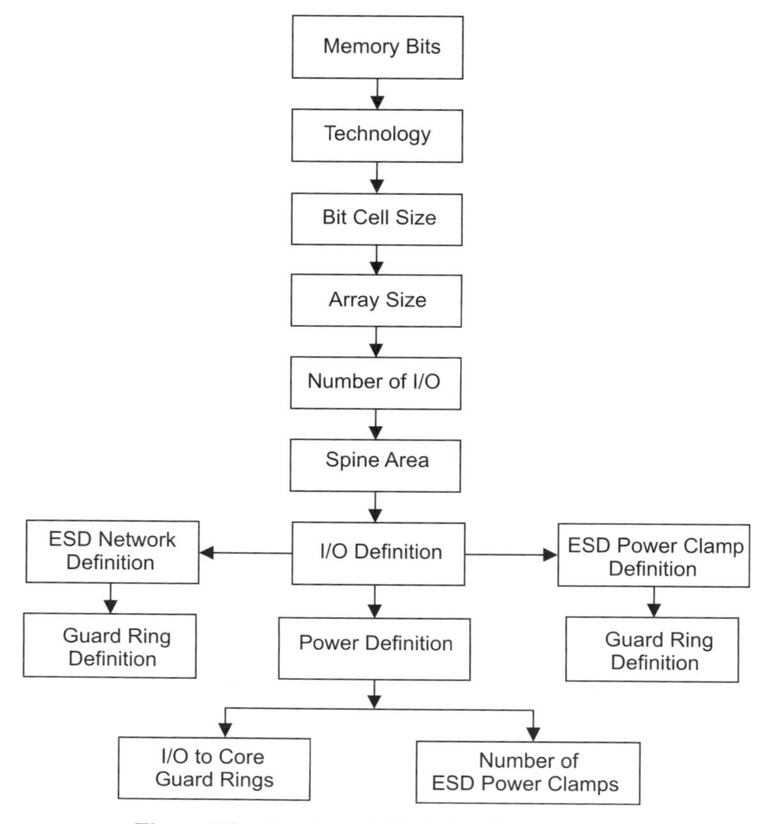

Figure 1.3 Top-down ESD design flow – memory

1.2 ESD DESIGN – THE SIGNAL PATH AND THE ALTERNATE CURRENT PATH

In semiconductor chip design, the role of a semiconductor chip is to receive a signal, process the signal, and transmit the signal.

In ESD design synthesis, the role of the ESD network solution is to establish an alternate current path to avoid damage along the signal path that impacts its function or operation characteristics [7,11]. As a result, simplistically, the ESD network must transmit the ESD current out of the sensitive signal path to an alternative path or current loop. This is achieved by diverting the ESD current to the power grid, or the ground plane. The fundamental requirements along the alternative current path are as follows:

- An alternative current path must exist between any signal pin and any grounded reference (e.g., signal pin, power pin, ground pin).
- An ESD element must divert the ESD current to the power plane or ground plane.

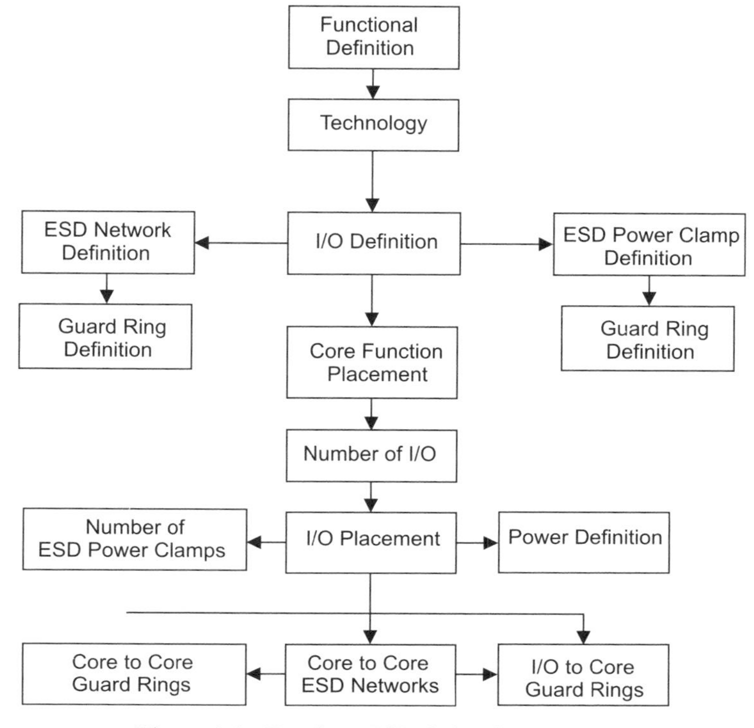

Figure 1.4 Top-down ESD design flow – ASICs

- An ESD element must be able to transmit the ESD current to the power rail or ground rail without damage (to some specification level).
- The power rail and ground rail must be able to source the ESD current without damage (to some specification level).
- The alternative current path must achieve the ESD current discharge to the grounded reference to some specification level prior to damage along the signal path.

To achieve this objective, there are some conditions on the alternative current path:

- ESD networks are required to address both positive and negative polarity events.
- The ESD network must have low turn-on voltage and low resistance prior to destruction of the circuitry along the signal path.
- The power grid and the ground rail resistance must be sufficiently low to avoid IR voltage drops.
- Bi-directional electrical connectivity must exist, providing an alternative current path between all independent rails through ESD networks, or other means (e.g., circuitry, inductors, bond wires, packaging, etc.).

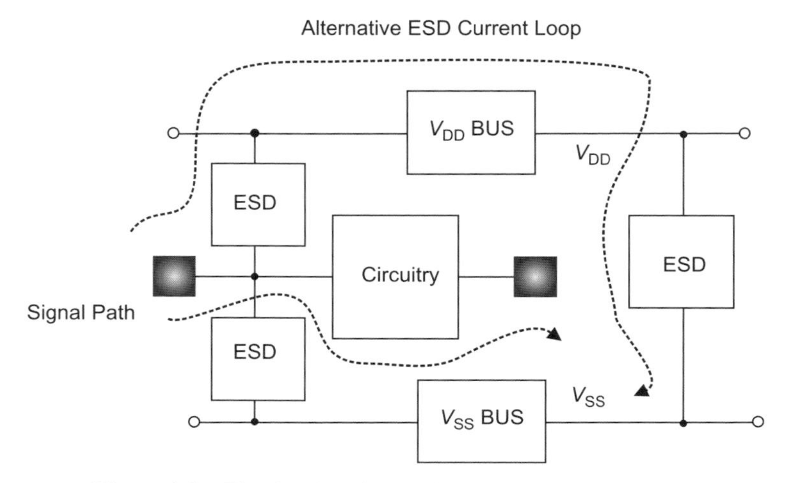

Figure 1.5 The signal path and alternative ESD current path

Figure 1.5 shows an example of a semiconductor high-level schematic of the chip architecture. The figure highlights the signal path and the alternative ESD current path created by the ESD networks.

1.3 ESD ELECTRICAL CIRCUIT AND SCHEMATIC ARCHITECTURE CONCEPTS

In this section, discussion of ESD from a chip architecture, and the electrical schematic viewpoint will be shown. What are the ideal characteristics that we are looking for from an ESD network? What are the ideal characteristics from a frequency domain perspective? How is the chip architecture related to the testing procedure and the events that occur in a real chip?

1.3.1 The Ideal ESD Network and the Current–Voltage DC Design Window

The DC *I–V* characteristics may determine the "on" and "off" characteristics of the ESD network during functional operation, and its ESD effectiveness as an ESD network to protect other circuitry. An ideal ESD network has the following characteristics [7,11]:

- The ESD device, circuit, or network is "off" during the DC functional regime between signal levels between the most negative power supply voltage and the most positive power supply (associated with the signal pin).
- The ESD network has an "infinite resistance" when in the "off" state, which can be expressed as

$$\left.\frac{dI}{dV}\right|_{\text{off}} = \frac{1}{R} = 0$$

The ESD network is "on" during voltage excursions that undershoot below the most negative power supply, or voltage excursions that overshoot the most positive power supply (during ESD testing).

- The ESD network has a "zero resistance" when

$$\left.\frac{dI}{dV}\right|_{\text{on}} = \frac{1}{R} = \infty$$

The ESD network operation extends beyond the "electrical safe-operation area" (electrical SOA) in DC voltage level or DC current level [26–30].

- The ESD network operation does not extend beyond a "thermal safe-operation area" (thermal SOA) in DC voltage level or DC current level [26–30].
- The ESD network operation does not reach the current-to-failure, voltage-to-failure, or power-to-failure prior to the ESD specification level objective [15–30].

ESD networks can consist of *I–V* characteristics of the following form:

- Step function *I–V* characteristics.
- S-type *I–V* characteristics.
- N-type *I–V* characteristics.

Step function *I–V* characteristics have a single "off" state as the structure is biased. At some voltage value, the device is "on." For example, a diode element has a step function *I–V* characteristic and is suitable for ESD protection. In the case of a diode element, the ideality is a function of the on-resistance of the diode element.

1.3.2 The ESD Design Window

In the defining of an ESD network, there is a desired range of operation. The "ESD design window" is the region of desired operation on a current–voltage (*I–V*) plot (Figure 1.6). On the *I–V* plot, there is a region defined for functional operation of the semiconductor chip. The application voltage is designated as from a voltage of $V=0$ to $V=V_{DD}$. On the *x*-axis, there is an absolute maximum voltage (also known as ABS MAX). On the *y*-axis, there is an operational current and an absolute current magnitude which the application must remain below without damage. The operational current–voltage range forms a rectangular region on the *I–V* plot. The ESD network must operate between the V_{DD} power supply and the absolute maximum voltage. On the current axis, the ESD network must discharge as high as possible to avoid the failure of the semiconductor component. The ESD current discharge should achieve the ESD specification levels. Hence, there is a region in which the ESD network is to operate without interfering with functional operation, but must discharge enough current prior to destruction of the semiconductor chip. In addition, the current

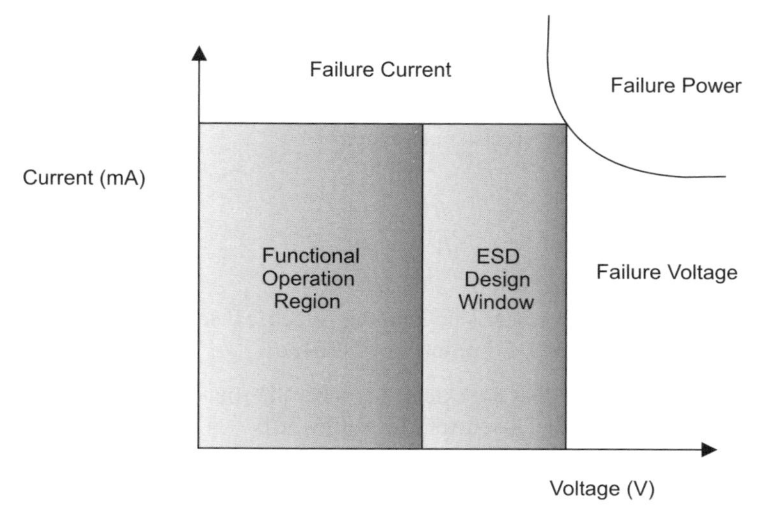

Figure 1.6 ESD design window and SOA

magnitude must exceed the latchup current criteria for voltages lower than the power supply voltage.

Figure 1.7 shows an example of a diode in the ESD design window. Because of the non-ideality of the diode element, there is a region where the DC voltage of the semiconductor devices in the technology are exceeded.

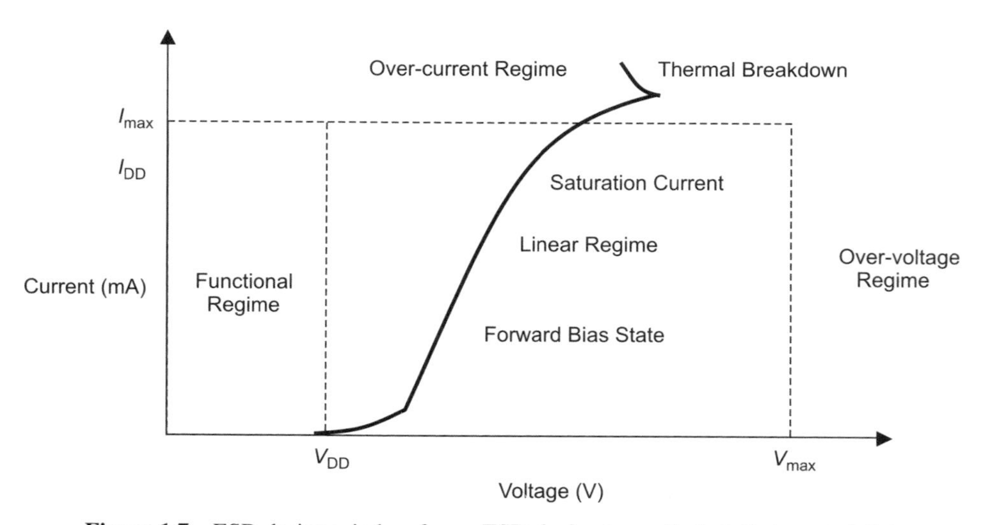

Figure 1.7 ESD design window for an ESD device (e.g., diode *I–V* characteristic)

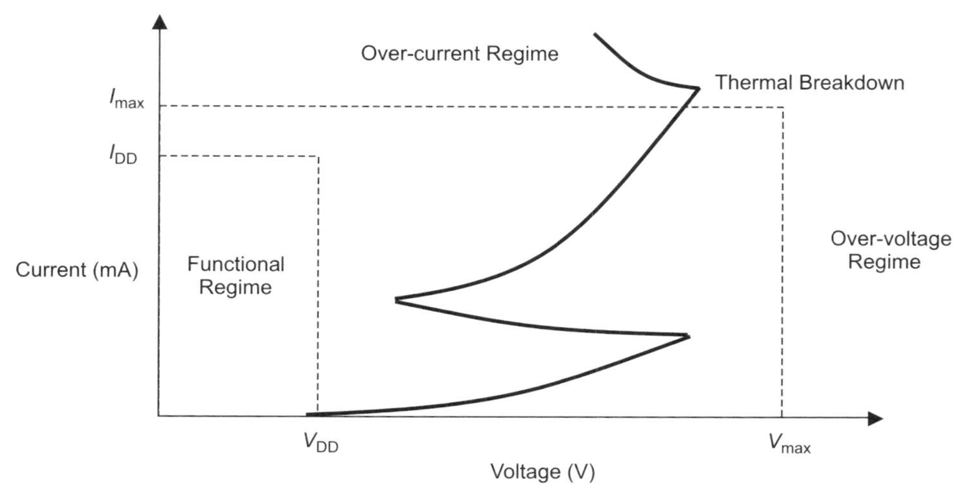

Figure 1.8 ESD design window for an S-type *I–V* characteristic ESD device

S-type characteristics are semiconductor devices or circuits that have two current states for a given voltage state. For example, an n-channel MOSFET or silicon-controlled rectifier (e.g., pnpn device) has an S-type *I–V* characteristic. Figure 1.8 shows an example of an n-channel MOSFET in a MOSFET drain-to-source configuration in an ESD design window. To utilize the MOSFET as an ESD network, the MOSFET snapback must occur within the current–voltage window of the technology limits of its safe operation area (SOA) of the other structures in the technology.

Figure 1.9 shows the ESD design window as a function of the technology generation. As observed, as the power supply voltage is reduced, the ESD design window decreases for successive technology generations.

1.3.3 The Ideal ESD Networks in the Frequency Domain Design Window

From an RF ESD design perspective, the characteristics of an RF ESD design are focused on its RF characteristics at the RF application frequency [8]. Figure 1.10 shows an example of ESD phenomenon frequencies, and RF application frequencies. RF applications are now faster than ESD phenomena for applications that exceed 5 GHz. This opens opportunities for unique RF ESD design implementations [8].

From an RF perspective, the ideal RF ESD network has the following features [8]:

- An ideal RF ESD network would have zero impedance during an ESD pulse.
- An ideal RF ESD network has infinite impedance during RF functional applications.

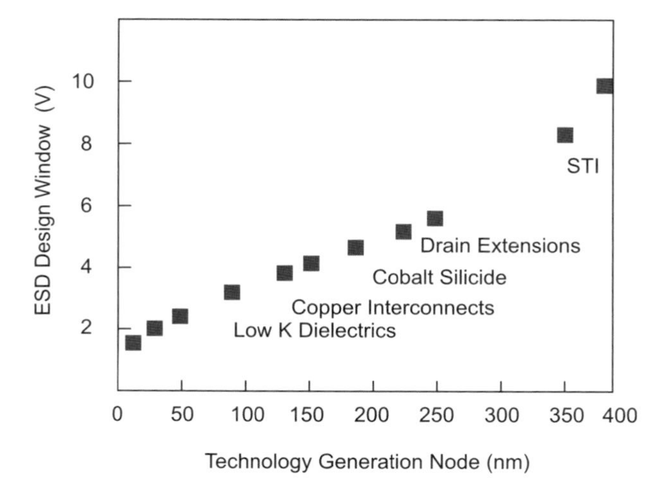

Figure 1.9 ESD design window as a function of technology generation

- An ideal RF ESD network during RF functional operation would not be a function of the current or voltage conditions.
- An ideal RF ESD network during RF functional operation is not temperature dependent.

Hence, from the frequency design window for RF ESD design, it is desirable to have an RF ESD network with zero impedance at low frequencies (e.g., HBM, MM, and CDM phenomenon regime below 5 GHz), and high impedance in the application frequency (e.g., RF application frequency regime).

Figure 1.11 shows the RF ESD frequency design window, and an ideal impedance characteristic imposed on the window. In a frequency regime below the CDM phenomenon, the ideal ESD impedance would be high during DC functional response, with zero impedance during an ESD event. At high frequencies, in the range of the RF application frequency, the impedance would have a high value (Figure 1.12).

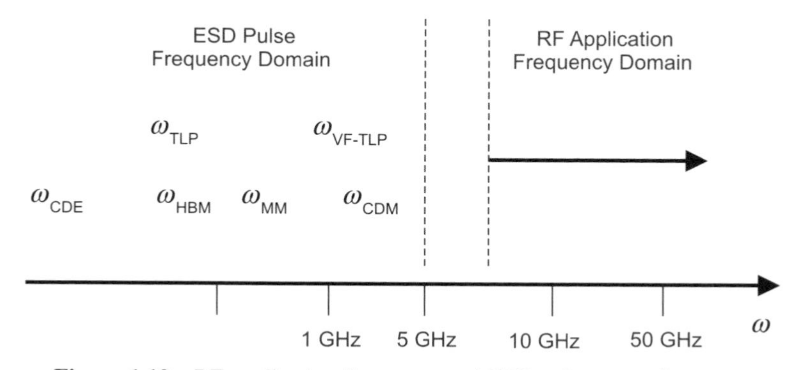

Figure 1.10 RF application frequency and ESD pulse event frequency

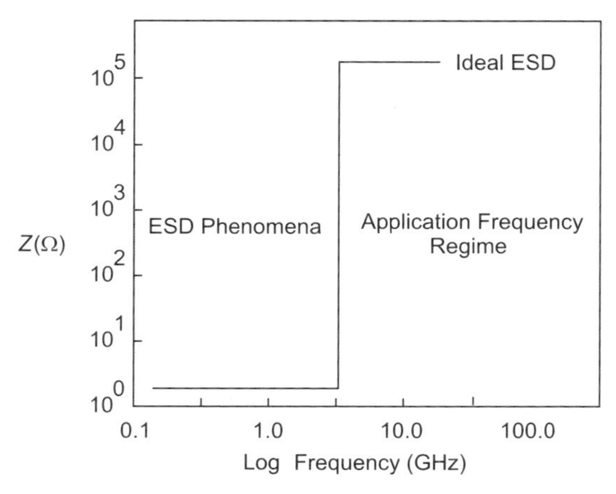

Figure 1.11 RF ESD frequency domain design window and the RF ESD device impedance

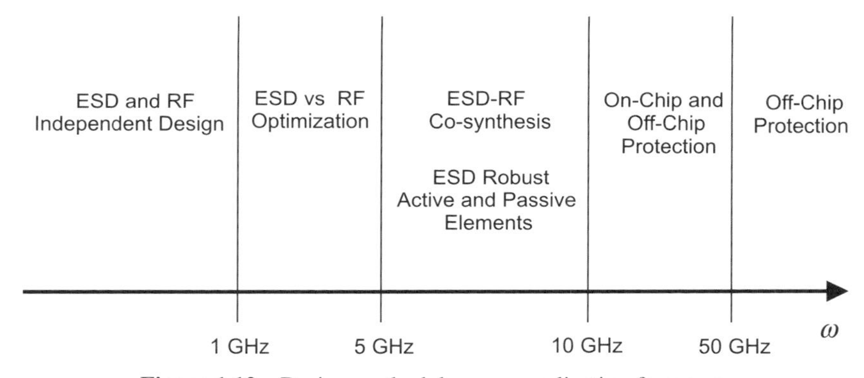

Figure 1.12 Design methodology vs. application frequency

1.4 MAPPING SEMICONDUCTOR CHIPS AND ESD DESIGNS

In semiconductor chip design, products are mapped from technology-to-technology for improved productivity. Designs are mapped from one foundry to another, and from technology generation to technology generation. Designs undergo re-mapping from bipolar to CMOS [6–8,39–41], and CMOS to SOI processes [6–8,11,41–44]. Additionally, there are both generational changes as well as productivity "shrinks." In this process the ESD design synthesis is influenced.

1.4.1 Mapping Across Semiconductor Fabricators

In semiconductor development, it is not uncommon to implement the same semiconductor chip design in multiple semiconductor fabrication facilities in today's business environments. Mapping of a semiconductor chip can occur in the following manner:

- Multiple semiconductor fabricators using the same technology.
- Multiple semiconductor fabricators using different technologies.

In the case of a utilization of multiple semiconductor fabricators, two different manufacturing lines can satisfy the same electrical specifications but in fact have unique features that influence the ESD semiconductor design. Semiconductor variations that influence ESD results can be as follows:

- Incoming wafer specifications for doping concentration.
- Incoming wafer specifications for epitaxy thickness.
- Photolithography tool bias conditions.
- Etch bias conditions.

In many manufacturing processes, the specifications contain only a lower-bound condition for some values. In other cases, although there is a range for a specification, there is variation within the specification window that can influence ESD results.

As an example, incoming wafers may be supplied from different vendors to different manufacturing facilities. Although the wafer specification is equivalent, the wafers may be "centered" differently within the specification window. In a p-/p++ substrate wafer, the epitaxial thickness may vary within the specification, as well as the measurement technique to control epitaxial thickness. In one case, a CMOS semiconductor chip was mapped into three different facilities on a p-/p++ wafer in a single-well semiconductor process. The ESD protection solution was a silicon controlled rectifier (SCR), whose triggering means the n-well-to-substrate breakdown. In the three different facilities, the breakdown voltages were different, leading to a different triggering voltage for the same semiconductor design. As a result, only one facility achieved the HBM ESD objective, whereas the other facilities did not.

To guarantee ESD equivalent results from fabricator-to-fabricator for a given technology, all electrical parameters that influence the ESD networks should be verified, and controlled to equivalent parameters. Secondly, ESD testing should be performed to determine the ESD pin distribution of the semiconductor product; this can be achieved by testing all signal and power pins to failure.

In the case of mapping across semiconductor fabricators of different semiconductor processes, the ESD robustness of all processes is not equivalent. Semiconductor process differences can influence the ESD circuits, and the functional circuits. All process variables can influence the ESD protection levels; which process features influence the product ESD robustness is dependent on the ESD circuit type used, and worst case ESD failure mechanism. To guarantee ESD equivalent results from fabricator-to-fabricator for a different technology, all electrical parameters that influence the ESD networks should be verified. Transmission line pulse (TLP) and human body model (HBM) measurements of the semiconductor device library and the ESD circuit library may be used to verify equivalency. Secondly, ESD testing should be performed to determine the ESD pin distribution of the semiconductor product; this can be achieved by testing all signal and power pins to failure.

Given that the semiconductor fabricators provide a set of ESD benchmark structures to quantify the semiconductor technology, and ESD TLP measurements, design adjustments can be made to the ESD circuit design and functional circuitry to compensate for differences between the semiconductor fabricators.

1.4.2 ESD Design Mapping Across Technology Generations

In semiconductor development, semiconductor chip designs are mapped from one technology generation to another for performance improvements, chip size reduction, and cost.

In the ESD design synthesis of mapping across technology generations, the semiconductor design can undergo different design and architectural conditions. In the re-mapping from one technology generation to another technology generation, the type of mapping can influence the ESD design. Two mapping processes are as follows:

- Lower power supply voltage for both peripheral I/O and core circuitry.
- Maintain the power supply voltage on the peripheral I/O, but reduced core circuitry supply voltage.

In the first case, the reduction of the power supply voltage for the entire semiconductor chip can lead to both circuit topology and technology differences that influence the ESD results. With the lowering of the power supply voltage, given that the first generation contained elements in series, these can be reduced with the re-mapping process. With a reduction of the series elements in the ESD networks, ESD improvement can be obtained.

With the technology scaling, the physical dimensions of the devices are reduced. Dimensional scaling of the technology can influence the ESD results as follows:

- Lower dielectric breakdown voltage.
- Lower MOSFET and bipolar snapback voltage.
- Different current-to-failure.
- Different power-to-failure.
- Different voltage-to-failure.

With the technology scaling, the changes in the ESD robustness can be lower or higher, depending on the decisions of the doping concentration and film thickness. In the case where the ESD robustness is reduced, design modifications can be made to improve the product ESD results. The design modification can be the physical size of the ESD networks, innovation, or novelty.

In the second case, the product must maintain the same external or application voltages in the scaled technology. To obtain the density advantage of the scaled technology, the semiconductor chip design synthesis may take various directions:

- Single power supply supporting both mixed voltage interface circuitry.
- Single power supply, and regulated voltage for core circuitry.
- Multiple power supply pins.

In the case of many product applications, the peripheral circuitry is required to receive or transmit signals to semiconductor chips different from the native voltage of semiconductor technology. The ESD architecture and design synthesis are highly influenced by these conditions in both digital, analog, and power designs.

1.4.3 Mapping from Bipolar Technology to CMOS Technology

In the mapping of a design from bipolar technology to CMOS technology, the ESD design must be adjusted since the bipolar transistors differ significantly from CMOS MOSFET transistors [6–8,14]. Bipolar transistors operate as vertical bulk devices, whereas MOSFET transistors operate as lateral surface devices. As a result, bipolar transistors provide lower current densities, and are more suitable for power and ESD implementations. Bipolar transistors can also be utilized as forward biased elements, or reverse breakdown devices, providing flexibility in the different types of implementations.

In the bipolar implementations, bipolar transistors are utilized for ESD protection on the signal pins, and the power rails. Bipolar transistors are typically used in collector-to-emitter configuration. Bipolar transistors are also used in diode configuration (e.g., base-collector, or emitter-base). In the mapping of a schematic design from bipolar technology to CMOS technology, bipolar diodes can be mapped directly as CMOS-based p–n diode elements formed out of diffusions and wells [8]. ESD inputs that are grounded base bipolar circuits (e.g., collector connected to input, and emitter connected to ground) can be mapped into grounded gate MOSFETs. ESD power clamps between power rails can be mapped from bipolar collector–emitter configured networks to grounded-gate triggered MOSFET power clamps or RC-triggered MOSFET power clamps.

One of the critical differences is that the size and layout of the implementations are significantly different between the bipolar elements and the CMOS elements (Figure 1.13). Since the bipolar transistors are bulk devices, and MOSFETs are surface devices, the perimeter and width of the MOSFET ESD devices are increased to achieve the same levels of ESD robustness.

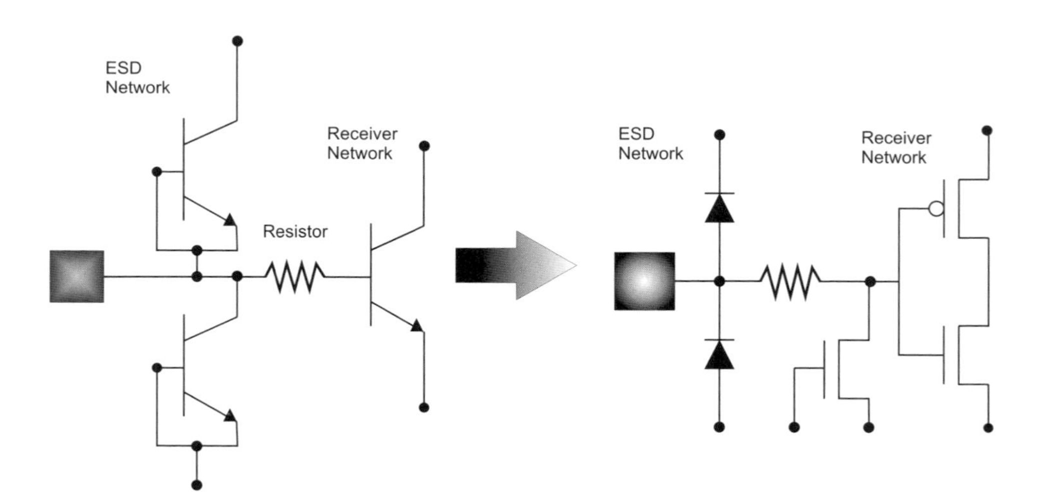

Figure 1.13 Mapping of bipolar to CMOS technology

1.4.4 Mapping from Digital CMOS Technology to Mixed Signal Analog–Digital CMOS Technology

In the mapping of a semiconductor chip from a digital CMOS technology to a digital–analog technology, the ESD design synthesis, architecture, and physical elements may require adjustments [14]. With the introduction of an analog domain, modifications are needed in the architecture of the semiconductor chip. Digital and analog domains are required to be established, as well as ESD protection for each independent power domain. In addition, added ESD protection is needed for signal lines that cross both domains. Digital designs also typically have simple connections, whereas analog networks typically have a higher number of electrical connections to many different sub-functions.

Another key difference is that the floorplanning of the semiconductor chip into separated analog and digital domains influences the placement of the physical elements and the electrical connections. In addition, the analog design layout may consist of different analog design layout considerations of both active and passive elements; these practices will be discussed in later sections of the text.

1.4.5 Mapping from Bulk CMOS Technology to Silicon on Insulator (SOI)

For performance enhancements, CMOS designs can be mapped from bulk CMOS to silicon on insulator (SOI) technology [6,7,11,43,44]. In the mapping from a bulk CMOS technology to an SOI technology, modifications are required in both the functional circuitry and the ESD designs. Mapping of the ESD design will be dependent on the ESD device choice, the thickness of the silicon film, the isolation, the technology features, and technology type. Various forms of SOI technology exist:

- Thin film partially depleted SOI (PD-SOI) MOSFET.
- Thick body partially depleted SOI (PD-SOI) MOSFET.
- Thin body fully depleted SOI (FD-SOI) MOSFET.
- SOI FinFET technology.
- Bipolar CMOS–DMOS SOI.

In partially depleted SOI (PD-SOI) technology, the MOSFET junctions extend to the buried oxide (BOX) region [6,7]. The silicon film on the BOX region is on the scale of the MOSFET junction depth. Isolation is formed, which extends from the device surface to the BOX film. The introduction of the BOX film, and integration with the isolation technology, leads to the elimination of all desired and undesired parasitic elements under the isolation structure. From an ESD perspective, all undesired parasitic elements are eliminated. For I/O and ESD elements, isolating guard rings are no longer required. External latchup and internal latchup is no longer an issue in thin body PD-SOI.

In the re-mapping process from CMOS to PD-SOI, the diode elements and SCRs can not conduct current vertically, or under the isolation region; this leads to modification of the ESD

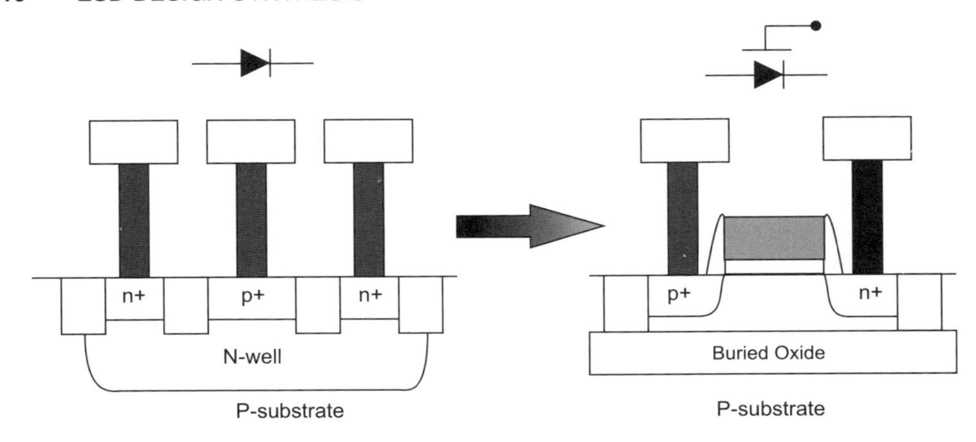

Figure 1.14 Mapping of bulk CMOS to SOI technology

networks to allow a conduction path. Lateral elements will be required for ESD protection due to the elimination of any active device under the isolation structure. As a result, new ESD structures will be required.

For thick body PD-SOI, the buried oxide film extends under the well or buried layer depths and does not fully isolate the p-channel MOSFET and n-channel MOSFET devices. The isolation structure may extend to the buried oxide film, allowing separation of tubs and floating relative to the substrate wafer. As a result, in the re-mapping from CMOS to thick body SOI, ESD networks that utilize surface devices may not require design modification.

For bulk bipolar CMOS–DMOS (BCD) technology, the depth of the BOX region separates the devices from the substrate wafer, and provides isolation of the different device types. The devices that extend to the BOX region will be influenced in the mapping process. The BOX region will influence both the breakdown voltages and the current conduction. As a result, ESD networks will require modifications.

In the mapping from bulk CMOS to FinFET technology, the layout and design of all devices in the technology are modified [11]. ESD networks will also be required to be modified due to the means of current distribution in the FinFET structure (Figure 1.14).

1.4.6 ESD Design – Mapping CMOS to RF CMOS Technology

Today, many designs are being defined from a CMOS base to an RF CMOS base technology [8,11,39–41]. CMOS-based designs are typically developed as a custom layout, including the ESD networks. In the CMOS ESD designs, the designs are not scalable elements and typically do not have scalable models.

Ironically, today many applications are using RF CMOS technology for the usage of the scalable models and AC models. In many application spaces, RF CMOS foundry technologies are being used for the higher-quality models. For analog design, small and large signal models are needed in the linear and the saturation region of the MOSFET devices, as well as in high-quality passive elements with good matching characteristics.

In the mapping of a digital or analog design into an RF CMOS technology, the ability to have better-quality AC models is available. For ESD design, it is also possible to have ESD designs with scalable elements. ESD designs can be constructed from parameterized cell elements. The primitive ESD design elements can be hierarchically defined, forming hierarchical parameterized cell ESD networks. Such elements can be constructed in a Cadence™ design environment [8]. By using an RF CMOS technology, the design synthesis environment allows for this ESD design strategy.

In the mapping of CMOS to RF CMOS, the circuit topology of the RF networks will introduce passive elements, such as capacitors and inductors [8]. Inductors and capacitors are used for DC isolation, input and output matching, filters, and matching. Inductors and capacitors are preferred resistor elements. As a result, in the re-mapping of circuits to RF CMOS, resistor elements are replaced by both capacitor and inductor elements. In ESD networks, typical CMOS ESD solutions (such as resistor ballasting and series resistor elements) are replaced with inductive and inductive–capacitive techniques.

In addition, in the ESD design synthesis, the RF circuits and ESD networks must be co-synthesized [8]. ESD-RF design co-synthesis is needed to achieve the RF performance objectives and matching characteristics.

1.5 ESD CHIP ARCHITECTURE, AND ESD TEST STANDARDS

The ESD standards are written to test the ESD robustness between possible "events." These events contain two features. The first feature is a given waveform. The second feature is the possible interactions that can occur within a semiconductor chip. The possible interaction is the case that a given pin receives an ESD event, and a second pin is the reference ground. The first feature tests the nature of the pulse, and the second feature is evaluating the current paths. In the ESD standards, different procedures exist in application of the ESD pulse to the pin or package, as well as the "pin combinations."

1.5.1 ESD Chip Architecture and ESD Testing

The pin combinations are evaluating the ESD robustness of the possible paths, which in essence is testing the existence of a suitable ESD current path to discharge the ESD current. As a result, the chip architecture must be constructed to establish current paths, alternate current paths, and "elements" within the path that are suitable to discharge the ESD current. The elements in the path are the devices, circuits, and electrical interconnects within the path.

A key point is that for some of the ESD standards, the semiconductor chip architecture is structured to satisfy the ESD pin combinations of the ESD test standard.

1.6 ESD TESTING

In this section, a brief discussion of ESD testing will be given, with relevance to ESD chip architecture. As stated in the last section, the architecture of the chip is structured to pass the ESD testing procedure. As a supplier, or customer, it is an expectation that a certain number of ESD test standards are performed, and achieve given objectives.

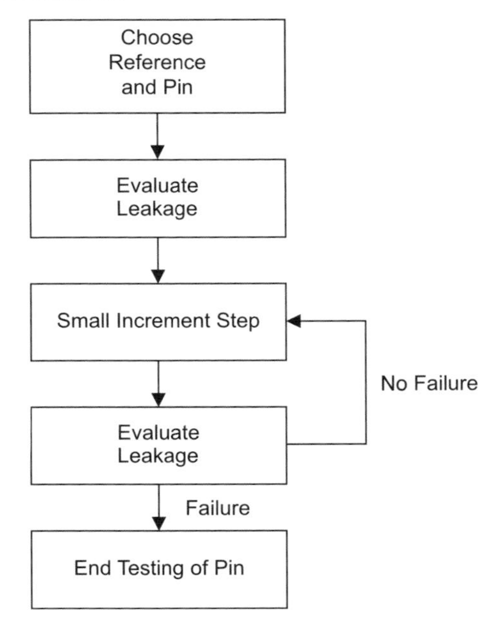

Figure 1.15 ESD test flow

1.6.1 ESD Qualification Testing

Qualification testing to demonstrate ESD robustness is performed by the supplier, or customer. ESD testing is typically completed in the environment that it is shipped from the supplier to the customer. Figure 1.15 is an example of the ESD test flow. The ESD testing can be completed on wafer, bare die, or packaged components. The majority of the ESD testing standards are defined and structured to address the packaged component.

1.6.2 ESD Test Models

ESD specifications contain different waveforms and pin combinations depending on the intent of the ESD event that the procedure is to simulate [47–82]. Component-level ESD test models include the following:

- Human body model (HBM) [47–53].
- Machine model (MM) [54].
- Charged device model (CDM) [55].
- Socketed device model (SDM).
- Human metal model (HMM) [79–82].

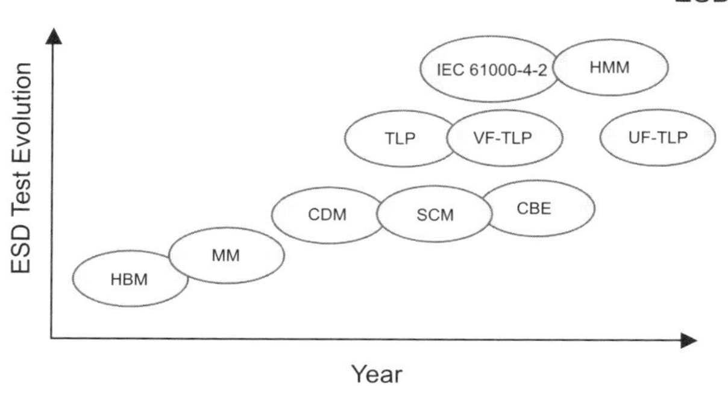

Figure 1.16 ESD testing evolution

ESD system-level tests, presently being performed on chips, sub-systems, and systems, are as follows [56–64,76–82]:

- Charged board event (CBE).
- Cable discharge event (CDE) [56–64].
- IEC 61000-4-2 ESD pulse [76–79].
- Human metal model [79–82].

Figure 1.16 is a chart demonstrating the evolution of testing in the semiconductor industry. The HBM test model was the first developed ESD specification for qualification of semiconductor devices. This was followed by the MM test specification, and CDM test specification. Today, there are additional test specifications at various phases of becoming ESD test standards.

1.6.3 ESD Characterization Testing

ESD characterization tests that are being performed on wafers and on packages include the following [47–54,65–75]:

- Human body model (HBM) [47–53].
- Machine model (MM) [54].
- Transmission line pulse (TLP) [65–68].
- Very fast transmission line pulse (VF-TLP) [69–74].
- Ultra-fast transmission line pulse (UF-TLP) [75].

Wafer-level HBM and MM testing can be performed as a two-pin test.

1.6.4 TLP Testing

ESD characterization tests that are being performed on wafers and on packages include the following [65–75]:

- Transmission line pulse (TLP) [65–68].

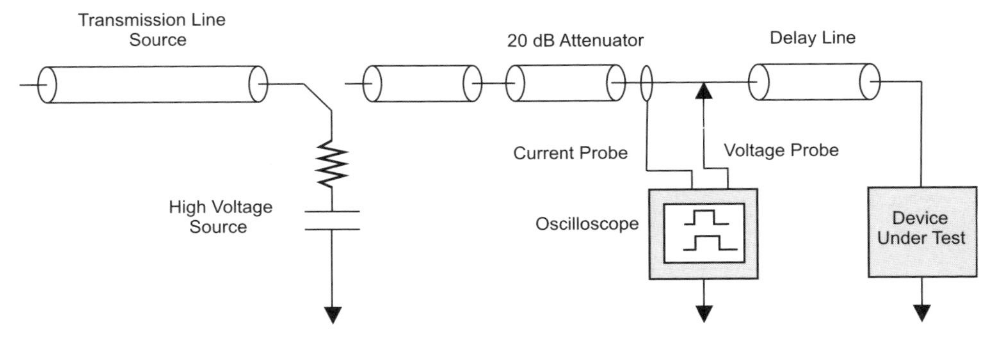

Figure 1.17 TLP test system

- Very fast transmission line pulse (VF-TLP) [69–74].
- Ultra-fast transmission line pulse (UF-TLP) [75].

In Figure 1.17, an example of a TLP system is shown. In a TLP system, a transmission line (TL) cable is charged by a high-voltage source. A switch is closed to apply the stress to the device-under-test (DUT). The TL pulse width is defined by the length of the cable in a given test system. The voltage and current are measured across the device under test as the pulse propagates through the device. There are various configurations of the TLP system. In some TLP systems, the configurations are such as to measure the reflected and transmitted signal through the device. A current and voltage value is determined in the device through averaging of the data in a "measurement window" within the pulse event.

A pulsed I–V characteristic can be constructed by increasing the voltage on the cable, and extracting the (I,V) data points in this successive step stress. Figure 1.18 shows an

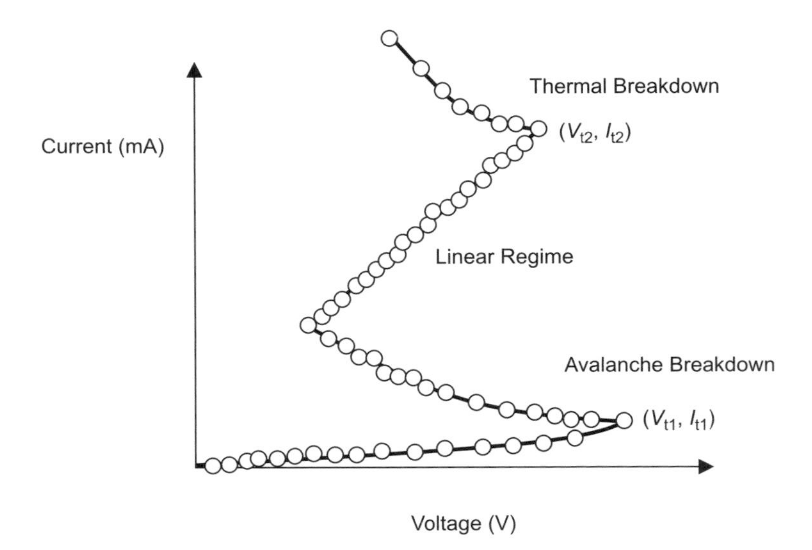

Figure 1.18 TLP pulsed I–V characteristic

example of a TLP pulsed *I–V* characteristic. Each point represents a single pulse in the step stress, where the (*I*,*V*) point is an averaged measurement extracted from the measurement window.

1.7 ESD CHIP ARCHITECTURE AND ESD ALTERNATIVE CURRENT PATHS

In the ESD design synthesis of a semiconductor chip, both the electrical characteristics and the physical placement are key to good full-chip ESD design. In the following section, a brief introduction is provided on the chip architecture of the I/O circuits, cores, and ESD elements.

1.7.1 ESD Circuits, I/O, and Cores

In the design synthesis of a semiconductor chip, the physical placement and chip architecture of the I/O circuits, the ESD elements and the core circuitry are key to the planning process, as well as the success of the implementation in achieving ESD robust designs [1–14].

I/O circuitry physical placements are found for a semiconductor chip as follows:

- **Peripheral I/O:** Periphery of the semiconductor chip.
- **Internal I/O Banks:** Large I/O clusters (also referred to as "banks") within the semiconductor chip interior.
- **Small Internal I/O Banks:** Small I/O clusters (e.g., "nibble" architecture) within the semiconductor chip interior.
- **Array I/O:** Individual "array" I/O distributed within the semiconductor core.

In each of these different design implementations, the architectures introduce new challenges to the ESD design synthesis implementation. The placement of the ESD signal pin devices, within domain ESD circuits (e.g., ESD power clamps) and domain-to-domain ESD circuits (e.g., ground-to-ground) are highly influenced by these architectures. In future sections, these will be discussed in great detail.

1.7.2 ESD Signal Pin Circuits

ESD circuits are placed on signal pins to re-direct the ESD current from the signal path to an alternative current path [7]. The ESD signal pin circuitry establishes connectivity to at least one power rail. ESD signal pin networks can be classified as follows:

- Single power rail connectivity.
- Dual power rail connectivity.
- Triple power rail connectivity.

Single power rail signal pin ESD networks are used in NMOS, CMOS, bipolar, DMOS, and BCD technologies. Examples of single power rail signal pin ESD networks are as follows:

- NMOS – thick oxide MOSFET – resistor – thin oxide MOSFET.
- CMOS – grounded gate NMOS (GGNMOS).
- CMOS – SCRs.
- Bipolar – grounded base bipolar in collector–emitter configuration.
- DMOS – grounded gate LDMOS transistor.

The advantage of a single power rail connected ESD signal pin network is that it is typically not connected to the power supply voltage, but connected to a ground power rail. This avoids issues associated with power-up and power-down conditions, power rail sequencing, power loss, and fail-safe conditions. The disadvantage is that during the reference grounding of the power rail V_{DD}, there is no current path established. These structures provide one mode of operation in a forward bias mode, and a second in reverse bias mode.

Dual power rail signal pin ESD networks are used in CMOS, bipolar, DMOS, and BCD technologies. Examples of dual power rail signal pin ESD networks are:

- CMOS – dual diode ESD networks.
- CMOS – series diode ESD networks.
- CMOS – dual silicon controlled rectifiers (DSCRs).
- LDMOS – dual diode configured LDMOS transistors.

The advantage of a dual power rail connected ESD signal pin network is that it is connected to the power supply voltage and ground power rail. Figure 1.19 shows a dual power rail connected ESD network. This establishes two alternative forward bias current paths and electrical connectivity to both power rails for positive or negative ESD pulse events. In these networks, the voltage across the structure is significantly less due to the low turn-on voltage, leading to less power density in the metallurgical junctions.

Triple rail ESD signal pin networks are also used in applications with multiple power or ground connections. An example is as follows:

- CMOS – DRAM triple rail ESD networks.

The advantage of triple rail ESD networks is that they establish direct connectivity to separate power supplies, or separate ground connections. Triple rail ESD networks also save significant space and have low capacitance.

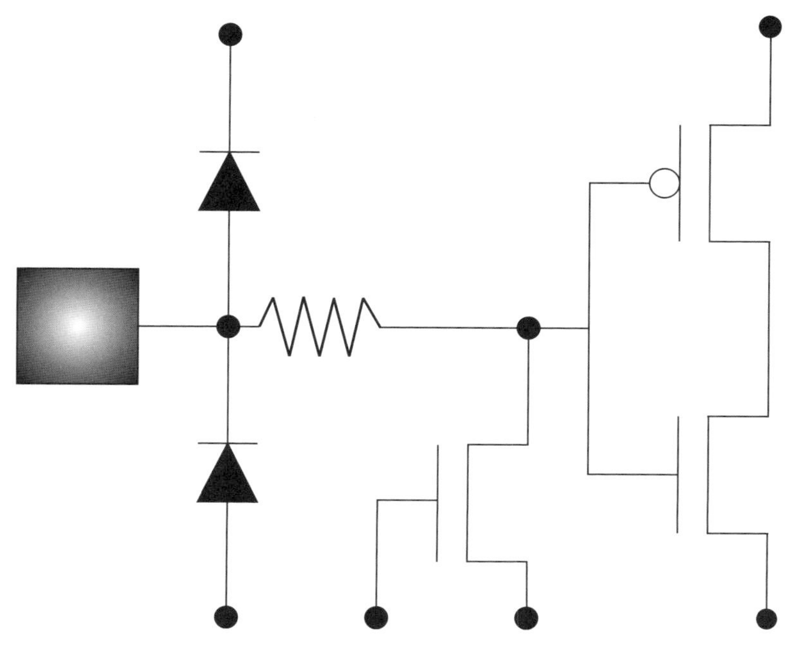

Figure 1.19 ESD input circuit

1.7.3 ESD Power Clamp Networks

In the ESD design synthesis, within a given circuit power domain, ESD networks are placed between the power and ground rails to improve the ESD robustness of the semiconductor chip [1–14].

ESD power clamps serve several purposes in the ESD design methodology:

- **Power Rail ESD Protection:** ESD power clamps provide ESD protection between the power rails (e.g., V_{DD}) and ground rails (e.g., V_{SS}).
- **Over-voltage Protection:** ESD power clamps provide over-voltage and over-current protection to the functional circuits in parallel with the ESD power clamp.
- **Alternative Current Loop:** ESD power clamps provide closure of an alternative current loop between signal pins and its power and ground rails.
- **Bi-directionality:** ESD power clamps establish bi-directional current paths between the power rail and the ground rails.
- **Signal Pin to Signal Pin:** ESD power clamps provide pin-to-pin ESD protection by establishment of a bi-directional current path (in conjunction with the ESD signal pin devices).
- **Low Impedance:** ESD power clamps lower the impedance of the semiconductor chip from the power rail to its associated ground rail in the alternative current loop.

- **Impedance Independence for Chip Capacitance:** ESD power clamps provide independence of the impedance of the semiconductor chip in the ESD protection process.

ESD power clamps can be classified as follows:

- Voltage-initiated forward bias ESD clamps.
- Voltage-initiated reverse bias breakdown ESD clamps.
- Active circuit-initiated ESD power clamps.
- Frequency-triggered ESD power clamps.
- Frequency and voltage-triggered ESD power clamps.

ESD power clamps can serve native-voltage, mixed-voltage, or high-voltage semiconductor chip sectors. ESD power clamps use both passive and active elements. For CMOS technology, ESD power clamps are formed using MOSFETs, diode, resistor, and capacitor elements. For bipolar technology, the ESD power clamps are constructed of bipolar transistors, varactors, and resistor elements. In a BiCMOS technology, both CMOS-based and bipolar-based ESD power clamps can be utilized. In addition, hybrid BiCMOS ESD power clamps can be formed that utilize both bipolar and MOSFET transistor elements in the same ESD power clamp circuit. In a power technology, the ESD power clamps can utilize PDMOS and NDMOS transistor elements. Figure 1.20 shows an example of an ESD power clamp.

1.7.4 ESD Rail-to-Rail Circuits

In the architecture of a semiconductor chip, the power grid is separated into different power domains. Power domains are separated for the following reasons:

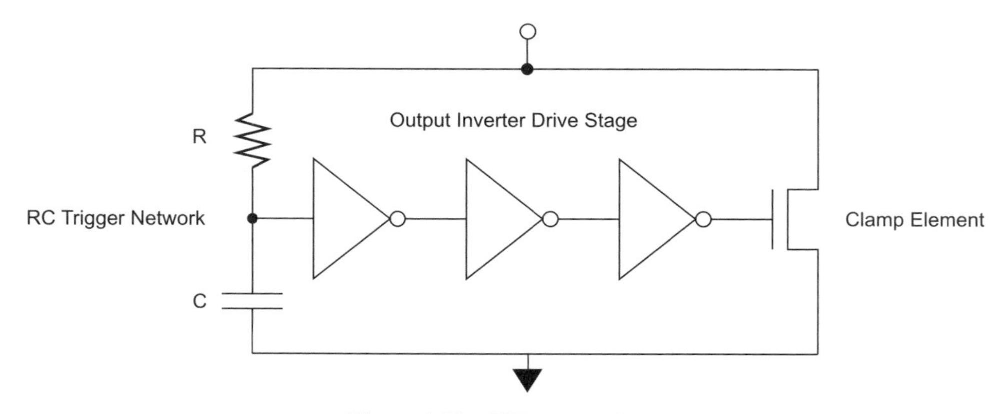

Figure 1.20 ESD power clamp

- Noise isolation between grounds.
- Different power supply voltage requirements.
- Different circuit functions.

ESD failure mechanisms can occur with the separation of ground rails. This can occur in the following applications:

- ESD failure between peripheral I/O circuit ground (e.g., V_{SS}(I/O)) and chip substrate (V_{SUB}).
- ESD failure between peripheral I/O circuit ground and core ground.
- ESD failure between digital ground (e.g., V_{SS}) and analog ground (e.g., AV_{SS}).
- ESD failure between digital ground (e.g., V_{SS}) and radio frequency ground (RFV_{SS}).

ESD networks are needed between the power rails to establish a current path between the power rails. Establishment of a current path between the power rails provides both ESD protection between the two power rails, and ESD protection between a signal pin to power rail.

In the architecture of a semiconductor chip, the voltage of each ground can be of the same potential ($V_{SS} = AV_{SS} = 0$) or different values ($V_{SS} = 0$, and $V_{EE} = -3$ V).

ESD failure mechanisms can occur with the separation of V_{DD} power rails. This can occur in the following applications:

- ESD failure between peripheral I/O circuit (e.g., V_{DD}(I/O)) and chip power (V_{DD}).
- ESD failure between peripheral I/O circuit (e.g., V_{DD}(I/O)) and core (V_{DD}) with a regulator between the two power rails.
- ESD failure between digital power (e.g., V_{DD}) and analog power (e.g., AV_{DD}).
- ESD failure between digital power (e.g., V_{DD}) and radio frequency ground (RFV_{CC}).

ESD networks are needed between the power rails to establish a current path between the power rails. Establishment of a current path between the power rails provides both ESD protection between the two power rails, and ESD protection between a signal pin to power rail.

In the architecture of a semiconductor chip, the voltage of each power can be of the same potential ($V_{DD} = AV_{DD}$) or different values ($V_{CC} = 5$ V, and $V_{DD} = 3.3$ V).

A common rail-to-rail network is the usage of a bi-directional series diode string (Figure 1.21). Bi-directional diode strings can consist of CMOS diodes, bipolar transistors, or diode-configured CMOS elements.

1.7.5 ESD Design and Noise

In semiconductor chips, the switching of circuitry can lead to noise generation that can influence circuit functions. In the architecture of a semiconductor chip, different domains are separated due to noise generation. Noise is generated from undershoot and overshoot

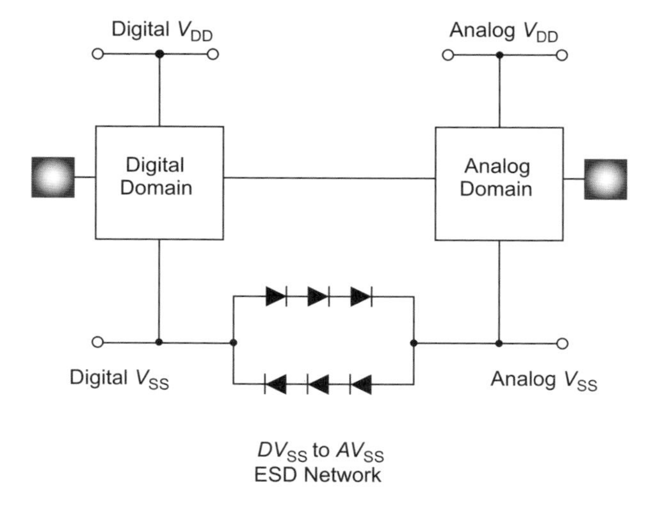

Figure 1.21 Bi-directional rail-to-rail ESD network

phenomena of signals leading to substrate injection. Noise can also be generated from switching of off-chip drivers circuitry, and large power devices.

In a semiconductor chip, there is circuitry which is sensitive to the noise generation. To avoid noise from impacting functionality of the sensitive circuitry, separate power domains are created on a common substrate. Examples of separation of power domains and semiconductor chip functions are as follows:

- Separation of peripheral circuitry from core circuitry.
- Separation of peripheral circuitry from core memory regions.
- Separation of digital and analog functions.
- Separation of digital, analog, and RF functions.
- Separation of power, digital, and analog functions.

In ESD design, the separation of different circuit domains for noise isolation can lead to ESD failures. ESD failures can occur due to the following situations:

- Lack of a forward bias current path between a first power rail and a second power rail.
- Lack of a forward bias current path between a pin of a first domain to a power rail of a second domain.
- Lack of a forward bias current path between a pin of a first domain to a pin of a second domain.

As a result, in the ESD design synthesis, the architecture of the semiconductor chip which is separated for noise, must allow for current to flow from domain to domain. As a result, there is a tradeoff between the noise isolation and the requirement of allowance of current flow between pins and power rails of different domains. Various means have been utilized to achieve this. Examples of solutions between independent power domains are as follows:

- Symmetric and asymmetric bi-directional ESD networks between domains of $V_{DD}(i)$ and $V_{DD}(j)$.
- Symmetric and asymmetric bi-directional ESD networks between domains of $V_{DD}(i)$ and $V_{SS}(j)$.
- Symmetric and asymmetric bi-directional ESD networks between domains of $V_{SS}(i)$ and $V_{SS}(j)$.

1.7.6 Internal Signal Path ESD Networks

In ESD design, the separation of different circuit domains for noise isolation can lead to ESD failures internal to the semiconductor chip [11]. Traditionally, the majority of ESD failures occur near the signal pads, and not in internal signal lines. But, signal lines that cross power domains, where there is no current path through the power grid, leads to voltage stress internal to the semiconductor chip. Figure 1.22 is an example of a signal line that crosses the power domain from digital to analog. The solutions are as follows:

- Internal signal line ESD networks.
- Low-voltage differential domain-to-domain ESD networks.

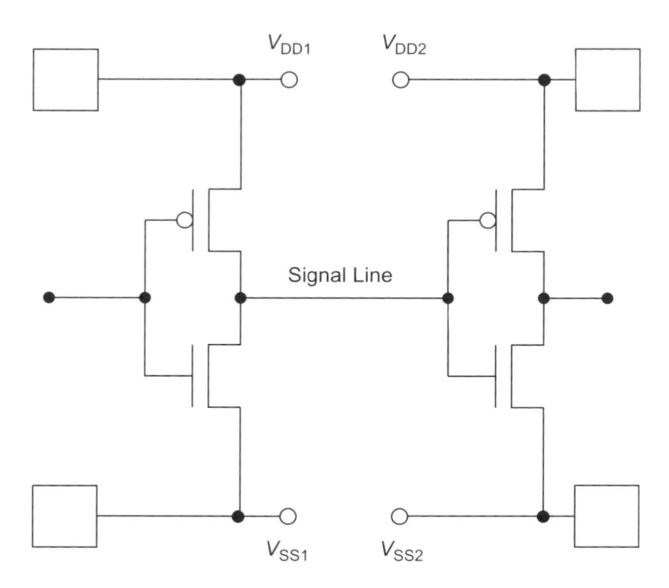

Figure 1.22 Internal signal path concern

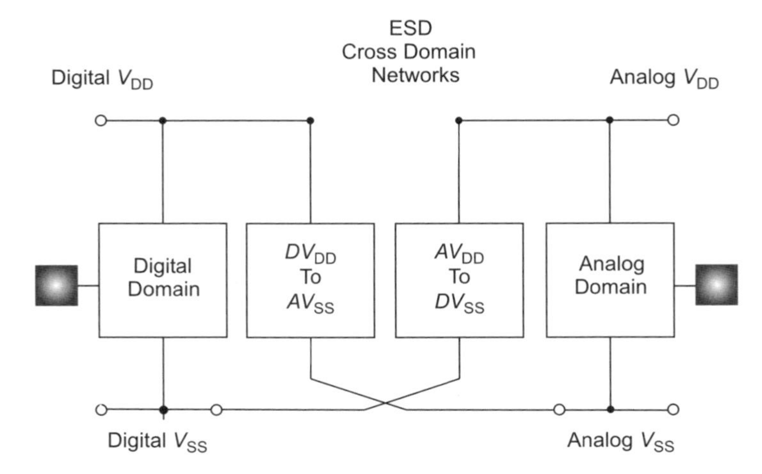

Figure 1.23 Cross-domain ESD power clamp

- Cross-domain ESD networks.
- Third-party networks between the power domains.

1.7.7 Cross-Domain ESD Networks

With the separation of power domains, the lack of a bi-directional current path through the power grid from domain to domain can lead to ESD failures. In addition, ESD failures can occur due to internal signal lines that cross domains. A solution to address this issue is to create ESD power clamp networks that cross the domains. Figure 1.23 shows an example of a cross-domain ESD network. Cross-domain ESD networks establish an electrical connection of:

- Symmetric and asymmetric bi-directional ESD networks between domains of $V_{DD}(i)$ and $V_{SS}(j)$.
- Symmetric and asymmetric bi-directional ESD networks between domains of $V_{DD}(j)$ and $V_{SS}(i)$.

1.8 ESD NETWORKS, SEQUENCING, AND CHIP ARCHITECTURE

In product applications, the sequence of power supplies and signal pins is an issue for ESD design synthesis. In some products the sequence of power to the power rails and signal pins is defined, and in others it is not. The decision whether a semiconductor chip is sequence dependent or sequence independent can influence the type of ESD networks that are used in the implementation [7].

In sequence-dependent applications, where the sequence of pins and power supplies is dependent, ESD circuits on signal pins can include the following:

- Diodes.
- Series diodes.
- Triple-well series diodes.
- MOSFETs.
- Series cascode MOSFETs.
- Silicon controlled rectifiers.

In sequence-dependent applications, the sequence of power supplies can be specified allowing usage of ESD diodes between power supplies. ESD series diodes can be placed between power supplies with different voltage levels or the same voltage level when the application specifies the procedure for power-up and power-down. ESD diodes can be placed on signal pins when the power is supplied to the power rails prior to signal pins.

In sequence-independent applications, where the sequence of pins and power supplies is not specified, ESD circuits on signal pins can include the following:

- MOSFETs.
- Series cascode MOSFETs.
- Silicon controlled rectifiers.

Sequence-independent networks that use control networks can be utilized on the power supplies and signal pins ESD circuits. Control networks and switch networks can be utilized to sense the voltage state of the signal pin or power supply to prevent forward biasing of ESD networks. Control networks can be utilized for both non-isolated and isolated epitaxial and well regions.

1.9 ESD DESIGN SYNTHESIS – LATCHUP-FREE ESD NETWORKS

In ESD design synthesis, ESD networks can lead to CMOS latchup. Latchup can be initiated in a powered state of the functional semiconductor chip [9,12].

In ESD design, it is an objective to provide an ESD network that is "activated" when an ESD event occurs, and disabled during functional operation. Functional operation leads to a powered state which includes the following:

- Power-up.
- Power-down.

- Sequencing of the independent power supplies.
- Direct current (DC) operation.
- Alternating current (AC) operation.

Additional conditions also exist that are of concern:

- System-level transient excursions.
- Board-level transient excursions.
- Reliability stress conditions.
- Latchup simulation stress testing.

During the release of a product, the ESD network must not latchup during any of these conditions. ESD networks that have a tendency to latchup are the following [9,12,14]:

- High-voltage silicon controlled rectifiers (HV-SCR).
- Medium-voltage silicon controlled rectifiers.
- Low-voltage triggered silicon controlled rectifiers (LVTSCR).
- Grounded gate silicon controlled rectifiers (GG-SCR).
- PNP-triggered SCR (PNP-SCR).
- PMOS-triggered SCR (PMOS-SCR).

To avoid CMOS latchup, in the ESD network design synthesis, the following requirements must exist:

- **Minimum Voltage Condition:** The ESD SCR network must have a "trigger voltage" condition above the worst case power supply condition, and above the reliability stress voltage requirements.
- **Maximum Voltage Condition:** The ESD SCR network must have a "trigger voltage" condition below the absolute maximum voltage condition (maximum condition without electrical damage to the component).
- **Minimum Current Condition:** The ESD SCR network must have a "holding current" condition above the latchup specification current condition (e.g., 100 mA).
- **Maximum Current Condition:** The ESD SCR network must have a "holding current" condition below the maximum current condition (maximum current condition without electrical damage to the component).

Figure 1.24 shows an example of an ESD design box for an ESD network SCR that demonstrates the above voltage and current requirements. In the figure, the trigger voltage of the ESD element exceeds the latchup specification current magnitude.

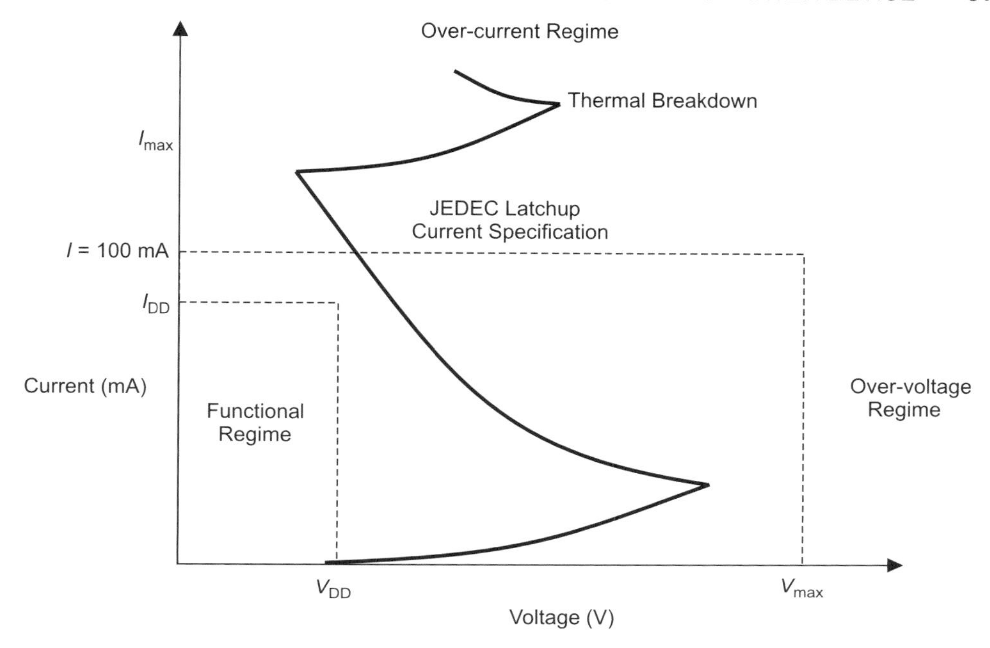

Figure 1.24 ESD design box for latchup-free SCR ESD network

1.10 ESD DESIGN CONCEPTS – BUFFERING – INTER-DEVICE

In the semiconductor chip architecture, a technique to avoid current flow to the sensitive circuits is to integrate "buffering" into the circuit net [7]. Adding series impedance in the signal path provides margin to transfer the current to the alternative current loop established by the ESD networks. In digital and analog circuits, adding resistance in series is a well-established practice. In RF circuits, capacitive and inductive elements can be introduced to avoid current flow to the sensitive active elements.

Buffering elements can serve the following roles:

- **Current:** Lower the current flow through the signal path.
- **Voltage:** Lower the voltage at the signal pin node with formation of a "resistor divider" network.
- **Impedance Matching:** Impedance matching of signal pins.

Buffering elements can be resistors that are inherent in the signal pin, or added resistance. The resistor elements can be as follows:

- Metal layer interconnects (e.g., aluminum or copper).
- Local metal interconnects (e.g., tungsten "M0" wiring layer).
- Polysilicon resistor.

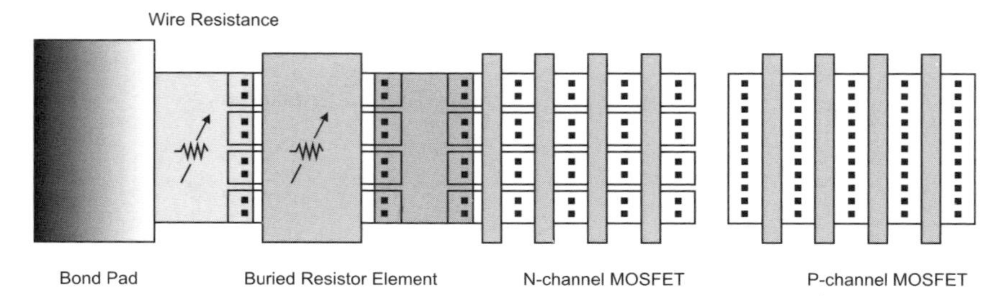

Figure 1.25 ESD design synthesis of buffering and impedance matching resistor element

- Diffused resistors (e.g., n-diffusion, n-well, buried resistors).
- Active elements (e.g., MOSFET or bipolar transistor).

The series resistor "buffering element" can be integrated inside or outside the ESD network. It is common practice to integrate the buffering element in a multi-stage ESD network. With integration of the resistor into a multi-stage ESD network, the resistor element can also reduce the voltage at the signal pin node by formation of a "resistor divider" with an ESD shunt element.

The integration of buffering can also be placed in sophisticated design methodologies to address impedance matching and ESD protection (Figure 1.25). A high-performance design methodology is to impedance match all identical pins independent of the wiring to the specific pin. In an "array I/O" architecture, each pin may be located in a different location. Hence, it is a design methodology to place a resistor that is adjustable so that the sum of the wire interconnect resistance and the series resistance is a constant. In this fashion, the impedance of all identical pins is the same. This design methodology is also satisfactory for ESD development as well, since it will reduce the ESD variation based on pin location.

1.11 ESD DESIGN CONCEPTS – BALLASTING – INTER-DEVICE

In the ESD design synthesis techniques, ballasting is utilized to both limit and distribute the ESD current within a circuit [7]. In a circuit with parallel circuit elements, ballasting is utilized to prevent a single element undergoing over-voltage or thermal runaway. By limiting the current flow through a given element, the current will re-distribute through the parallel elements.

Ballasting is utilized in both functional circuits and ESD networks [7]. In semiconductor technology, ballasting is a common practice within output circuits and power circuits. Ballasting techniques are used in an output circuit that has multiple parallel elements to improve the current distribution through the elements. Ballasting is used in output circuits, such as off-chip driver (OCD) elements. Ballasting techniques can be used in both bipolar transistors and MOSFET transistors. Ballasting is also utilized in LDMOS technology, to improve the power distribution in large power devices [14,26–30].

Ballasting is also utilized within an ESD network. Inter-device ballasting is commonly used in the following ESD networks:

- Grounded gate n-channel MOSFET networks.
- Grounded base bipolar collector-to-emitter networks.

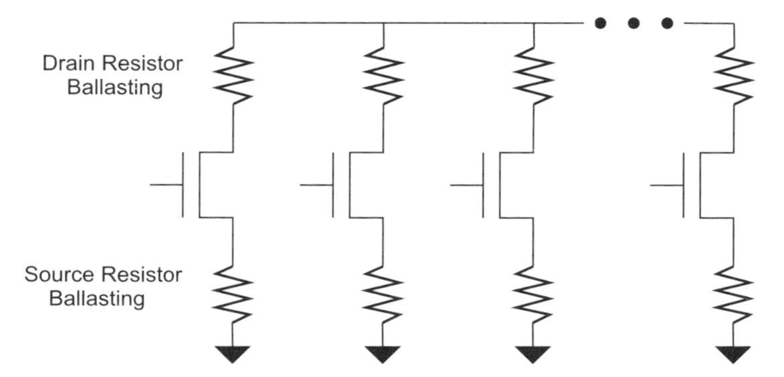

Figure 1.26 Multi-finger MOSFET with resistor ballasting

Ballasting is achieved using series resistor elements. These resistor elements can be independent resistor elements, or integrated with the bipolar or MOSFET element. Resistor elements used for ballasting are as follows:

- N-well resistors.
- Buried resistor (BR) elements.
- N-diffusion resistor MOSFET source/drain with silicide block mask.
- P-diffusion resistor MOSFET source/drain with silicide block mask.
- Polysilicon resistor.
- Bipolar base resistor.

Figure 1.26 shows an example of a MOSFET circuit with introduction of ballasting. Ballast resistors are introduced in series with the MOSFET element. The ballast resistor can be placed in series with the MOSFET drain or source region. It is common to introduce the MOSFET ballast resistor in the drain region.

Figure 1.27 shows an example of a bipolar circuit with the introduction of ballasting [8]. Ballast resistors are introduced in series with the bipolar element. The ballast resistor can be

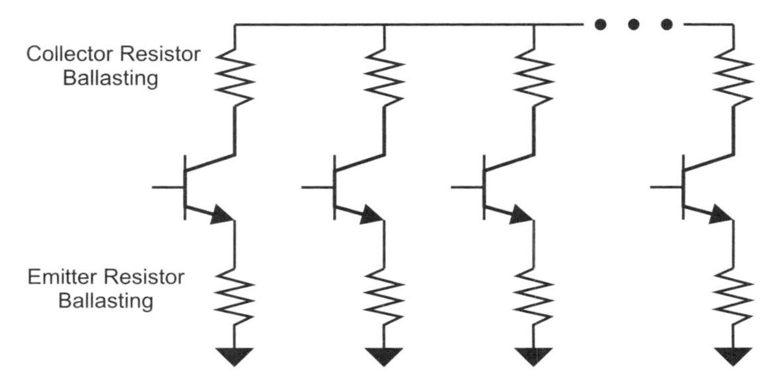

Figure 1.27 Bipolar circuit with resistor ballasting

placed in series with the bipolar base, collector, or emitter region. It is common to introduce the bipolar ballast resistor in the emitter region to provide ballasting between the bipolar collector and emitter in off-chip driver circuitry.

1.12 ESD DESIGN CONCEPTS – BALLASTING – INTRA-DEVICE

In the ESD design synthesis techniques, ballasting can be introduced into a single device element to improve current distribution along the width of the physical structure [7,8]. Ballasting is utilized to both limit and distribute the ESD current within a circuit element. Within a single device element, ballasting can be achieved through semiconductor process features or physical layout.

In a circuit with parallel circuit elements, ballasting is utilized to prevent a single element undergoing over-voltage or thermal runaway. Ballasting within a single element prevents electro-current constriction within a given physical element and re-distributes the current spatially within that physical element.

Ballasting can be introduced by semiconductor process solutions. Physical regions which can lead to improvement in ballasting are as follows:

- N-well sheet resistance.
- Buried-layer sheet resistance.
- Sub-collector sheet resistance.

Lateral resistance can be improved by the introduction of silicide blocking masks. Silicide blocking masks can be used as follows:

- MOSFET source/drain region.
- Bipolar extrinsic base region.

1.13 ESD DESIGN CONCEPTS – DISTRIBUTED LOAD TECHNIQUES

In semiconductor chip design, a concern of circuit designs is the loading capacitance of an ESD circuit and its impact on circuit performance on signal pins. Most ESD elements contain silicon junctions, leading to an increase in the capacitance on a signal pin. An ESD design practice of using a distributed ESD structure leads to a reduction in the loading effect on the signal pins [8]. This can be achieved using a multi-stage ESD network; this ESD design practice can be implemented in both high-speed digital and RF applications. Using series resistor elements, the ESD network can form a resistor–capacitor (RC) transmission line. Using a series inductor, the ESD network can form an inductor–capacitor (LC) transmission line. Figure 1.28 provides an example of a multi-stage network. Some examples of multi-stage ESD networks are as follows:

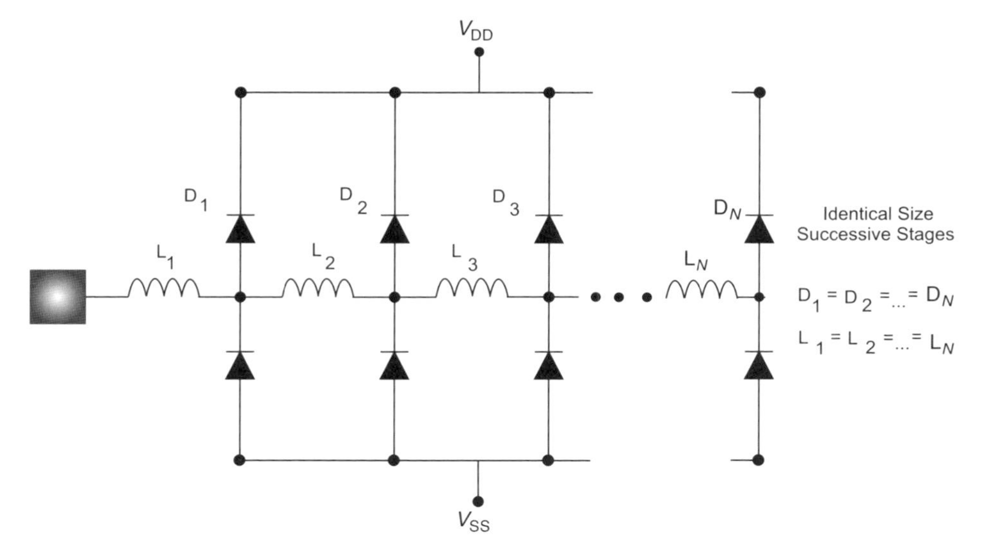

Figure 1.28 ESD distributed load techniques

- Primary dual-diode network, series resistor, and secondary dual diode network, ad infinitum.
- Primary dual-diode network, series inductor, and secondary dual-diode network, ad infinitum.
- Grounded gate MOSFET, series inductor, grounded gate MOSFET, ad infinitum.

1.14 ESD DESIGN CONCEPTS – DUMMY CIRCUITS

An ESD design discipline technique is the usage of auxiliary circuits for improved ESD robustness [7,8,11]. Auxiliary circuits, or "dummy circuits," can be used in ESD design for the following applications:

- De-coupling from signal pins.
- Disabling of feedback networks on signal pins.
- Disabling of well and tub structures.
- De-coupling from power supply rails.
- Disabling of feedback networks connected to power rails.
- Connections to "floating" structures.
- Impedance matching to an active circuit.

Auxiliary circuits can be used to de-couple elements or sub-circuits from a signal pin during electrical overstress or ESD events. De-coupling can be achieved through resistor elements, capacitor elements, or active ESD dummy networks. In addition, dummy circuits can disable or de-couple feedback networks that can be "pinned" during ESD stress.

Auxiliary circuits can be used to de-couple elements or sub-circuits from a power rail during electrical overstress or ESD events. De-coupling can be achieved through passive elements or active ESD dummy networks.

Dummy circuits can be used to connect to "floating" structures that can undergo ESD stress. Bond wire pads, metal structures, MOSFET gate structures, and wafer substrates are left "floating" in some semiconductor applications that lead to ESD failures. An ESD design practice is to connect these structures to passive elements, active circuits, or ESD networks to avoid electrical overstress.

Dummy circuits are also used to provide "impedance matching" between parallel circuit elements. This ESD design discipline practice is used in OCD circuitry.

1.15 ESD DESIGN CONCEPTS – POWER SUPPLY DE-COUPLING

An ESD design concept is the usage of a means to de-couple the ESD power supply [7,8,11]. This is achieved by de-coupling of elements in the ESD current path. Circuit elements can be introduced which lead to the avoidance of current flow to those physical elements. The addition of "ESD de-coupling switches" can be used to de-couple sensitive circuits as well as to avoid the current flow to these networks or sections of a semiconductor chip. ESD de-coupling elements can be used to allow elements to undergo "open" or "floating" states during ESD events. This can be achieved within the ESD network, or within the architecture of a semiconductor chip.

De-coupling of sensitive elements or de-coupling of current loops can be initiated by the addition of elements that allow the current loop to "open" during ESD events [7]. During ESD testing, power rails and ground rails are set as references. The de-coupling of nodes, elements, or current loops relative to the grounded reference prevents over-voltage states in devices, and eliminates current paths. These de-coupling elements can avoid "pinning" of electrical nodes. Hence, integration of devices, circuit elements, or circuit functions that introduce de-coupling of electrical connections to ground references, and power supply references, is a key unique ESD design practice.

During ESD testing, it is a requirement to ground every independent power supply reference. As a result, unanticipated current paths are created to the reference ground power supply rail, which can lead to ESD failures.

1.16 ESD DESIGN CONCEPTS – FEEDBACK LOOP DE-COUPLING

During ESD testing, feedback can lead to unanticipated ESD failures. An ESD design concept is the usage of a means to de-couple feedback networks [7,8,11]. Feedback loops can lead to unique ESD failures and lower ESD results significantly. The de-coupling of nodes, elements, or current loops relative to the grounded reference prevents over-voltage states in devices, and eliminates current paths initiated by the feedback elements. These de-coupling elements can avoid "pinning" of electrical nodes. Hence, integration of devices, circuit elements, or circuit functions that introduce de-coupling of electrical connections to ground references, and power supply references of the feedback elements during ESD testing, is also a key unique ESD design practice.

1.17 ESD LAYOUT AND FLOORPLAN-RELATED CONCEPTS

1.17.1 Design Symmetry

Design symmetry is an ESD design practice to maximize the ESD robustness [45]. Design symmetry is also a design practice of analog and power technologies [31–38]. The capability of the ESD network to dissipate high-current pulse events is directly related to the network's topology and its design symmetry. The more uniform the current distribution is through the ESD network during a discharge, the better the utilization of the area of the structure, and as a consequence, the greater the robustness of the circuit design.

The distribution of current during an ESD event is dependent upon the design symmetry of the ESD network and its components.

From experience, to the degree that the design of the ESD network (or structure) on all levels of the integrated circuit departs from a symmetric configuration, the greater is the current localized or non-uniformities in the ESD network. With a symmetrical distribution of the current, the peak power-to-failure per unit area is lowered, producing superior results. Additionally, the more uniform the current distribution, the more uniform the thermal field as well. Since semiconductor element electrical and thermal parameters are temperature dependent (e.g., mobility, electrical conductivity, thermal conductivity), the more uniform the current distribution, the more symmetrical the temperature distribution within the device.

In integrated circuit design, a key ESD design concept is to maintain a high degree of design symmetry within a structure on all design levels. In both the ESD network and I/O driver circuit, an evaluation of the power distribution of an ESD event within the circuit is an indicator of the robustness of the integrated circuit. Hence, physical layout design symmetry can be used as a heuristic determination of the power distribution within a physical structure [45].

To evaluate ESD design symmetry, this can be done visually in a design review, or through means of an automated computer aided design (CAD) tool [45].

Integrated circuits are produced on a uniform substrate, which is the subject of numerous mask operations. The masks create, from lines and shapes, individual devices on the layers of the integrated circuits. Hence, the mask physical layout features can be used to quantify the ESD design symmetry. This can be done on each of the layout design levels of the ESD structure.

To define ESD design symmetry, an axis of symmetry can be defined in the ESD design. Semiconductor design layout is two-dimensional, allowing us to define an axis of symmetry in the x- and y-direction. In this fashion, "moments" can be defined about the axis of symmetry as a means of quantifying the degree of symmetry, and identify non-symmetric features.

Before manufacturing the integrated circuit in silicon, the data file which defines the lines and shapes of each mask is available for evaluating the design to be implemented in silicon. In this methodology, a method can define the symmetry which evaluates on a level-by-level basis (Figure 1.29).

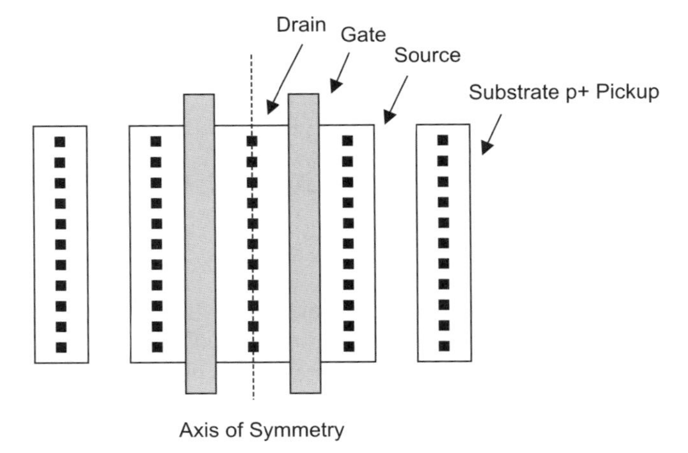

Figure 1.29 Design symmetry

In this design methodology, the method provides for evaluating the degree of design symmetry of the proposed semiconductor device, by considering various topological features of the design such as the directional flow of current into and out of the device, circuit element design symmetry, metal and contact symmetry, and other design features which reduce the robustness of an ESD protection network.

The semiconductor design can be "checked" before implementing in silicon by evaluating each ESD shape. In the event that any level fails the check, the level can be redesigned before implementing in silicon.

1.17.2 Design Segmentation

In the ESD design synthesis techniques, a form of ballasting is known as segmentation. Segmentation in essence is a means of ballasting where a device is "cut" into multiple parallel devices or elements [6–8]. The segmentation process may be initiated with a single element, or multiple elements, and then "cut" into more elements. The goal of the segmentation process is to improve current distribution along the width of the physical structure or multiple structures by physical separation using design layout techniques, which does not significantly increase the design area.

Segmentation can be utilized in any region of the semiconductor device that improves the current uniformity in the structure leading to high ESD robustness. In a MOSFET structure, segmentation can be introduced in the following regions:

- Introduction of isolation in the MOSFET source or drain regions.

In a silicon controlled rectifier, segmentation can be introduced in the following regions:

- Introduction of isolation in the anode region.
- Introduction of isolation in the cathode region.
- Introduction of masking in the sub-collector, body, well, and tub regions.

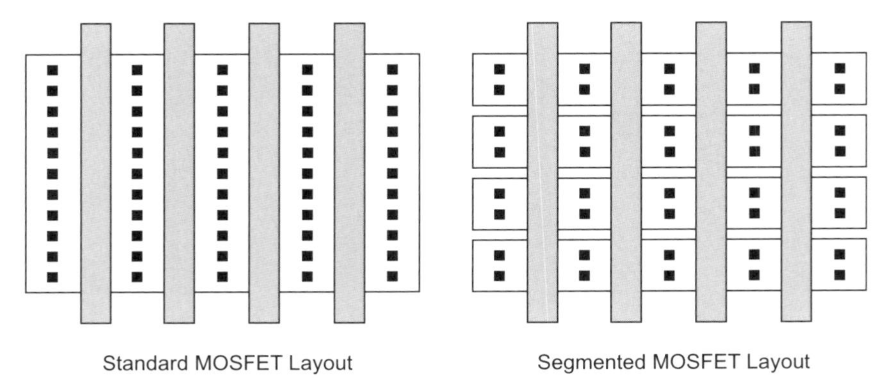

Figure 1.30 Multi-finger MOSFET structure with and without introduction of segmentation

In a bipolar transistor, segmentation can be introduced in the following regions:

- Introduction of isolation in the emitter.
- Introduction of isolation and separation in the collector or sub-collector regions.

Segmentation can be introduced in a single-finger MOSFET or multi-finger MOSFET. In a single-finger MOSFET, the segmentation will improve the current uniformity along the MOSFET width.

Figure 1.30 shows an example of a multiple-finger MOSFET structure with and without introduction of segmentation. The first device layout has a continuous MOSFET source and drain region. In the second device layout, the MOSFET source and drain regions were separated into four segments; this is an example of segmentation of a multiple-finger element.

In essence, the concept of segmentation is the reduction of a device into multiple devices to improve the ESD robustness. In order for this ESD design methodology to be successful, it is important to maintain layout symmetry.

1.17.3 ESD Design Concepts – Utilization of Empty Space

In ESD design synthesis, sections of the semiconductor chip area have no physical structures in the area. These “white space” regions exist in the semiconductor chips. White space regions in semiconductor chip design occur in the following areas:

- Area in the corners of the semiconductor chip.
- Area between the bond pads.
- Area under the signal bond pads.

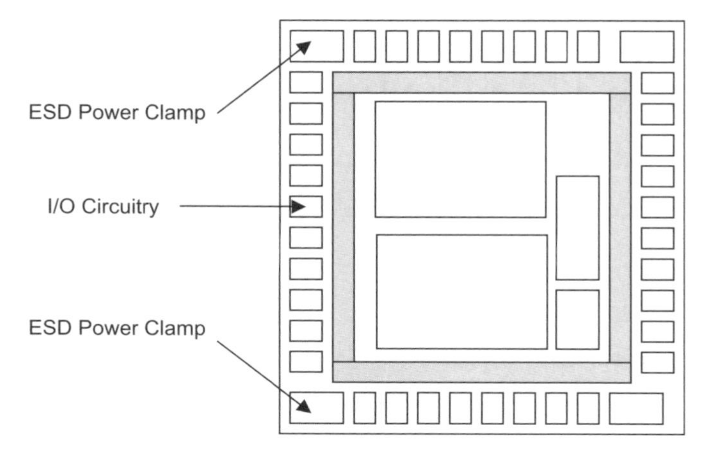

Figure 1.31 Utilization of semiconductor chip corners for ESD protection circuitry

- Area under power bond pads.
- Area under buses or "wiring bays."
- Area between power domains.

It is common practice in ESD design synthesis to utilize the area not used for functionality. Utilization of the area can be used for ESD protection in the following fashion:

- Placement of ESD power clamps in the corners of the semiconductor chip.
- Placement of ESD signal pad ESD elements or ESD power clamps in the area between the bond pads.
- Placement of the ESD signal pad ESD elements or ESD power clamps under the bond pads.
- Placement of de-coupling capacitors to increase chip capacitance.
- Placement of "back-up" ESD designs or experimental ESD designs in the unused area.

Figure 1.31 shows the usage of the semiconductor chip corner for ESD protection. This is a common practice in semiconductor design synthesis.

1.17.4 ESD Design Synthesis – Across Chip Line Width Variation (ACLV)

In semiconductor development, semiconductor process variation can introduce structural and dimensional non-uniformity. Photolithography and etch tools can introduce these non-uniformities that exist on a local and global design level. These variations can manifest themselves by introducing variations in both active and passive elements. For MOSFET transistors, variation in the MOSFET channel length in single-finger and multiple-finger MOSFET layouts can lead to non-uniform "turn-on"; this effect can influence both active functional circuits and ESD networks [46]. In bipolar transistors, the line width variation can lead to different sizes in emitter structures, leading to non-uniform current distribution in multi-finger bipolar transistors. For resistor elements, resistor elements that are utilized for ballasting in multi-finger structures can also lead to non-uniform current in the different fingers in the structure.

Design factors that influence the lack of variation are the following semiconductor process and design variables:

- Line width.
- Line-to-line space.
- "Nested-to-isolated" ratio.
- Orientation.
- Physical spacing between identical circuits.

It is a circuit design practice and an ESD design synthesis practice to provide a line width which is well controlled. For line-to-line space, in an array of lines, the spacing is maintained to provide maximum matching between adjacent lines. For example, in a multi-finger MOSFET structure, the spacing between the polysilicon lines is equal, to provide the maximum matched characteristics.

Given any array of parallel lines, the characteristics of the "end" or edges of the array can have different characteristics from the other lines. In an array of lines, whereas one edge is adjacent to another line, the other edge is not; this leads to one line-to-line edge space appearing "nested" and the outside line-to-line edge space appearing "semi-infinite" or "isolated." To address the problem of poorly matched edge lines, the following semiconductor process and ESD design solutions are used:

- **Process:** Cancellation technique of photolithography and etch biases.
- **Design:** Use of dummy edge lines.
- **Circuit:** Use of "gate-driven" circuitry.

Orientation can also influence the line width of identical circuits, both locally and globally. An ESD design practice is to maintain the same x–y orientation of ESD circuits in a semiconductor chip to minimize variation pin-to-pin. This is not always possible in a peripheral architecture where the ESD element is rotated on the four edges of the semiconductor chip. Note that in this case, the circuit itself (e.g., OCD) may also undergo an orientation effect. It is a good ESD design synthesis practice that addresses the orientation issue with compensation and matching issues for orientation of the ESD elements (in conjunction with the circuit it is protecting).

On a macroscopic full-chip scale, variations in the photolithography and etching can vary from the top to the bottom of a semiconductor chip. In the design of a semiconductor chip, these can be compensated with a pre-knowledge of the photolithography and etch variations of a technology.

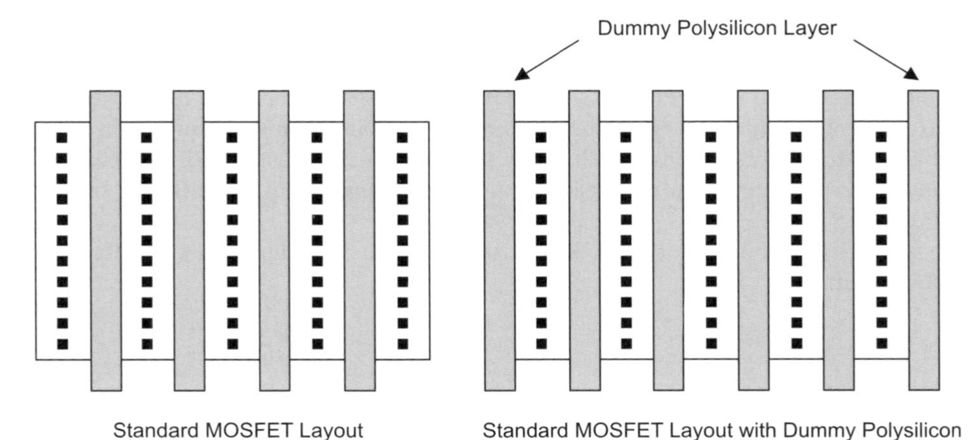

Figure 1.32 MOSFET with and without dummy shapes

1.17.5 ESD Design Concepts – Dummy Shapes

To improve uniformity in the current distribution within an output circuit (e.g., OCD network), or an ESD network, an ESD design concept is the introduction of "dummy shapes."

This technique is also utilized in advanced technologies (e.g., CMOS and bipolar), as well as circuit functions and applications that require a high degree of tolerance control or matching (e.g., analog, or high-speed digital circuits). Due to line width variations, there is a "nested" to "isolated" offset due to both photolithography and etch variations.

An ESD design concept is the introduction of "dummy shapes" adjacent to the ends of an array of active shapes. In a MOSFET structure, the polysilicon MOSFET gate line width influences the MOSFET "snapback." MOSFET "snapback" is a function of the MOSFET channel length. In a multi-finger structure, the line width variation within a given circuit can lead to non-uniform current distribution.

Figure 1.32 shows a multi-finger MOSFET structure with and without "dummy shapes." The dummy shapes are placed on the ends of the array at the same pitch as the active MOSFET fingers. With the introduction of the dummy fingers, the line width tolerance of the edge lines will improve, leading to an improved current distribution of the ESD current during MOSFET snapback.

1.17.6 ESD Design Concepts – Dummy Masks

In the ESD design synthesis, "dummy masks" are used to provide ESD design features desired in ESD design. Dummy masks used in ESD design synthesis are as follows:

- Introduction of masks to block silicide formation.
- Introduction of masks to block implants to produce non-uniform doping concentration.

Silicide "masking" can be introduced by the following processes:

- MOSFET gate structure.
- Removable MOSFET gate structure.
- Silicide block mask.
- Bipolar emitter structure masks.

MOSFET gate structures can be used in semiconductors for silicide blocking in the following devices and structures needed for ESD protection:

- Lateral "gated diodes" in bulk CMOS or SOI technologies.
- Lateral pnpn structures in bulk or SOI technology.
- Lateral buried resistor elements.

MOSFET gate structures can be removed after formation of the implants. In the above list of elements, the gate structure can be removed to avoid ESD dielectric failure mechanisms.

1.17.7 ESD Design Concepts – Adjacency

An ESD design concept is the issue of adjacency [31–33]. In the physical layout design of ESD structures, the adjacency of structures internal and external to the ESD element is a concern.

Structures adjacent to ESD structures can lead to both ESD failure mechanisms and latchup. Adjacent structures can lead to parasitic device elements not contained within the circuit schematics. Parasitic npn, pnp, and pnpn are not uncommon in the design of ESD structures. These parasitic bipolar transistor elements can occur between the ESD structure and the guard rings around the ESD structure. These parasitic elements can occur in single-well, dual-well, and triple-well CMOS, DMOS, bipolar, BiCMOS, and BCD technologies.

Figure 1.33 shows an example of two adjacent resistor elements for a differential pair circuit. For matching purposes, the resistor layouts are symmetrical about a common axis, and adjacent to each other. The figure shows a differential pair with input IN(+) and IN(−) for matching. In the case of ESD testing between the two signal pins, a lateral npn can form between the two n+ resistor elements at the input.

1.18 ESD DESIGN CONCEPTS – ANALOG CIRCUIT TECHNIQUES

In analog design, unique design practices are used to improve the functional characteristics of analog circuitry [14,31–33]. In the ESD design synthesis of analog circuitry, the ESD design practices must be suitable and consistent with the needs and requirements of analog circuitry. Fortunately, many of the analog design practices are aligned with ESD design practices.

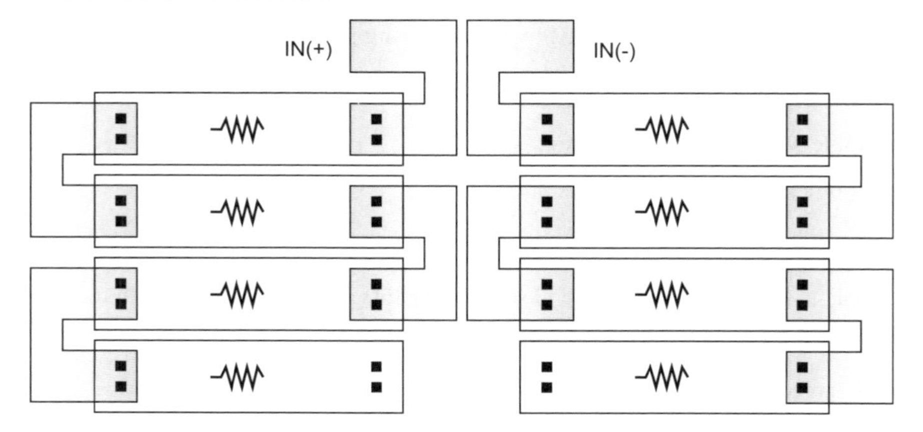

Figure 1.33 Adjacency in an RF differential pair

In the analog design discipline, there are many design techniques to improve tolerance of analog circuits. Analog design techniques include the following:

- **Local Matching:** Placement of elements close together for improved tolerance.
- **Global Matching:** Placement in the semiconductor die.
- **Thermal Symmetry:** Design symmetry.

A key analog circuit design requirement is matching [31–33]. To avoid semiconductor process variations, matching is optimized by the local placement. Placement within the die location is also an analog concern due to mechanical stress effects. In analog design, there is a concern over the temperature field within the die, and the effect of temperature distribution within the die.

Many of the analog design synthesis and practices are also good ESD design practices. The design practice of matching and design symmetry are also suitable practices for electrostatic discharge design. But, there are some design practices where a tradeoff exists between the analog tolerance and ESD; this occurs when parasitic devices are formed between the different analog elements within a given circuit, or circuit-to-circuit. Figure 1.33 demonstrates both design symmetry and matching characteristics that are suitable for RF or analog design implementations.

1.19 ESD DESIGN CONCEPTS – WIRE BONDS

In the ESD design synthesis, one must consider the packaging and wire bonds in the ESD design.

Wire bonds typically connect from the package to the bond pads. In multi-chip die, wire bonds can connect from one silicon die to another. The inductance of the wire bond is an ESD consideration. Wire bond inductance can influence the nature of the ESD event. In ESD events which have low resistance (e.g., MM events), the inductance of the wire bond influences the impinging current waveform.

An ESD design concept is the usage of the wire bond for ESD design co-synthesis. An example of this in RF applications can serve as a matching filter, and co-integrated with ESD structures. ESD networks in RF applications integrate both capacitor and inductor elements for ESD–RF co-synthesis or as ESD networks. For RF ESD design, the wire bond inductor lead can be integrated into a T-match network comprising a first inductor, a capacitive element, and a second inductor.

1.20 DESIGN RULES

In the ESD design, a critical part of the ESD business strategy is to have ESD checking and verification methods. ESD checking and verification can evaluate both physical layout, schematic, and electrical resistances; for a successful whole-chip implementation, it is key to evaluate both layout and schematic as well as electrical parameters.

1.20.1 ESD Design Rule Checking (DRC)

As part of the process of ESD design synthesis, the semiconductor chip design layout and schematics are checked to be in conformance with the technology rules, as well as the specific ESD design rules. ESD design rule checking (DRC) is used to evaluate the layout of circuits, and ESD circuits. The ESD DRC rule sets are not standardized across the semiconductor industry. ESD DRC rule sets can evaluate the following:

- ESD network type.
- Special ESD design layers.
- ESD element layout dimensional parameters (e.g., width, length, spacing).
- Interconnects (e.g., wire and via).
- Guard rings.

ESD DRC rules are evaluated in the submission process of a semiconductor design.

1.20.2 ESD Layout vs. Schematic (LVS)

As part of the process of ESD design synthesis, the semiconductor chip design schematics are checked to be in conformance with the technology, as well as the specific ESD design rules. ESD layout vs. schematic (LVS) design checking and verification is used to evaluate the ESD schematics and circuits. ESD LVS rule sets are not standardized across the semiconductor industry. ESD LVS rule sets can evaluate the following:

- Existence of an ESD network on a given signal or power pin.
- ESD network type.

- ESD network connectivity.
- Circuit-to-ESD network compatibility.
- ESD element layout design parameters (e.g., width, length, perimeter, area).

ESD LVS rules are evaluated in the submission process of a semiconductor design.

1.20.3 Electrical Resistance Checking (ERC)

Evaluation of resistance is critical for a full-chip ESD implemenation. As part of the process of ESD design synthesis, evaluation of resistance in the alternative current loop is key to success. Electrical resistance checking (ERC) plays a key role in ESD design synthesis. ERC evaluation is used for the following:

- Bond pad to ESD series resistance.
- Signal pin series resistance.
- ESD signal pin to power rail resistance.
- ESD power rail resistance.
- ESD power rail of signal pin to the nearest ESD power clamp.
- Power rail to ESD power clamp resistance.

1.21 SUMMARY AND CLOSING COMMENTS

In this chapter, ESD concepts are introduced from layout, circuits, to design rule checking. A "sampler" of concepts is laid out for the reader, to begin viewing the ESD design synthesis from a broader perspective. ESD design synthesis extends from the smallest contact, to full-chip integration. With this awareness, it is possible to realize the extent of the ESD design discipline in semiconductor design.

In Chapter 2, the topic of ESD architecture and floorplan concepts will be highlighted. The chapter focus is on "peripheral I/O" and "array I/O" architectures, and how they influence the placement of the various elements for the whole-chip design integration. Chapter 2 also focuses on native-voltage, mixed-voltage, and mixed-signal chip integration, from digital, analog, to RF applications.

PROBLEMS

1.1. In mapping from a 5 V CMOS technology to a 3.3 V technology, what adjustments must be made in the ESD protection strategy for the ESD input circuit and a voltage-triggered ESD power clamp when the ESD input network is a grounded gate MOSFET, and the voltage-triggered MOSFET clamp contains a forward biased trigger circuit? What is the change in the power bus or ground rail (if any)?

1.2. In mapping from a 5 V CMOS technology to a 3.3 V technology, what adjustments must be made in the ESD protection strategy for the ESD input circuit and an RC-triggered ESD power clamp when the ESD input network is a series of p+/n-well diodes to V_{DD}, and n+ diffusion to ground?

1.3. Draw the SOA for a bipolar transistor. Explain each domain on the *I–V* plot. What are the limits?

1.4. Draw the SOA for a MOSFET transistor. Explain each domain on the *I–V* plot. What are the limits?

1.5. Explain the difference between the electrical safe operating area (E-SOA) and the thermal safe-operating area (T-SOA). How do these regions compare to the understanding of the functional regime of circuits? How do these relate to the ESD results? How do these relate to latent failures and hard failures?

1.6. Why is symmetry important in ESD networks? How does symmetry play a role in voltage distribution? Current distribution? Power distribution?

1.7. Why is adjacency important for matching in analog applications? Why is adjacency a problem for ESD protection? Draw matched elements of a differential receiver and draw the parasitic elements between the devices.

1.8. Draw a semiconductor chip that is 20 mm by 20 mm. Assume the bond pads for the circuits are 100 μm × 100 μm and 50 μm spacing between bond pads. Assuming a mosaic (or array) of bond pads, what is the maximum number of bond pads that can be placed on the semiconductor chip. Generalize an equation for the number of bond pads and the chip size relationship.

1.9. Draw a semiconductor chip that is 20 mm by 20 mm. Assume the bond pads for the circuits are 100 μm × 100 μm and 50 μm spacing between bond pads. Assuming the bond pads are on the perimeter, what is the maximum number of bond pads that can be placed on the semiconductor chip? Generalize an equation for the number of bond pads and the chip size relationship. If the ESD devices for each bond pad are as large as the bond pad, what is the percentage of chip area needed for ESD protection?

1.10. Draw a semiconductor chip that is 20 mm by 20 mm. Assume the bond pads for the circuits are 100 μm × 100 μm and 50 μm spacing between bond pads. Assuming a mosaic (or array) of bond pads, what is the maximum number of bond pads that can be placed on the semiconductor chip? Generalize an equation for the number of bond pads and the chip size relationship. If the ESD devices for each bond pad are as large as the bond pad, what is the percentage of chip area needed for ESD protection?

REFERENCES

1. S. Dabral and T. J. Maloney. *Basic ESD and I/O Design*. Chichester, UK: John Wiley and Sons, Ltd, 1998.
2. A.Z. H. Wang, *On Chip ESD Protection for Integrated Circuits*. New York: Kluwer Academic Publications, 2002.

3. A. Amerasekera and C. Duvvury. *ESD in Silicon Integrated Circuits*. 2nd Edition. Chichester, UK: John Wiley and Sons, Ltd, 2002.
4. J. Vinson, J. Bernier, G. Croft, and J.J. Liou. *ESD Design and Analysis Handbook*. New York: Kluwer Academic Publications, 2002.
5. H. Gossner, K. Esmark, and W. Stadler. *Advanced Simulation Methods for ESD Protection Development*. Elsevier Science, Amsterdam, 2003.
6. S. Voldman. *ESD: Physics and Devices*. Chichester, UK: John Wiley and Sons, Ltd., 2004.
7. S. Voldman. *ESD: Circuits and Devices*. Chichester, UK: John Wiley and Sons, Ltd., 2005.
8. S. Voldman. *ESD: RF Circuits and Technology*. Chichester, UK: John Wiley and Sons, Ltd., 2006.
9. S. Voldman. *Latchup*. Chichester, UK: John Wiley and Sons, Ltd., 2007.
10. S. Voldman. *ESD: Circuits and Devices*. Beijing, China: Publishing House of Electronic Industry (PHEI), 2008.
11. S. Voldman. *ESD: Failure Mechanisms and Models*. Chichester, UK: John Wiley and Sons, Ltd., 2009.
12. M.D. Ker and S.F. Hsu. *Transient Induced Latchup in CMOS Integrated Circuits*. Singapore: John Wiley and Sons, (Asia) Pte Ltd., 2009.
13. M. Mardiquan. Electrostatic Discharge: *Understand, Simulate, and Fix ESD Problems*. New York: John Wiley and Sons, Inc., 2009.
14. V. Vashchenko and A. Shibkov. *ESD Design in Analog Circuits*. New York: Springer, 2010.
15. D. M. Tasca. Pulse power failure modes in semiconductors. *IEEE Transactions on Nuclear Science*, **NS-17**, (6), December 1970; 346–372.
16. D.C. Wunsch and R.R. Bell. Determination Of threshold voltage levels of semiconductor diodes and transistors due to pulsed voltages. *IEEE Transactions on Nuclear Science*, **NS-15**, (6), December 1968; 244–259.
17. J.S. Smith and W.R. Littau. Prediction of thin-film resistor burn-out. *Proceedings of the Electrical Overstress/Electrostatic Discharge (EOS/ESD) Symposium*, 1981; 192–197.
18. M. Ash. Semiconductor junction non-linear failure power thresholds: Wunsch–Bell revisited. *Proceedings of the Electrical Overstress/Electrostatic Discharge (EOS/ESD) Symposium*, 1983; 122–127.
19. V.I. Arkihpov, E. R. Astvatsaturyan, V.I. Godovosyn, and A.I. Rudenko, *Plasma accelerator with closed electron drift. International Journal of Electronics*, **55**, 1983; 135–145.
20. V. A. Vlasov and V. F. Sinkevitch, Elektronnaya Technika, No. 4, 1971; 68–75.
21. V. M. Dwyer, A. J. Franklin, and D.S. Campbell. Thermal failure in semiconductor devices. *Solid State Electronics*, **33**, 1989; 553–560.
22. W.D. Brown. Semiconductor device degradation by high amplitude current pulses. *IEEE Transactions on Nuclear Science*, **NS-19**, December 1972; 68–75.
23. D.R. Alexander and E.W. Enlow. Predicting lower bounds on failure power distributions of silicon npn transistors. *IEEE Transactions on Nuclear Science*, **NS-28**, (6), Dec. 1981; 4305–4310.
24. E.N. Enlow. Determining an emitter-base failure threshold density of npn transistors. *Proceedings of the Electrical Overstress/Electrostatic Discharge (EOS/ESD) Symposium*, 1981; 145–150.
25. D. Pierce and R. Mason. A probabilistic estimator for bounding transistor emitter-base junction transient-induced failures. *Proceedings of the Electrical Overstress/Electrostatic Discharge (EOS/ESD) Symposium*, 1982; 82–90.
26. P. Antognetti. *Power Integrated Circuits: Physics, Design, and Applications*. New York: McGraw-Hill, 1986.
27. P.L. Hower and P.K. Govil. Comparison of one- and two-dimensional models of transistor thermal instability. *IEEE Transactions of Electron Devices*, **ED-21**, (10), 1974; 617–623.
28. P. Hower, J.Lin, S. Haynie, S. Paiva, R. Shaw, and N. Hepfinger, Safe operating area considerations in LDMOS transistor. *International Symposium on Power Semiconductors and ICs (ISPSD),* 1999; 55–84.

29. P. Hower and S. Pendeharker, Short and long-term safe operating area considerations in LDMOS transistors. *Proceedings of the International Reliability Physics Symposium (IRPS)*, 2005; 545–550.
30. J.J. Liou, S. Malobabic, D.F. Ellis, J.A. Salcedo, J.J. Hajar, and Y.Z. Zhou. Transient safe operating area (TSOA) definition for ESD applications. *Proceedings of the Electrical Overstress/Electrostatic Discharge (EOS/ESD) Symposium*, 2009; 17–27.
31. P. Gray, Hurst, Lewis, and Meyer. *Analysis and Design of Analog Integrated Circuits. Fifth Edition*. New York: John Wiley and Sons, Inc., 2009.
32. W.M.C. Sansen. *Analog Design Essentials*. Netherlands: Springer, 2006.
33. A. Hastings. *The Art of Analog Layout*. Second Edition. New Jersey: Pearson Prentice Hall, 2006.
34. S. K. Ghandi. *Semiconductor Power Devices*, New York: John Wiley and Sons, Inc., 1977.
35. P. Antognetti. *Power Integrated Circuits: Physics, Design, and Applications*. New York: McGraw-Hill, 1986.
36. B. J. Baliga. *High Voltage Integrated Circuits*. New York: IEEE Press, 1988.
37. B.J. Baliga. *Modern Power Devices*. New York: John Wiley and Sons, Inc. 1987.
38. V. Vashchenko, M. Ter Beek, W. Kindt, P. Hopper. ESD protection of high voltage tolerant pins in low voltage BiCMOS processes. *Proceedings of the Bipolar Circuits Technology Meeting (BCTM)*, 2004; 277–280.
39. S. Voldman. The state of the art of electrostatic discharge protection: Physics, technology, circuits, designs, simulation and scaling. *Invited Talk, Bipolar/BiCMOS Circuits and Technology Meeting (BCTM) Symposium*, 1998; 19–31.
40. R. Singh, D. Harame, and M. Oprysko, *Silicon Germanium: Technology, Modeling and Design*, New York, John Wiley and Sons, Inc., 2004.
41. S. Voldman. The impact of MOSFET technology evolution and scaling on electrostatic discharge protection. *Journal of Microelectronics Reliability*, 1998, **38**; 1649–1668.
42. S. Voldman. The impact of technology evolution and scaling on electrostatic discharge (ESD) protection in high-pin-count high-performance microprocessors. *Proceedings of the International Solid State Circuits Conference (ISSCC)*, Session 21, WA 21.4, February 1999; 366–367.
43. S. Voldman. Electrostatic discharge (ESD) protection in silicon-on-insulator (SOI) CMOS technology with aluminum and copper interconnects in advanced microprocessor semiconductor chips. *Proceedings of the Electrical Overstress/Electrostatic Discharge (EOS/ESD) Symposium*, 1999; 105–115.
44. S. Voldman, D. Hui, L. Warriner, D. Young, R. Williams, J. Howard, V. Gross, W. Rausch, E. Leobangdung, M. Sherony, N. Rohrer, C. Akrout, F. Assaderaghi, and G. Shahidi. Electrostatic discharge protection in silicon-on-insulator technology. *Proceedings of the IEEE International Silicon on Insulator (SOI) Conference*, 1999; 68–72.
45. S. Voldman, Method for evaluating circuit design for ESD electrostatic discharge robustness. U.S. Patent No. 6,526,548 Feb 25th 2003.
46. S. Voldman, J. Never, S. Holmes, and J. Adkisson. Linewidth control effects on MOSFET ESD robustness. *Proceedings of the Electrical Overstress/Electrostatic Discharge (EOS/ESD) Symposium*, 1996; 101–109.
47. ANSI/ESD ESD-STM 5.1 – 2007. ESD Association Standard Test Method for the Protection of Electrostatic Discharge Sensitive Items – Electrostatic Discharge Sensitivity Testing – Human Body Model (HBM) Testing Component Level. Standard Test Method (STM) document, 2007.
48. ANSI/ESD SP 5.1.2-2006 ESD Association Standard Practice for the Protection of Electrostatic Discharge Sensitive Items – Human Body Model (HBM) and Machine Model (MM) Alternative Test Method: Split Signal Pin – Component Level, 2006.
49. T. Meuse, R. Barrett, D. Bennett, M. Hopkins, J. Leiserson, J. Schichl, L. Ting, R. Cline, C. Duvvury, H. Kunz, and R. Steinhoff. Formation and suppression of a newly discovered secondary EOS event in HBM test systems. *Proceedings of the Electrical Overstress/Electrostatic Discharge (EOS/ESD) Symposium*, 2004; 141–145.

50. R.A. Ashton, B. E. Weir, G. Weiss, and T. Meuse. Voltages before and after HBM stress and their effect on dynamically triggered power supply clamps. *Proceedings of the Electrical Overstress/Electrostatic Discharge (EOS/ESD) Symposium*, 2004; 153–159.
51. J. Barth and J. Richner. Voltages before and after current in HBM testers and Real HBM. *Proceedings of the Electrical Overstress/Electrostatic Discharge (EOS/ESD) Symposium*, 2005; 141–151.
52. R. Gaertner, R. Aburano, T. Brodbeck, H. Gossner, J. Schaafhausen, W. Stadler, F. Zaengl. Partitioned HBM test – A new method to perform HBM test on complex tests. *Proceedings of the Electrical Overstress/Electrostatic Discharge (EOS/ESD) Symposium*, 2005; 178–183.
53. T. Brodbeck and R. Gaertner. Experience in HBM ESD testing of high pin count devices. *Proceedings of the Electrical Overstress/Electrostatic Discharge (EOS/ESD) Symposium*, 2005; 184–189.
54. ANSI/ESD ESD-STM 5.2 – 1999. ESD Association Standard Test Method for the Protection of Electrostatic Discharge Sensitive Items – Electrostatic Discharge Sensitivity Testing – Machine Model (MM) Testing Component Level. Standard Test Method (STM) document, 1999.
55. ANSI/ESD ESD-STM 5.3.1 – 1999. ESD Association Standard Test Method for the Protection of Electrostatic Discharge Sensitive Items – Electrostatic Discharge Sensitivity Testing – Charged Device Model (CDM) Testing Component Level. Standard Test Method (STM) document, 1999.
56. Intel Corporation. Cable discharge event in local area network environment. White Paper, Order No: 249812-001, July 2001.
57. R. Brooks. A simple model for the cable discharge event. *IEEE802.3 Cable-Discharge Ad-hoc Committee*, March 2001.
58. Telecommunications Industry Association (TIA). Category 6 Cabling: Static discharge between LAN cabling and data terminal equipment, *Category 6 Consortium*, December 2002.
59. J. Deatherage, D. Jones. Multiple factors trigger discharge events in Ethernet LANs. *Electronic Design*, Vol. 48, (25), 2000; 111–116.
60. W. Stadler, T. Brodbeck, R. Gartner, and H. Gossner. Cable discharges into communication interfaces. *Proceedings of the Electrical Overstress/Electrostatic Discharge (EOS/ESD) Symposium*, 2006; 144–151.
61. ESD Association. DSP 14.1-2003. ESD Association Standard Practice for the Protection of Electrostatic Discharge Sensitive Items – System Level Electrostatic Discharge Simulator Verification Standard Practice. Standard Practice (SP) document, 2003.
62. ESD Association. DSP 14.3-2006. ESD Association Standard Practice for the Protection of Electrostatic Discharge Sensitive Items – System Level Cable Discharge Measurements Standard Practice. Standard Practice (SP) document, 2006.
63. ESD Association. DSP 14.4-2007. ESD Association Standard Practice for the Protection of Electrostatic Discharge Sensitive Items – System Level Cable Discharge Test Standard Practice. Standard Practice (SP) document, 2007.
64. H. Geski. DVI compliant ESD protection to IEC 61000-4-2 level 4 standard. *Conformity*, September 2004; 12–17.
65. S. Voldman, R. Ashton, J. Barth, D. Bennett, J. Bernier, M. Chaine, J. Daughton, E. Grund, M. Farris, H. Gieser, L. G. Henry, M. Hopkins, H. Hyatt, M.I. Natarajan, P. Juliano, T. J. Maloney, B. McCaffrey, L. Ting, and E. Worley. Standardization of the transmission line pulse (TLP) methodology for electrostatic discharge (ESD). *Proceedings of the Electrical Overstress/Electrostatic Discharge (EOS/ESD) Symposium*, 2003; 372–381.
66. ANSI/ESD Association. ESD-SP 5.5.1-2004. ESD Association Standard Practice for the Protection of Electrostatic Discharge Sensitive Items – Electrostatic Discharge Sensitivity Testing – Transmission Line Pulse (TLP) Testing Component Level. Standard Practice (SP) document, 2004.
67. ANSI/ESD Association. ESD-STM 5.5.1-2008. ESD Association Standard Test Method for the Protection of Electrostatic Discharge Sensitive Items – Electrostatic Discharge Sensitivity Testing –

Transmission Line Pulse (TLP) Testing Component Level. Standard Test Method (STM) document, 2008.

68. ANSI/ESD STM5.5.1-2008 Electrostatic Discharge Sensitivity Testing – Transmission Line Pulse (TLP) – Component Level, 2008.
69. ANSI/ESD STM5.5.2-2007, Electrostatic Discharge Sensitivity Testing – Very Fast Transmission Line Pulse (VF-TLP) – Component Level, 2007.
70. ESD Association. ESD-SP 5.5.2. ESD Association Standard Practice for the Protection of Electrostatic Discharge Sensitive Items – Electrostatic Discharge Sensitivity Testing Very Fast Transmission Line Pulse (VF-TLP) Testing Component Level. Standard Practice (SP) document, 2007.
71. ANSI/ESD Association. ESD-SP 5.5.2-2007. ESD Association Standard Practice for the Protection of Electrostatic Discharge Sensitive Items – Electrostatic Discharge Sensitivity Testing – Very Fast Transmission Line Pulse (VF-TLP) Testing Component Level. Standard Practice (SP) document, 2007.
72. ESD Association. ESD-STM 5.5.2. ESD Association Standard Test Method for the Protection of Electrostatic Discharge Sensitive Items – Electrostatic Discharge Sensitivity Testing Very Fast Transmission Line Pulse (VF-TLP) Testing Component Level. Standard Test Method (STM) document, 2009.
73. K. Muhonen, R. Ashton, J. Barth, M. Chaine, H. Gieser, E. Grund, L.G. Henry, T. Meuse, N. Peachey, T. Prass, W. Stadler, and S. Voldman. VF-TLP round robin study, analysis, and results. *Proceedings of the Electrical Overstress/Electrostatic Discharge (EOS/ESD) Symposium,* 2008; 40–49.
74. ANSI/ESD Association. ESD-STM 5.5.1-2008. ESD Association Standard Test Method for the Protection of Electrostatic Discharge Sensitive Items – Electrostatic Discharge Sensitivity Testing – Very Fast Transmission Line Pulse (VF-TLP) Testing Component Level. Standard Practice (SP) document, 2009.
75. T.W. Chen, C. Ito, T. Maloney, W. Loh, and R.W. Dutton. Gate oxide reliability characterization in the 100 ps regime with ultra-fast transmission line pulsing system. *Proceedings of the Electrical Overstress/Electrostatic Discharge (EOS/ESD) Symposium,* 2007; 16–21.
76. International Electro-technical Commission (IEC). IEC 61000-4-2 Electromagnetic Compatibility (EMC): Testing and Measurement Techniques – Electrostatic Discharge Immunity Test, 2001.
77. E. Grund, K. Muhonen, and N. Peachey. Delivering IEC 61000-4-2 current pulses through transmission lines at 100 and 330 ohm system impedances. *Proceedings of the Electrical Overstress/Electrostatic Discharge (EOS/ESD) Symposium,* 2008; 132–141.
78. IEC 61000-4-2 Electromagnetic Compatibility (EMC) – Part 4-2:Testing and Measurement Techniques – Electrostatic Discharge Immunity Test, 2008.
79. R. Chundru, D. Pommerenke, K. Wang, T. Van Doren, F.P. Centola, J.S. Huang. Characterization of human metal ESD reference discharge event and correlation of generator parameters to failure levels – Part I: Reference Event. *IEEE Transactions on Electromagnetic Compatibility*, **46** (4), November 2004; 498–504.
80. K. Wang, D. Pommerenke, R. Chundru, T. Van Doren, F.P. Centola, and J.S. Huang. Characterization of human metal ESD reference discharge event and correlation of generator parameters to failure levels – Part II: Correlation of generator parameters to failure levels. *IEEE Transactions on Electromagnetic Compatibility*, **46** (4), November 2004; 505–511.
81. ESD Association. ESD-SP 5.6-2008. ESD Association Standard Practice for the Protection of Electrostatic Discharge Sensitive Items – Electrostatic Discharge Sensitivity Testing – Human Metal Model (HMM) Testing Component Level. Standard Practice (SP) document, 2008.
82. ANSI/ESD SP5.6-2009 Electrostatic Discharge Sensitivity Testing – Human Metal Model (HMM) – Component Level, 2009.

2 ESD Architecture and Floorplanning

2.1 ESD DESIGN FLOORPLAN

One of the most fundamental issues and challenges in the ESD design discipline is ESD architecture and floorplanning. The integration of the devices, circuits, sub-functions, and cores is critical to the success of EOS and ESD robust designs. In each application space, the floorplan and layout of a semiconductor chip may be different, leading to unique challenges for ESD protection design [1–6]. Whether it is DRAM [10–12], SRAM [1–3,6], NVRAM, microprocessor [1–4,12–20], CPU, ASICs [1–6], or semiconductor foundry, each has unique challenges for ESD design. Whether it is single voltage, mixed voltage or mixed signal, the ESD design strategy and architecture has to be modified. Additionally, CMOS, BiCMOS, and BCD technology produce digital [1–20], analog [21–24], power [25–29], and RF [30–32] applications with integration, layout, and design.

In this text, one of the goals is to teach how to construct a semiconductor chip to achieve an ESD robust implementation. Significant focus in publications address semiconductor device physics and ESD circuits, but the subject of how to integrate all the elements into a given product has limited exposure.

The ordering of materials in this text is constructed in the fashion that a semiconductor chip is assembled. In this chapter, we will begin by discussing the architecture and layout floorplanning for different chip architectures. The discussion will address both peripheral and "array" I/O configurations. For peripheral I/O architectures, architectures with aligned I/O to staggered I/O will be shown. For array I/O configurations, implementations of off-chip driver "banks," nibble architectures, to single I/O circuits in the chip will be discussed. In the chapter, issues of power bus architectures will also be shown. Mixed-voltage implementations, to mixed-signal architectures, and global floorplanning will be discussed [2–4]. In future chapters, each chapter will go into greater depth for the ESD circuits, the power bus, the guard rings, and chip integration.

ESD: Design and Synthesis, First Edition. Steven H. Voldman.

In semiconductor chip design, the circuit design team follows an application performance and specification objectives. The ESD specifications and objectives are part of the original design definition for the product. The core functions are defined, and the interface circuitry between the semiconductor chip and outside is also addressed. Once these objectives are established, the placement of the cores and the interface circuitry are defined. As part of the floorplanning process, the I/O interface circuitry placement and power bus are decided. At this time, the chapter will deviate to discuss the options for I/O placement, and how the ESD elements and power are integrated into the I/O floorplan.

2.2 PERIPHERAL I/O DESIGN

In semiconductor chip design, there are two fundamental classes of I/O configurations. These two different architectures have a significant effect on the ESD design architectures. We will refer to them as "peripheral I/O" and "array I/O" architectures [2,3].

In the "peripheral I/O" architecture, the signal and power bond pads are placed on the edge of the semiconductor chip. The primary reason is that, in the majority of cases, the I/O circuitry must interface with other chips, a system, or the outside environment. The packages use wire bonds between the silicon chip and the package. This decision is dependent on the cost, the package size, and the pin count of the semiconductor chip.

Figure 2.1 shows an example of the layout and floorplan for a peripheral I/O architecture. The width of the I/O design and spacing between the I/O cells is limited by the bond pad size,

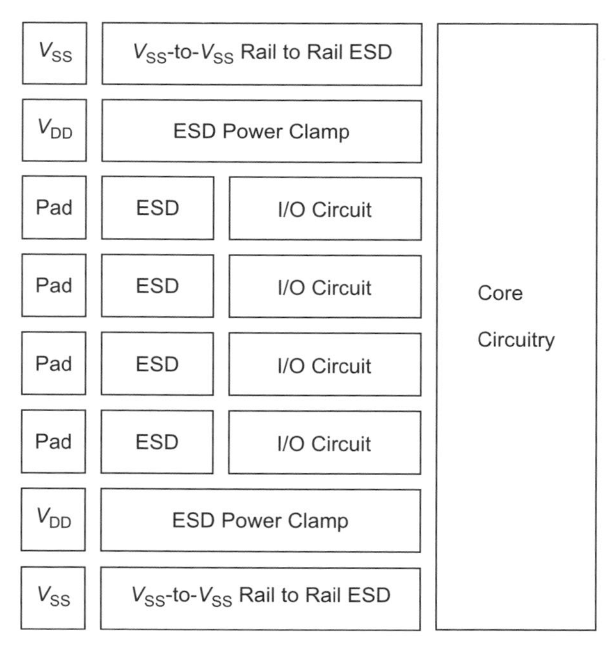

Figure 2.1 Peripheral I/O design

the bond pad design rules, and wire bond rules. In a "peripheral I/O limited design," the maximum number of I/O standard cells and power cells are defined according to the allowed rules of the technology and package, and technology definition rules.

In the peripheral I/O architecture, ESD elements are placed with the circuit in the custom or standard I/O cell. The peripheral I/O architecture must contain the input, output, and bi-directional circuitry, as well as service functions, and power pads. The ESD networks are inherently integrated with the I/O cells, and the power pads.

2.2.1 Pad-Limited Peripheral I/O Design Architecture

In semiconductor chips, the majority of products and applications encountered are designs with peripheral I/O circuitry. Product application types that are typically peripheral I/O can vary from memory, logic, digital, analog, mixed signal, mixed voltage, and power applications [1–30]. Some applications can not have bond pads, wiring, and other circuits in the center of the chip, such as CMOS imaging applications. In other applications, there is a limit on how far one can wire bond into a semiconductor chip. There are different cases of peripheral I/O chips of interest – one is the pad-limited peripheral I/O architecture.

In peripheral I/O design, the wire bonds and packaging requirements limit the number of I/O placed on the periphery of the semiconductor chip. Limitations to the number of I/O on the periphery of a chip are as follows:

- Bond pad dimensions (width and length).
- Bond pad to bond pad spacing.
- Bond pad to active circuitry requirements.

Figure 2.2 shows an example of a pad-limited peripheral I/O design architecture. From an ESD design synthesis perspective, the bond pad dimensions influence the following:

- Placement of the ESD network.
- Form factor of the ESD network.
- Electrical connections between the bond pad and the ESD network.
- Electrical connections between the ESD network and the power rails.
- Guard ring placement.
- ESD network interaction with I/O circuitry.
- Latchup considerations for the ESD network and adjacent elements.

In peripheral pad I/O design, ESD elements can be placed adjacent to the bond pad, or under the bond pad.

Figure 2.2 Pad-limited peripheral I/O design

2.2.2 Pad-Limited Peripheral I/O Design Architecture – Staggered I/O

In peripheral I/O design, the wire bonds and packaging requirements limit the number of I/O placed on the periphery of the semiconductor chip. To increase the number of I/O cells, one solution is peripheral I/O staggering of the cells. Figure 2.3 shows an example of a peripheral I/O design architecture with staggered I/O. With the staggering of the I/O cells, the bond wire to bond wire spacing is increased, allowing an increase in the number of I/O cells. This allows for the reduction of the pad size, and the pitch of the I/O circuitry. The I/O circuit standard cell is compensated by changing its aspect ratio, and the placement of the power buses.

In the staggered I/O configuration, ESD design synthesis and latchup considerations must address the following:

- Placement of the input location relative to the ESD element.
- Guard ring resistance between the power connection and the worst case resistance point of the guard ring.
- Standard cell to standard cell latchup (PFET of the first standard cell, and NFET of the second standard cell).

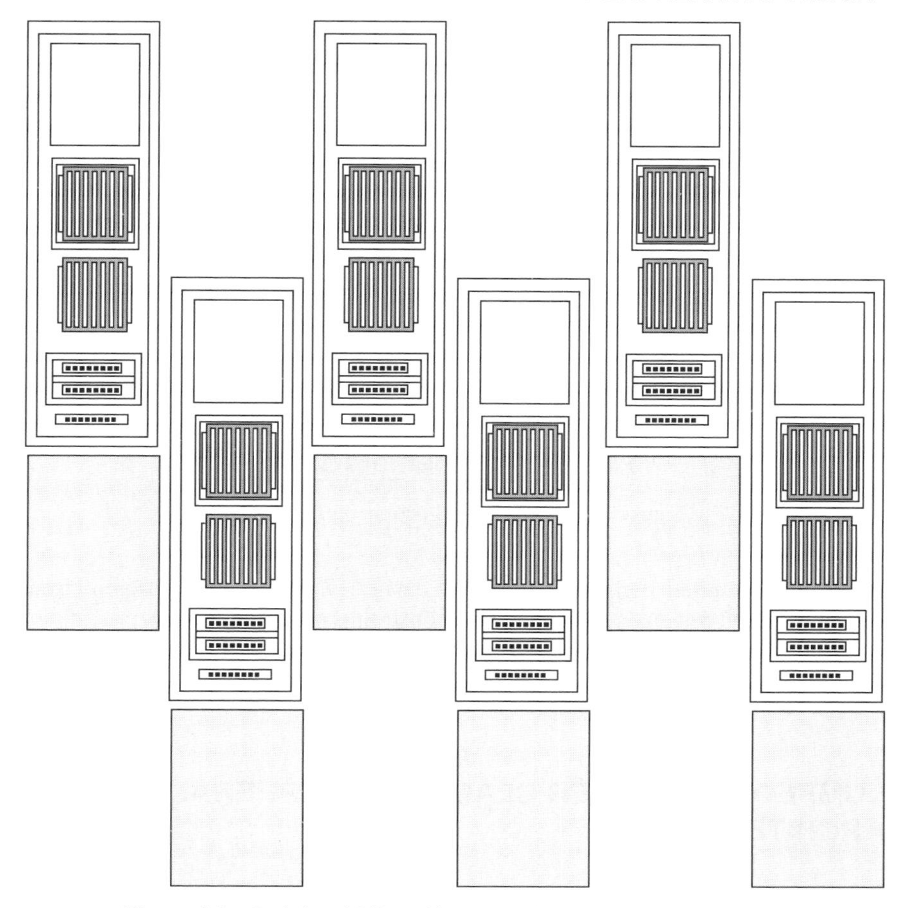

Figure 2.3 Peripheral I/O architecture with staggered I/O circuitry

2.2.3 Core-Limited Peripheral I/O Design Architecture

In semiconductor chips, some products and applications encountered are core-limited designs with peripheral I/O circuitry. Product application types that are typically peripheral I/O with core-limited applications are low pin-count chips that require large arrays, or large devices. These can be CMOS image processing chips, memory, mixed signal, mixed voltage, and power applications. For example, in a CMOS image processing chip, it can not have bond pads, wiring, and other circuits in the center of the chip; it is also desirable to have the size of the array as large as possible. In power electronics, there are applications with only a few pins, but large devices. In these architectures, the area on the periphery is not constrained to bond pad dimensions, bond pad to bond pad spacing, and other technology limits. They provide significant area in the periphery for support networks, logic, de-coupling capacitors, and ESD networks. Figure 2.4 shows an example of a core-limited peripheral I/O design

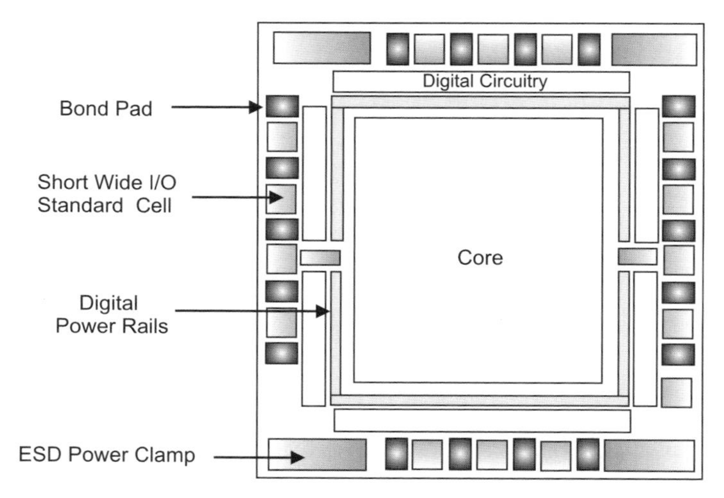

Figure 2.4 Core-limited peripheral I/O design

architecture. Core-limited peripheral I/O ESD design is significantly different from pad-limited peripheral I/O design due to many of the constraints not being present, providing design freedom in the architecture, floorplan, orientation, size, form factor, and footprint of the ESD networks.

2.3 LUMPED ESD POWER CLAMP IN PERIPHERAL I/O DESIGN ARCHITECTURE

In ESD design synthesis of a peripheral I/O architecture, ESD power clamps are placed in the "pad ring" with the signal bond pads, power bond pads, and power bus. The whole-chip ESD performance is dependent on how the ESD power clamps are distributed along the power rail. ESD power clamps can be integrated into the individual signal pad "cell," or more commonly placed as a "lumped" element in the semiconductor chip. In this section, the integration of lumped ESD power clamps is discussed.

2.3.1 Lumped ESD Power Clamp in Peripheral I/O Design Architecture in the Semiconductor Chip Corners

In ESD design synthesis, ESD power clamps are placed in the "pad ring" with the signal bond pads, power bond pads, and power buses. In many chip designs, the corners of the semiconductor chip are not utilized. The reasons for not using the corners are as follows:

- Restrictions on placing signal pins in the corner.
- Mechanical stress on the chip corners influencing the circuitry.

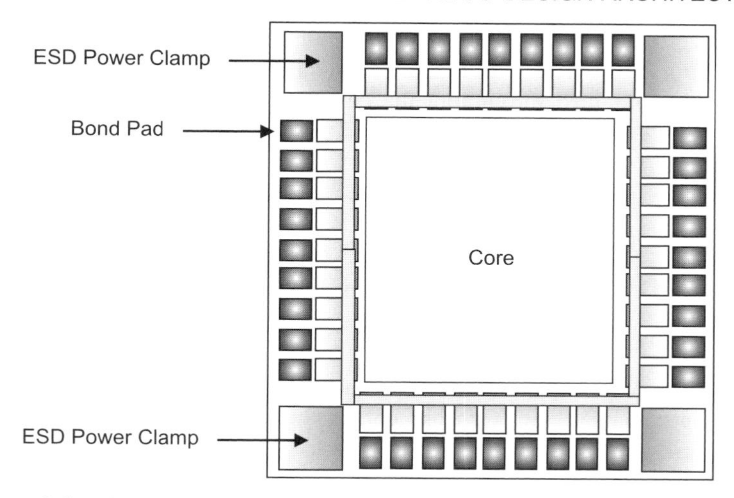

Figure 2.5 Floorplan with ESD power clamps in the semiconductor chip corners

- Photolithography control on the chip corners.
- Placement of identification markings.
- Corners are "white space" regions existing in the semiconductor chips.

It is common practice in ESD design synthesis to utilize this corner area for placement of the ESD power clamps between the power and ground circuitry. Figure 2.5 shows an example of a semiconductor chip with placement of the ESD power clamps in the corners. This ESD design synthesis is very common in very small semiconductor chips, below 5 mm × 5 mm. Placement of the ESD elements in the corners of large microprocessors or ASIC implementations is more limited due to the series resistance of the power bus.

2.3.2 Lumped ESD Power Clamp in Peripheral I/O Design Architecture – Power Pads

In peripheral I/O design, in very large semiconductor chips, or small semiconductor chips that require high ESD robustness, the ESD power clamps are placed at a higher spatial frequency. A natural placement of the ESD power clamps is in the peripheral "standard cell" regions where V_{DD} or V_{SS} power pins are required. In some ASICs, microprocessor or standard cell foundry methodologies, it is a requirement to place V_{DD} and V_{SS} power pins at a given frequency for a given number of I/O cells. For example, in some methodologies, it is a requirement to place a "power pin" adjacent to every fifth I/O standard cell. Placement of the ESD power clamps within the "power cell" or "power book" allows for the local placement of ESD networks within a given periodicity of every I/O signal pin. Additionally, the placement of the ESD power clamps can naturally be integrated into the design methodology as part of the power pin frequency requirements. In this system, the complete ESD power clamp network is contained throughout the periphery of the semiconductor chip design, in a given periodicity.

2.4 LUMPED ESD POWER CLAMP IN PERIPHERAL I/O DESIGN ARCHITECTURE – MASTER/SLAVE ESD POWER CLAMP SYSTEM

In the previous section, the complete ESD power clamp is placed in a given "power cell" or "power book" in a standard cell environment. In this methodology, it allows for the local placement of ESD networks within a given periodicity of every I/O signal pin. In this method, the initiation or "triggering" of each power clamp is independent of the ESD power clamp itself.

A different ESD methodology is to use a "master/slave" ESD design methodology. In a master/slave concept, a single ESD power clamp has the trigger or initiation circuit only in one ESD power clamp (e.g., the master ESD power clamp). Figure 2.6 shows a circuit schematic representation of the ESD master/slave implementation [24]. In the ESD network, a single trigger network is contained within the master ESD power clamp. The trigger network signal is sent to the slave ESD power clamp implementations.

Figure 2.7 shows an example floorplan of the master/slave system with the ESD power clamps in the corner of the chip. In the master/slave ESD system, the interconnection between the master trigger and the slave networks is shown. An advantage of the master/slave system is that a single trigger initiates the entire system. A second advantage is that the slave clamps can be placed at a high spatial frequency along the periphery. A disadvantage of the master/slave system is that the additional bus is required to transfer the signal from the master ESD power clamp to the slave clamps.

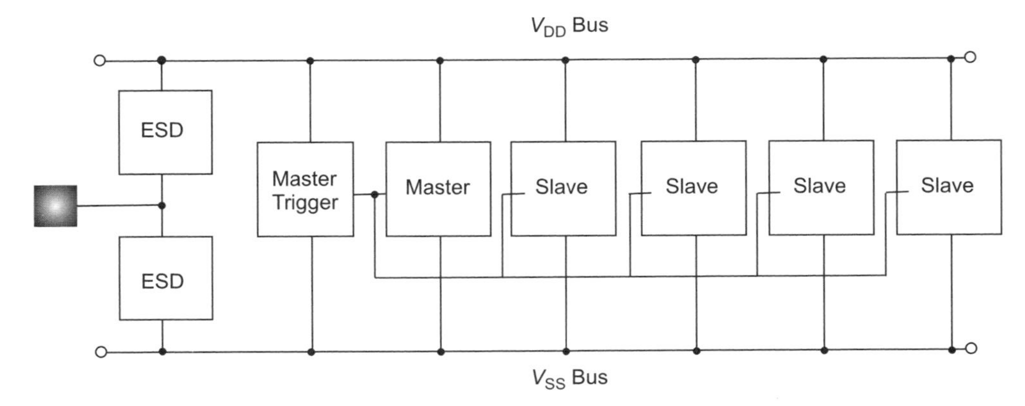

Figure 2.6 Master/slave ESD power clamps

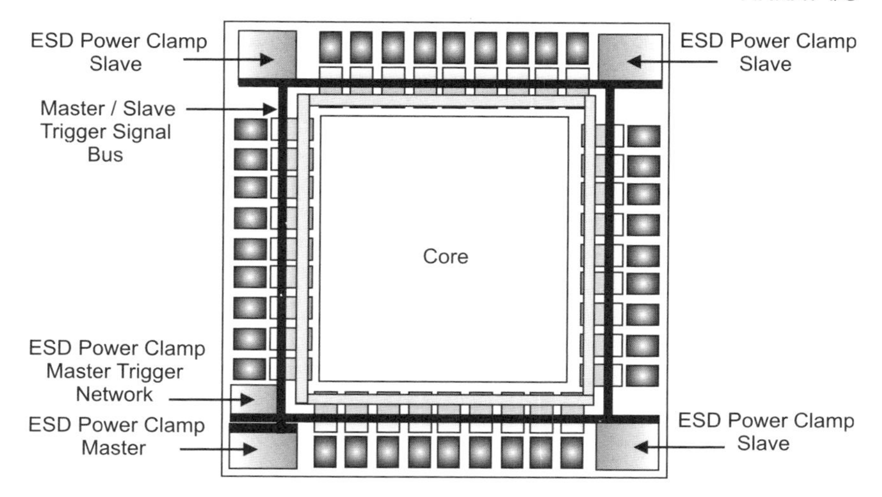

Figure 2.7 Master/slave ESD power clamps design layout and floorplan

2.5 ARRAY I/O

As the number of circuits increases on a semiconductor chip, the number of signal pins increases. For large semiconductor chips, floorplans are modified to "array I/O" where the inputs, outputs, and bi-directional circuitry is placed in the interior of a semiconductor chip. This architecture brings new challenges for ESD design synthesis.

In an array I/O semiconductor architecture, the package is designed in an array of solder balls equally spaced and arranged. For example, in a 1000-pin semiconductor chip, an array of 100×100 solder balls and bond pads is arranged on the top level. The metal layer below the bond pads is known as the "transfer wire" level, which is the method of connecting from the bond pads to the I/O circuitry in the semiconductor chip.

In this methodology, the bond pad is separated from the I/O circuitry and ESD network. In some cases, the bond pad is directly over the I/O cell; and in other cases, it is not. The I/O cell is placed in the "cores" surrounded by dense CMOS logic, and gate array circuitry (Figure 2.8). In this methodology, for impedance matching, the resistance in series must be matched for identical circuits. The design method addresses the matching by adjusting the series resistance using a resistor element, which compensates for the metal wire resistance. To have impedance matching of an off-chip driver (OCD), the impedance consists of the interconnect resistance, the corrective ballast resistance, and the OCD circuit impedance. The corrective ballast resistor would be adjusted to provide the matching. In most ESD architectures, the ESD ballast resistor is fixed; in this methodology, it is variable based on spatial placement of the I/O cell.

The ESD floorplan and issues are as follows:

- **Interconnects:** Metal interconnect width and via number limitations due to transfer wire.
- **Interconnects:** Wire width limitation for performance in high-speed input receivers.
- **Ballast Resistor:** Automated ballast resistor variations.

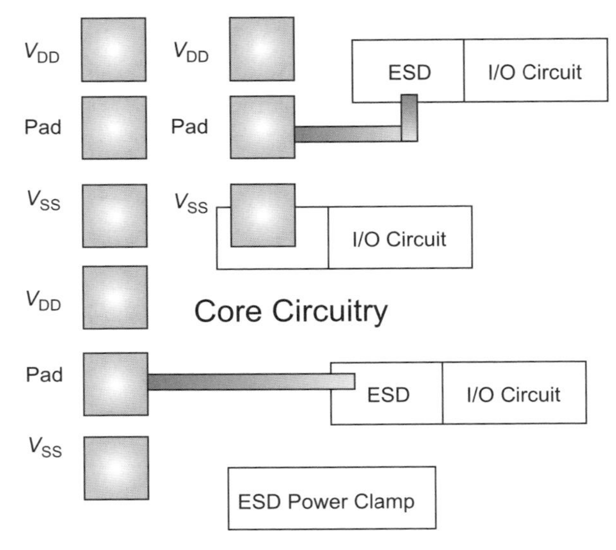

Figure 2.8 Array I/O architecture

- **ESD Power Clamp:** Placement of ESD power clamps.
- **Latchup:** Placement of I/O in dense logic.
- **I/O Footprint:** I/O cell aspect ratio may differ from the peripheral I/O standard cell.

2.5.1 Array I/O – Off-Chip Driver Banks

In the ESD design synthesis and floorplanning of semiconductor chips, for performance objectives, off-chip drivers are placed internally to the perimeter of the semiconductor chip. In many applications, it is efficient to have the off-chip drivers in "OCD banks" where groupings of four or eight circuits are placed internally to the semiconductor chip. Figure 2.9 shows an example of a semiconductor chip with internal OCD banks with service module.

With the placement of OCD banks within the chip perimeter, ESD protection issues occur in this architecture. Here are some of the issues associated with this architecture:

- **Local Power Bus:** Local power buses must be placed to support the OCD bank.
- **Local Service Module:** Local "service modules" are added to address needs of the OCD bank.
- **Wire Width:** Wire width is limited due to internal area constraints.
- **Guard Rings:** Guard rings must be placed to avoid interaction with adjacent circuitry or other circuit "cores" (e.g., memory, internal logic, de-coupling capacitors).

The following implementations are needed to ensure good ESD protection with this architecture:

- Low resistance of the local power rail in the OCD bank to avoid OCD functional and ESD roll-off effects at the ends of the local power rail.

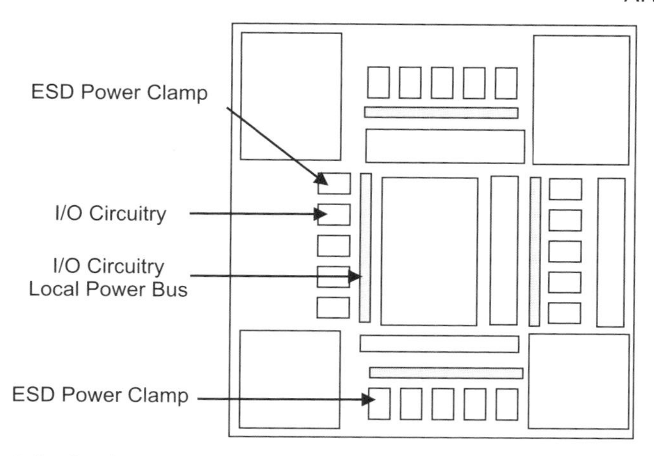

Figure 2.9 Semiconductor chip with internal OCD banks and service module

- Placement of an ESD power clamp network within the OCD bank "service module" between the local power bus and the local ground rail.
- ESD protection on each of the receiver and off-chip driver signal pins.
- Metal width and via number must be adequate between the bond pads and the signal pin ESD networks.
- An additional guard ring between the OCD circuitry, the ESD networks, and the service modules to avoid interaction with adjacent functional circuits.

Figure 2.10 is an example of an internal OCD bank architecture with ESD networks. One service module supports the internal OCD bank. This service module contains the ESD V_{DD}-to-V_{SS} power clamp.

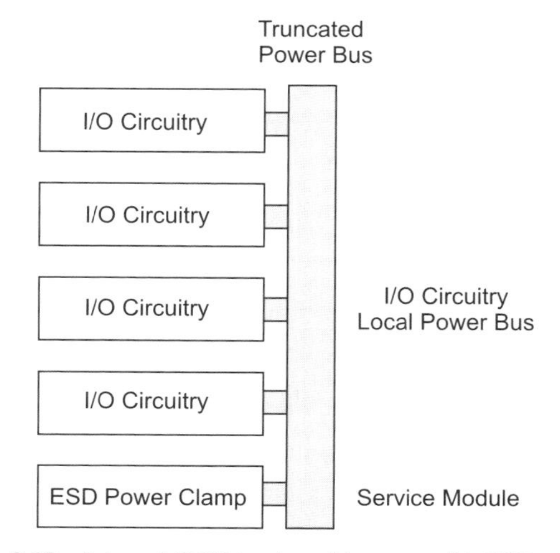

Figure 2.10 Internal OCD bank architecture with ESD networks

2.5.2 Array I/O Nibble Architecture

In the ESD design synthesis and floorplanning of semiconductor chips, for performance objectives, off-chip drivers are placed internally to the perimeter of the semiconductor chip. In many applications, it is efficient to have the off-chip drivers in small groupings of four I/O cells placed internally to the semiconductor chip. Figure 2.11 shows an example of a semiconductor chip with internal OCD circuits in a group of four, forming a "nibble."

With the placement of four I/O cells together within the chip perimeter, the following are some of the issues associated with this architecture:

- **Local Power Bus:** Local power buses must be placed to support the nibble grouping.
- **Interconnect Width:** Wire width is limited due to internal area constraints.
- **Guard Rings:** Guard rings must be placed to avoid interaction with adjacent circuitry or other circuit "cores" (e.g., memory, internal logic, de-coupling capacitors).

With a small nibble group, no service module is utilized. As a result, effective ESD design is established. In this example, the ESD solution utilized to provide good ESD protection incorporated the following:

- Sharing of ESD protection between the I/O cells.
- Sharing of ESD protection between two power rails.

With a small nibble group, elements can be shared locally to the nibble group for more compact design and improved ESD protection.

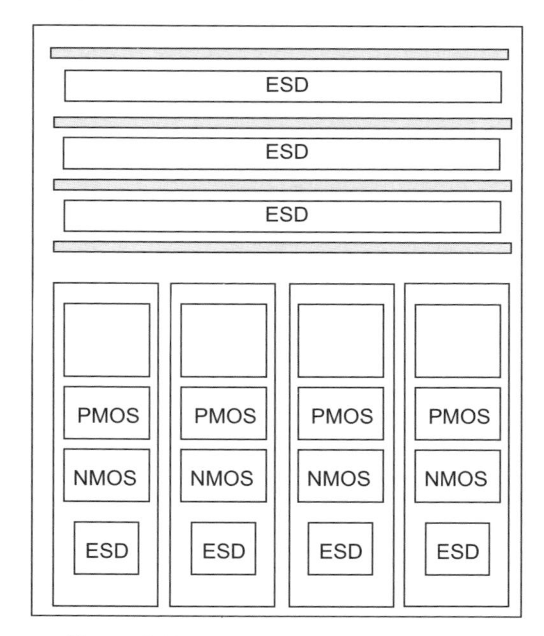

Figure 2.11 Array I/O nibble architecture

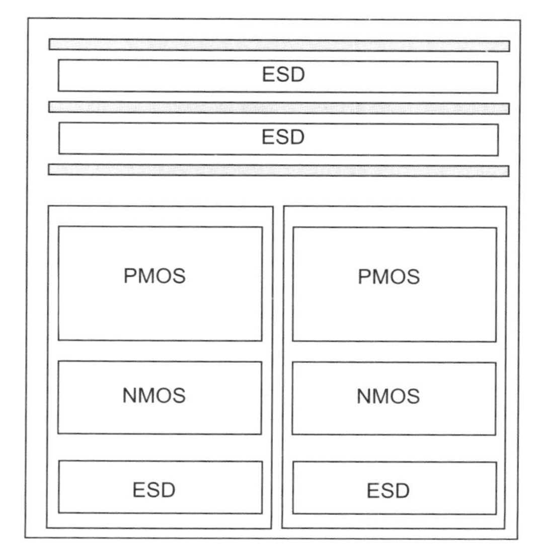

Figure 2.12 Array pair I/O architecture

2.5.3 Array I/O Pair Architecture

In some array architectures, instead of the individual external circuits being separated, they are placed in pairs. Figure 2.12 shows an example of an I/O pair placed within dense logic. In the case of array I/O pairs, the configuration can take advantage of sharing of guard rings, and wiring channels. I/O pair configurations can also be suitable for signals where matching is desired between two adjacent signals. I/O pairs are also suitable for differential receiver networks.

From an ESD perspective, all of the issues associated with array I/O are still present, with the additional issues of I/O to I/O interaction. ESD power clamps must be placed locally and/or embedded in the standard cell. One of the largest issues is the interaction of charge injection from the ESD elements to the adjacent circuitry and cores. Charge injection can disturb the surrounding circuitry or initiate CMOS latchup. Figure 2.13 shows an example of CMOS latchup in the adjacent circuitry near the array I/O pairs. Latchup was initiated by electron injection into the substrate by the ESD elements within the I/O cell [5].

2.5.4 Array I/O – Fully Distributed

In array I/O, the I/O can be arranged in groupings, or fully distributed. In a fully distributed array I/O, the ESD design synthesis must account for the spatial distribution of the I/O standard circuitry (Figure 2.14). The ESD floorplan and issues are as follows:

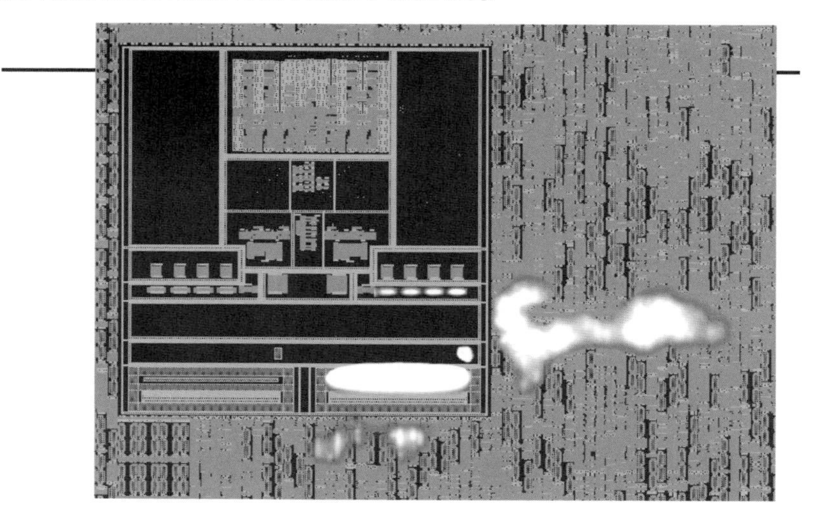

Figure 2.13 CMOS latchup in array pair I/O architecture

- **Interconnects:** Metal interconnect width and via number limitations due to transfer wire width.
- **Interconnects:** Wire width limitation for performance in high-speed input receivers.
- **Ballast Resistor:** Automated ballast resistor variations.

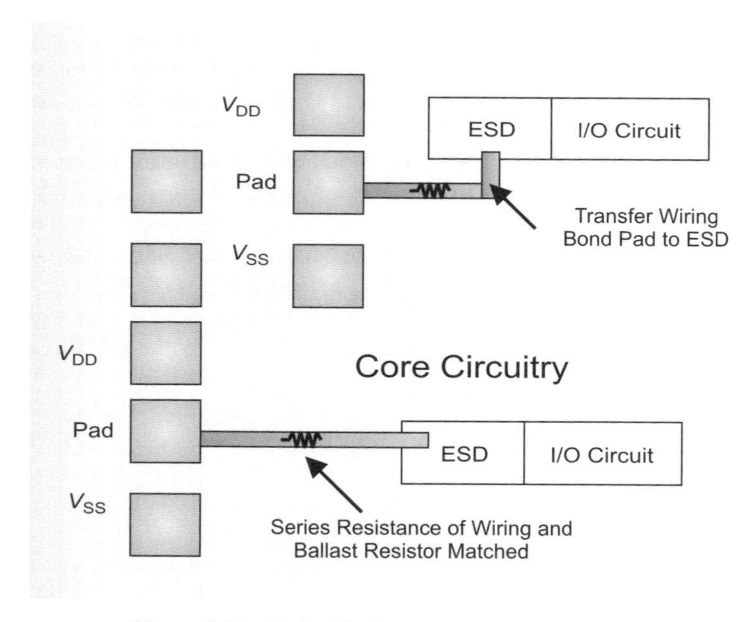

Figure 2.14 Fully distributed array I/O architecture

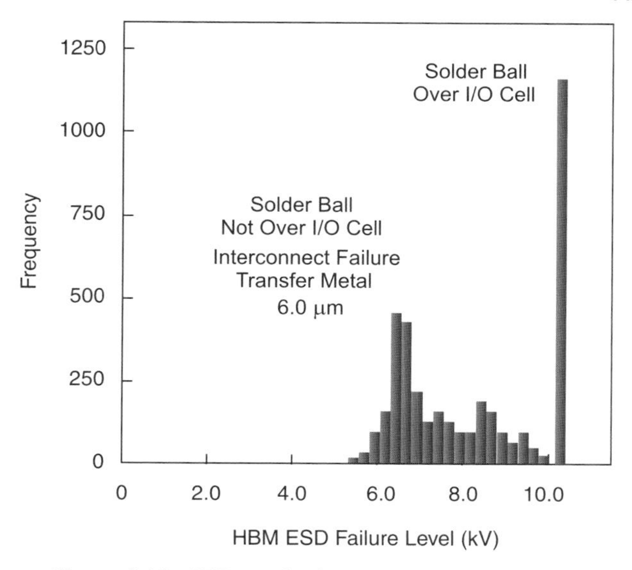

Figure 2.15 ESD results from array I/O architecture

- **ESD Power Clamp:** Placement of ESD power clamps.
- **Charge Injection:** Injection from ESD elements into adjacent dense logic.
- **Latchup:** Placement of I/O in dense logic can lead to external latchup [5].
- **I/O Footprint:** I/O cell aspect ratio may differ from the peripheral I/O standard cell.

Figure 2.15 shows an example of the ESD results from an array I/O semiconductor chip. In this technology generation, with array I/O, the ESD results were limited by the interconnects. The ESD results show two distributions of ESD failures. A Gaussian distribution of failures is observed at approximately 6.2 kV, and a second distribution is observed at 10 kV (e.g., ESD tester limitation) [1,2]. The first distribution has to do with the failure of the transfer wire metal of the ESD interconnects; the second distribution observes no failures. In the case where there was no transfer wire ESD failure, the bond pad and solder ball are placed directly over the array I/O cell. Hence, in this implementation, the transfer wire is the limitation of the array I/O. In the case of this technology, the metal layer was titanium/aluminum/titanium (Ti/Al/Ti), with a worst case failure of approximately 6000 V HBM. With the technology migration to copper interconnects, the semiconductor chip ESD results increased to 9000 V HBM levels. Since copper (Cu) interconnects have a higher critical current-to-failure, the ESD results increased.

Figure 2.16 shows a picture of the two cases of the solder ball and bond pad over the array I/O cell, and the case when the solder ball and bond pad were not placed over the array I/O cell.

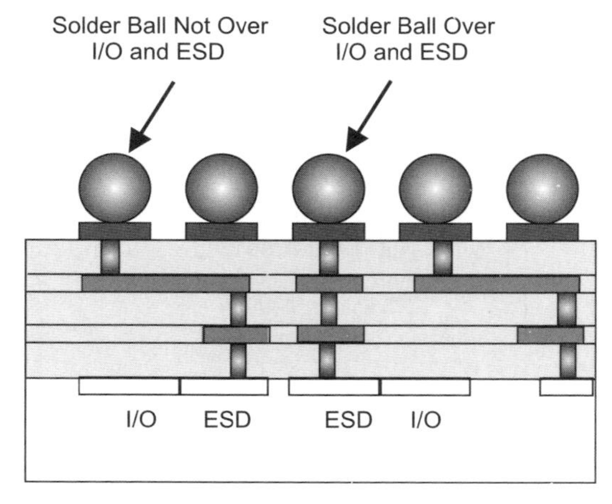

Figure 2.16 ESD failure distribution in array I/O architecture

In the case when the solder ball/bond pad structure was not over the array I/O cell, the failure distribution was associated with the "transfer wire" between the solder ball and the array I/O cell. In this case, the interconnect failed prior to the ESD network silicon failure.

In the case when the solder ball/bond pad structure was over the array I/O cell, the failure distribution was associated with the ESD failure of the ESD network, or the results exceeded the maximum test level of the test system.

Latchup and charge injection is also a concern in this array I/O floorplan [5]. Figure 2.17 shows an example of latchup from an array I/O to dense logic. In the figure, the center region is the ESD diode element. The ESD n-well diode element is surrounded by a guard ring of the array I/O standard cell. Outside the array I/O cell, the surrounding space is CMOS logic. The CMOS logic is a gate array configuration, with no guard ring structures between the n-well and p-well regions. In Figure 2.17, photon emissions are visible in the

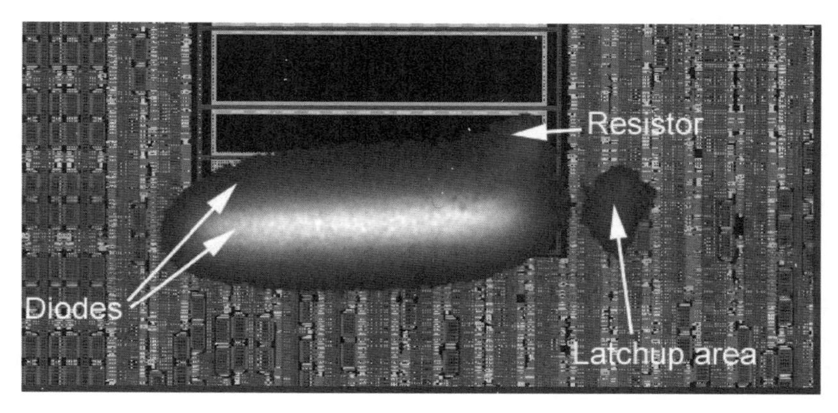

Figure 2.17 Latchup in array I/O architecture

ESD diode. In the dense logic, photons are evident in the dense logic area. The photons evident in the CMOS logic area are associated with forward biasing and the onset of CMOS latchup.

2.6 ESD ARCHITECTURE – DUMMY BUS ARCHITECTURES

In semiconductor chip architecture, there are advantages in establishing separate power buses to divert the current from the ESD signal pin network to the ESD power clamps. These "dummy ESD bus" structures play a key role in digital, analog, and RF applications.

2.6.1 ESD Architecture – Dummy V_{DD} Bus

In ESD architecture of a semiconductor chip, a "dummy" V_{DD} bus can be utilized that is distinct from the semiconductor chip power rail. ESD V_{DD} buses separate the ESD discharge from the chip power rail used by the functional circuits. Figure 2.18 shows an example of an ESD dummy V_{DD} bus. The ESD signal pin devices are electrically connected to the ESD dummy V_{DD} bus; this allows the ESD discharge to be directed to the ESD bus instead of the semiconductor V_{DD} power rail. Between the ESD dummy V_{DD} bus and ground is an ESD power clamp element. An additional ESD element can be placed between the V_{DD} power rail and the ESD dummy ground bus, and the semiconductor chip ground, V_{SS}. The usage of the ESD V_{DD} bus has the following ESD design synthesis advantages:

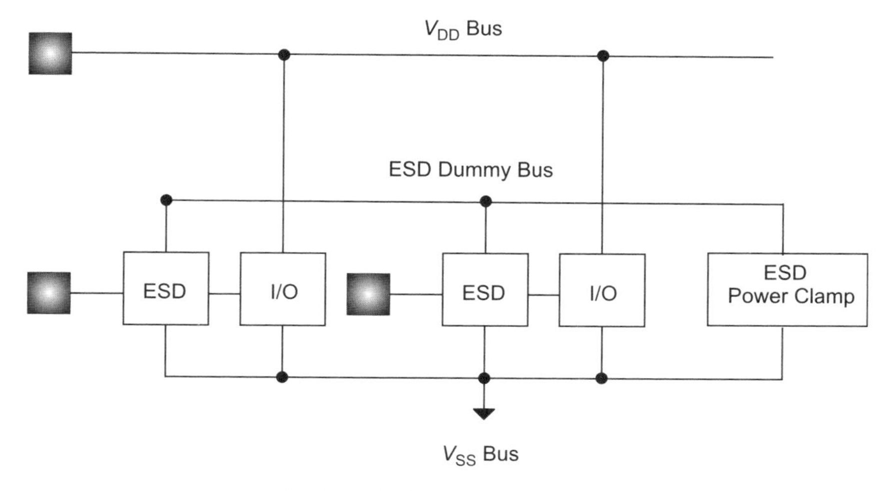

Figure 2.18 ESD dummy V_{DD} bus architecture

- **Full Chip Connections:** All ESD signal pins are connected to the independent bus.
- **Biasing Voltage:** The ESD dummy ground bus can be independently biased to another voltage, decreasing ESD capacitance on the input devices.
- **Voltage Tolerance:** A lower voltage tolerant level can be used for the ESD power clamp (due to the reduced voltage on the ESD dummy bus).
- **Floating Bus:** The ESD dummy bus can be left "floating."
- **Design Freedom:** ESD dummy bus design (e.g., width) can be set to address ESD concerns, independent of the semiconductor chip power requirements.

2.6.2 ESD Architecture – Dummy Ground (V_{SS}) Bus

In ESD architecture of a semiconductor chip, a "dummy" ground bus can be utilized that is distinct from the semiconductor chip ground. ESD ground buses separate the ESD discharge from the chip substrate ground bus used by the functional circuits. Figure 2.19 shows an example of an ESD dummy ground bus. The ESD signal pin devices are electrically connected to the ESD dummy ground bus; this allows the ESD discharge to be directed to the ESD bus instead of the semiconductor chip substrate. An additional ESD element is placed between the ESD dummy ground bus and the semiconductor chip ground, V_{SS}. The usage of the ESD ground bus has the following ESD design synthesis advantages:

- All ESD signal pins are connected to the independent bus.
- The ESD dummy ground bus can be independently biased to another voltage.

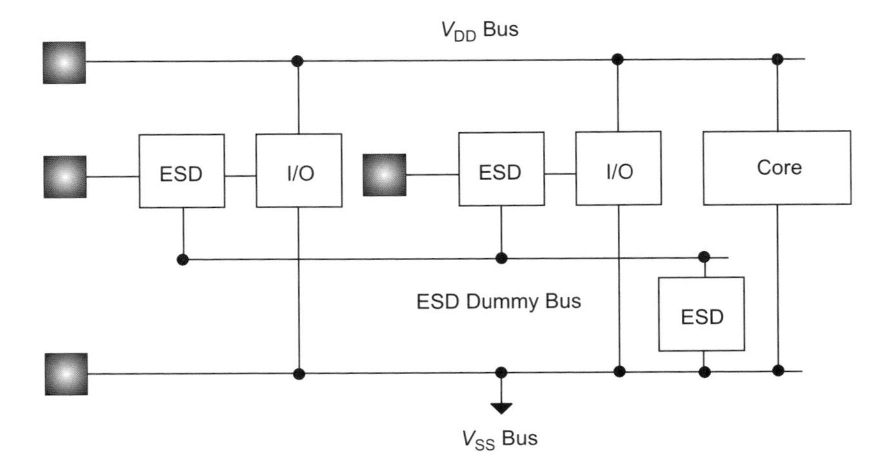

Figure 2.19 Dummy ESD ground bus architecture

- The ESD dummy ground bus can be left "floating."
- ESD dummy bus design and width can be set to address ESD concerns, independent of the semiconductor chip ground, V_{SS}.

2.7 NATIVE VOLTAGE POWER SUPPLY ARCHITECTURE

Semiconductor chips power supply architecture can be used for the native power supply voltage of the technology. The power supply architecture influences the type of ESD signal pin and the ESD power clamp choice. In the following sections, single and multiple power supply architectures will be discussed.

2.7.1 Single Power Supply Architecture

Semiconductor chips can interface with other semiconductor chips or systems with a common power supply voltage. Figure 2.20 shows an example of the semiconductor chip architecture. In the ESD design synthesis, an ESD network is connected to the bond pad. There are two typical architectures:

- The ESD network is connected to the V_{SS} ground rail only.
- The ESD network is connected to the V_{DD} power supply rail and the V_{SS} ground rail.

An example of the ESD signal pin network connected to the V_{SS} ground rail can be a grounded gate n-channel MOSFET (also referred to as GGNMOS). An example of an ESD signal pin network connected to both power rails is a dual diode (also referred to as a double diode).

To complete the alternative current loop, at least one ESD power clamp is placed in the semiconductor chip. An ESD power clamp is formed between the V_{DD} and the V_{SS} power rails.

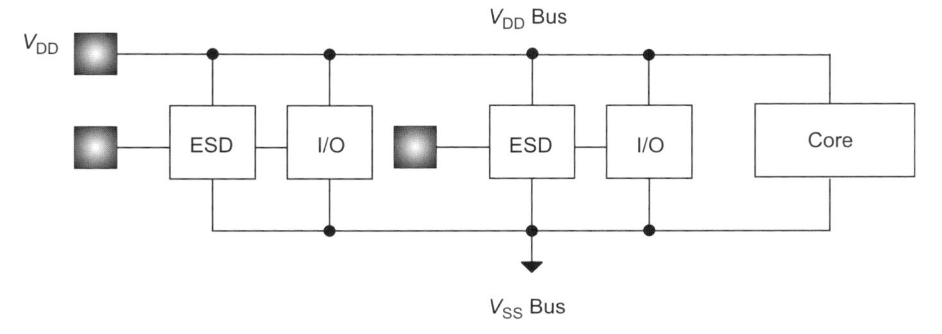

Figure 2.20 Single power supply architecture with native-voltage interface

2.8 MIXED-VOLTAGE ARCHITECTURE

In a system, different voltage levels exist for different semiconductor chips that interface with one another. The different voltage levels exist due to both different technology types and technology generations. This becomes an issue for both the interface circuitry, as well as semiconductor chip architectures. As a result, the ESD design synthesis must also address these configurations.

2.8.1 Mixed-Voltage Architecture – Single Power Supply

In a mixed-voltage environment, semiconductor chips must receive or transmit signal levels which are above the native voltage of the technology. To address this, some semiconductor chips have multiple power rails in the design to receive the higher voltage. In a single power supply architecture, only the native voltage of the technology is present on the semiconductor chip.

To address "receiving" or "driving" signal levels above the native voltage power supply, the circuits and ESD networks must have circuitry that is tolerant of the higher voltage levels. The circuit solution for this mixed-voltage interface requirement is addressed as follows:

- Mixed-voltage interface (MVI) OCDs contain series cascode n-channel MOSFET pull-down networks to avoid electrical overstress of the MOSFET [2–4].
- MVI OCDs contain "floating-well" p-channel MOSFET pull-up networks to avoid forward biasing of the p-channel MOSFET (also referred to as n-well-biased off-chip driver) [2,3,11–14].

Figure 2.21 shows an example of an n-well-biased OCD network.

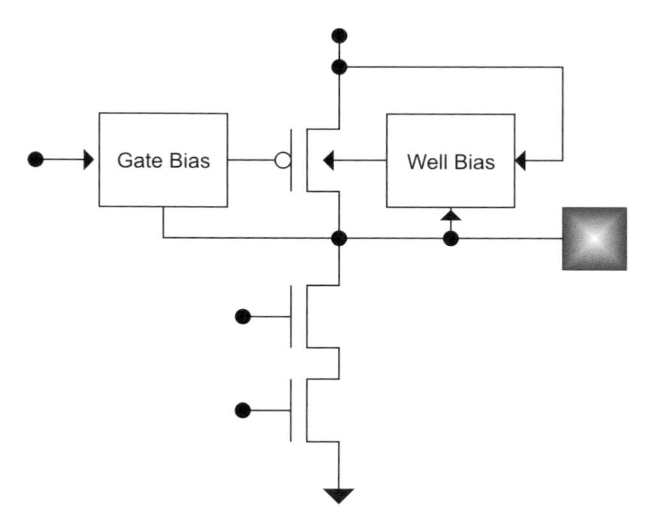

Figure 2.21 Mixed-voltage interface n-well-biased OCD

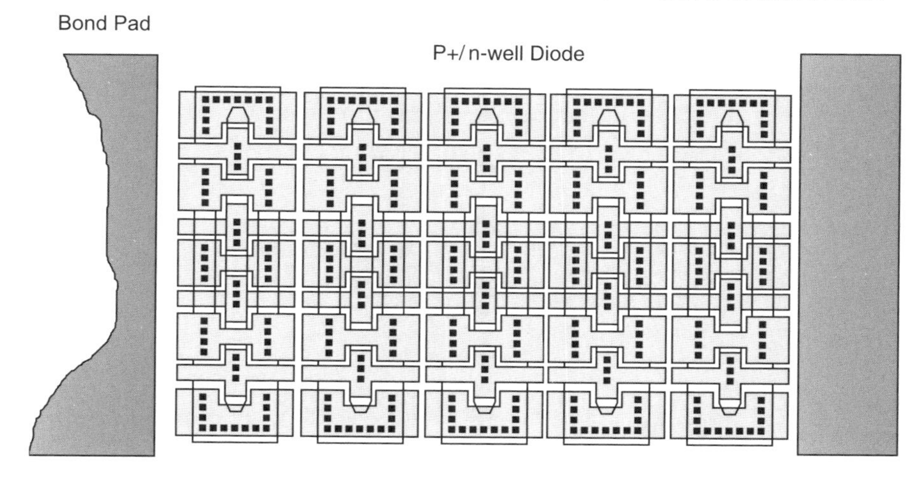

Figure 2.22 Mixed-voltage interface series diode ESD network layout

The ESD circuit solution for this mixed-voltage interface requirement is addressed as follows:

- No ESD element is connected to the V_{DD} power supply rail, to avoid forward biasing of the network.
- The MVI ESD network utilizes ESD series diodes to the power supply. ESD diodes consist of p-diffusion-based elements.
- The MVI ESD network contains a "floating-n-well" ESD network to avoid forward biasing of the p + /n-well diode [2].

Figures 2.22 and 2.23 show an example of an ESD network used for a 5 V to 3.3 V interface [2–4,13,14]. Five p+/n-well diodes are placed between the bond pad and the 3.3 V power supply of the semiconductor chip. The signal pad may switch from 0 to 5 V, yet must interface with the 3.3 V power supply. Five diodes are placed to prevent forward biasing of the ESD network below the worst case power supply conditions. In the first figure, the ESD design layout is shown between the bond pad and the V_{DD} power rail. In the second figure, a circuit schematic containing the ESD "diode string" network is shown.

2.8.2 Mixed-Voltage Architecture – Dual Power Supply

In systems, semiconductor chips are at different application voltages. Semiconductor chips receive signals above the native power supply voltage of the technology. One method to achieve this is to have a semiconductor chip with more than one power supply voltage; the external I/O

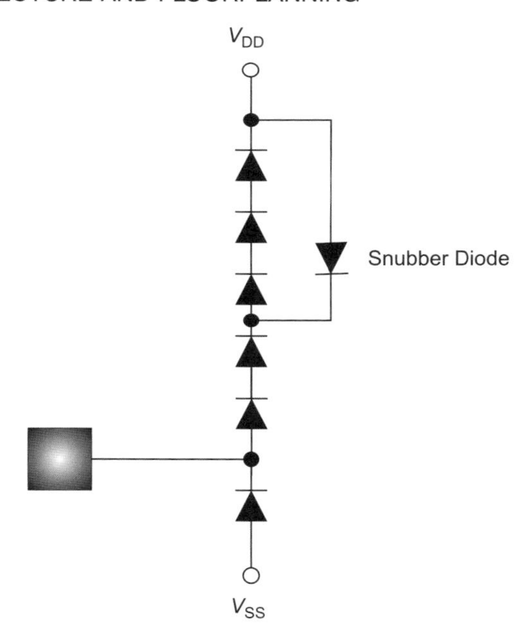

Figure 2.23 Mixed-voltage interface series diode ESD network

circuitry is power to the higher power supply voltage, and the internal core circuitry is powered at the native voltage.

Mixed-voltage interfaces also influence the ESD design synthesis and chip architecture. Figure 2.24 shows an example of a semiconductor chip with two power supply voltages. The external peripheral circuits are power to the incoming signal voltages, and the core circuitry is at the native voltage of the technology.

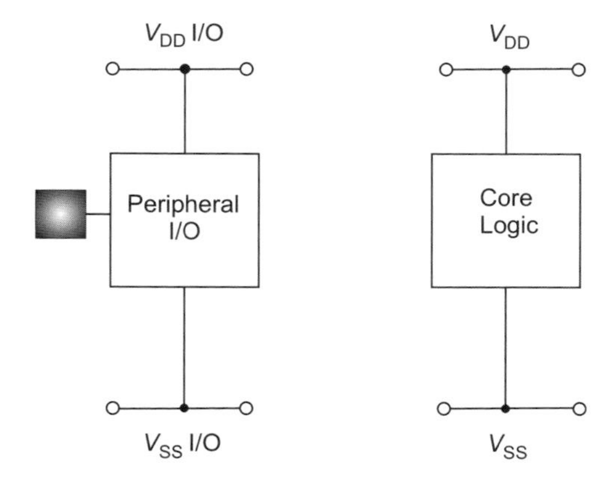

Figure 2.24 Dual power supply architecture

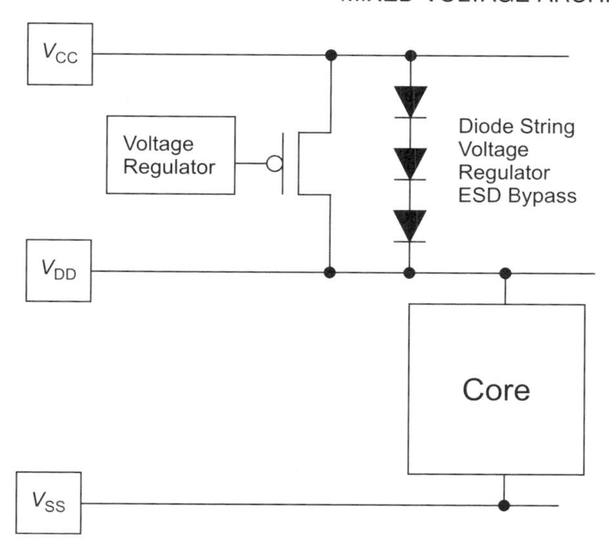

Figure 2.25 Dual power supply architecture with ESD network between the two V_{DD} power supply rails

There are multiple ESD architectures that can be established for this environment. The following architectures are examples:

- Sequence-dependent V_{DD}-to-V_{DD} ESD network.
- Sequence-independent V_{DD}-to-V_{DD} ESD network.
- ESD V_{DD}-to-V_{SS} power clamps in the peripheral I/O and core chip regions.

In the ESD design synthesis, a first architecture is to have an ESD network between the power supply rails. A bi-directional ESD network is placed between the two power supply rails. In semiconductor memory chips, the I/O capacitance is small compared to the core capacitance. In memory chips, to save area, the core capacitance can be used due to its low impedance. Figure 2.25 shows an example of a semiconductor memory chip that uses an ESD network between the power supplies. The ESD signal pin device transfers the charge to the external power rail, which then transfers the charge to the core region of the semiconductor chip. An alternative current loop is established from the bond pad to the core. The advantage of this architecture is that no additional ESD power clamps on the peripheral I/O are required, and the core chip capacitance is utilized. This architecture is suitable for large-memory chips, and microprocessors.

In another ESD design synthesis, ESD V_{DD}-to-V_{SS} power clamps are placed in the internal and external power domains. In this architecture, no ESD network is used between the V'_{DD} and V_{DD} power supply rails. Figure 2.26 shows an example of this architecture. The ESD signal pin device transfers the charge to the external power rail, which then transfers the charge to the peripheral ground rail of the semiconductor chip. In addition, there is an ESD network placed in the ground path (V'_{SS} to V_{SS}). The advantage of this architecture is that no ESD element exists between the two power domains. This avoids concerns over power sequencing, and noise

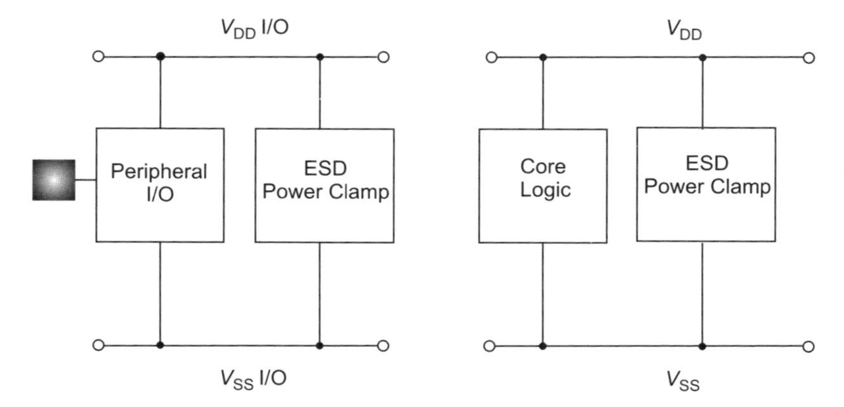

Figure 2.26 Dual power supply architecture with peripheral I/O and core ESD power clamp network

injection into the core circuitry. This implementation requires two different ESD V_{DD}-to-V_{SS} power clamps, due to the voltage tolerance of the internal circuits and external circuits.

2.9 MIXED-SIGNAL ARCHITECTURE

In a mixed-signal environment, semiconductor chips can include both digital and analog domains. For mixed-signal CMOS semiconductor chips, both the analog and digital circuitry are MOSFET device elements. For a BiCMOS technology, the digital and analog sectors can contain both CMOS and bipolar transistors.

In a semiconductor chip with bipolar transistors, the architecture of the semiconductor chip may have a V_{CC} power supply, a V_{SS} substrate ground power supply, and a negative-biased V_{EE} power rail. In this three-rail architecture, the ESD power clamps can have different topologies.

2.9.1 Mixed-Signal Architecture – Bipolar

In a mixed-signal environment, or pure analog applications, bipolar transistors are utilized. In a semiconductor chip with bipolar transistors, the architecture of the semiconductor chip may have a V_{CC} power supply, a V_{SS} substrate ground power supply, and a negative-biased V_{EE} power rail. In this three-rail architecture, the ESD power clamps can have different topologies.

In this architecture, the ESD design synthesis has a few options:

- ESD bipolar power clamps between V_{CC} and V_{SS}, and second ESD power clamp between V_{SS} and V_{EE}.
- ESD bipolar power clamps between V_{CC} and V_{SS}, and second ESD power clamp between V_{CC} and V_{EE}.

Figure 2.27 shows an example of the power supply architecture. In Figure 2.27, the ESD power clamps are stacked between the V_{CC} and V_{SS}, and a second one between V_{SS} and V_{EE}. In this architecture, the same ESD bipolar power clamp can be used for both cases.

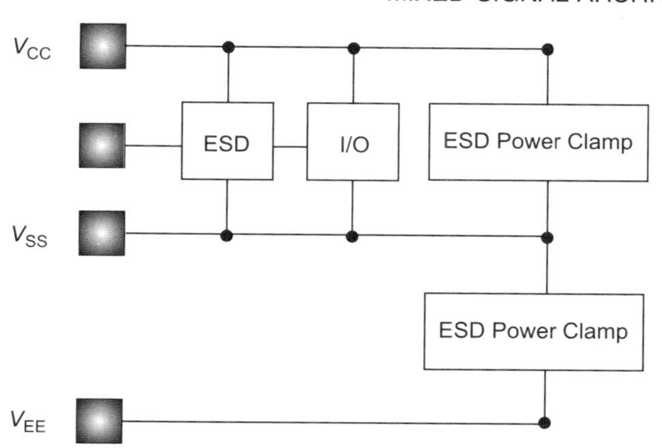

Figure 2.27 Mixed-signal architecture – bipolar ESD power clamps in a stacked fashion. ESD power clamps are between V_{CC} and V_{SS}, and between V_{SS} and V_{EE}

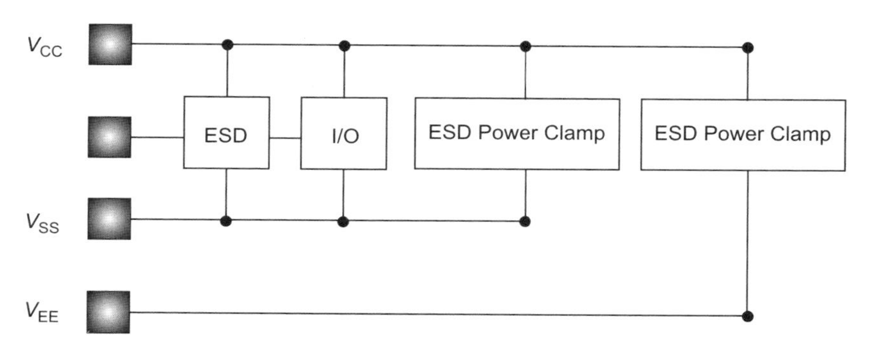

Figure 2.28 Mixed-signal architecture – bipolar ESD power clamps from V_{CC} to V_{SS}, and from V_{CC} to V_{EE}

In Figure 2.28, the ESD power clamps are stacked between the V_{CC} and V_{SS}, and a second one between V_{CC} and V_{EE}. In this architecture, the voltage tolerance across the ESD power clamps requires different bipolar transistors with higher breakdown voltages, or elements that are in series cascode within the ESD power clamp.

2.9.2 Mixed-Signal Architecture – CMOS

Mixed-signal semiconductor chips can be formed with a CMOS technology. In a mixed-signal chip with CMOS technology, the architecture is typically the same power supply voltage, with different power domains established for each chip function domain. Figure 2.29 is an example of a mixed-signal chip commonly found in the semiconductor industry.

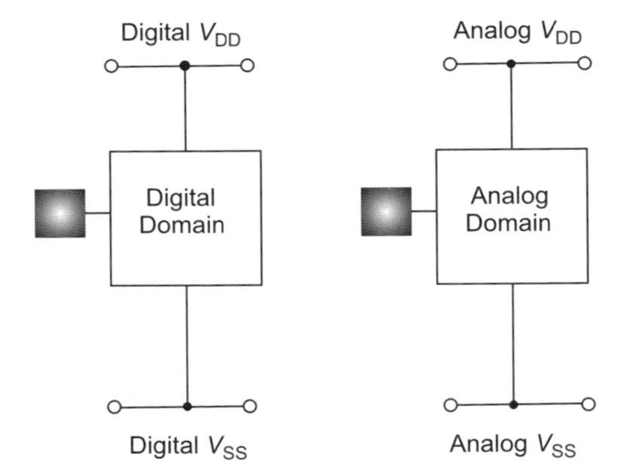

Figure 2.29 Mixed-signal architecture

2.10 MIXED-SYSTEM ARCHITECTURE – DIGITAL AND ANALOG CMOS

A common mixed-signal application is a semiconductor chip with digital and analog circuitry. In a mixed-signal (MS) architecture, typically the concern is the influence of the noise from the digital circuitry received at the analog circuitry; as a result, the digital and analog circuitry are separated into different power domains.

2.10.1 Digital and Analog CMOS Architecture

In a MS architecture, the digital and analog circuitry are separated into different power domains. Figure 2.30 shows an example of a semiconductor chip with a digital and analog domain. To avoid ESD failures in a MS semiconductor chip, ESD protection networks are

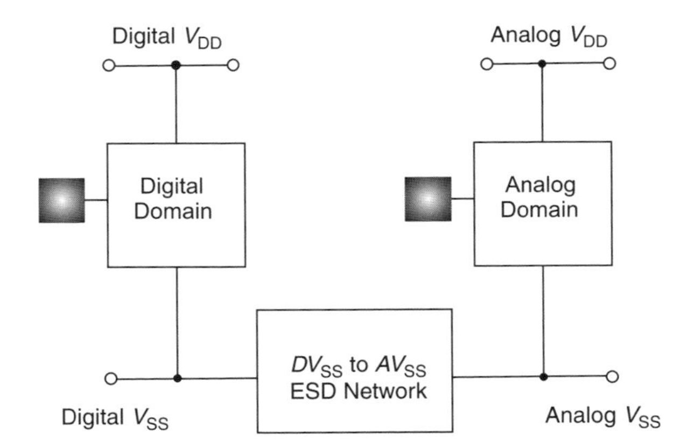

Figure 2.30 Mixed-signal architecture – digital and analog with DV_{SS}-to-AV_{SS} ESD network

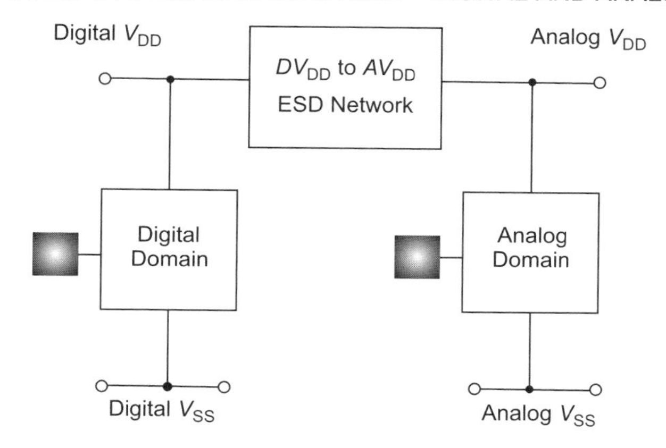

Figure 2.31 Mixed-signal architecture – digital and analog with DV_{DD}-to-AV_{DD} ESD network

placed between the analog ground (AV_{SS}) and the digital ground (V_{SS}). Typical architectures contain a separate ESD power clamp in each domain. An ESD power clamp exists in the digital domain, between V_{DD} and V_{SS}, and a second ESD power clamp exists in the analog domain, between analog V_{DD} (AV_{DD}) and analog ground (AV_{SS}).

Alternative architectures are as follows:

- **V_{DD}-to-AV_{DD}** ESD **Network:** ESD network between the digital power rail (V_{DD}) and the analog power rail (AV_{DD}) (Figure 2.31).
- **V_{DD}-to-AV_{SS} ESD Network:** ESD network between the digital power rail (V_{DD}) and the analog power rail (AV_{SS}) (Figure 2.32).

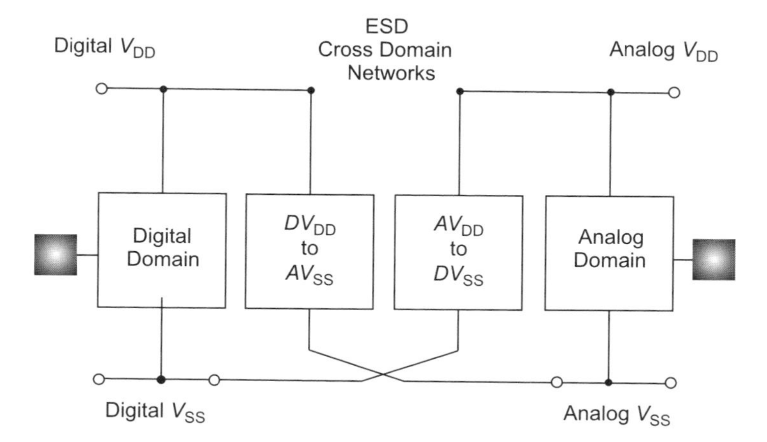

Figure 2.32 Mixed-signal architecture – digital and analog with cross-domain ESD network

2.10.2 Digital and Analog Floorplan – Placement of Analog Circuits

In analog design, many considerations are taken to provide good matching and non-uniform characteristics. Analog circuits avoid variations associated with process variations, temperature field, to mechanical stress. Many analog solutions introduce common centroid design practices to improve the matching. On a global level, analog circuits avoid placement next to digital and power domains. Additionally, analog circuits are not placed in the corners of a semiconductor die to wafer warpage, which can lead to mechanical stress variations of the devices.

Figure 2.33 shows an example of a semiconductor die, highlighting the regions where the analog circuits can not be placed [21–23]. Analog circuits can not be placed in the corners of a semiconductor die. For analog ESD design co-synthesis, the corners of the semiconductor die are ideal for ESD power clamps where the space is free to be utilized (Figure 2.34). For pure analog designs, or mixed-signal chips with digital and analog design, this is ideal for allowance of the ESD power clamps.

For analog design only, all four corners can be used for the analog power domain (e.g., AV_{DD} to AV_{SS}). Four ESD power clamps can be placed between the two power rails in a periphery design, using the area of the corners where analog circuitry is not allowed to be placed.

For mixed-signal designs, two of the corners can used for ESD power clamps for the digital domain (e.g., digital V_{DD} (DV_{DD}) and digital ground (DV_{SS})); and the other two corners can be used for the analog power domain ESD power clamps (e.g., AV_{DD} and AV_{SS}). In this architecture, "breaker cells" between the two power domains using ground to ground cells (AV_{SS} to V_{SS}) can be utilized. These breaker cells can be placed in the peripheral architecture design. It is typical in these designs that the digital circuits are separated from the analog

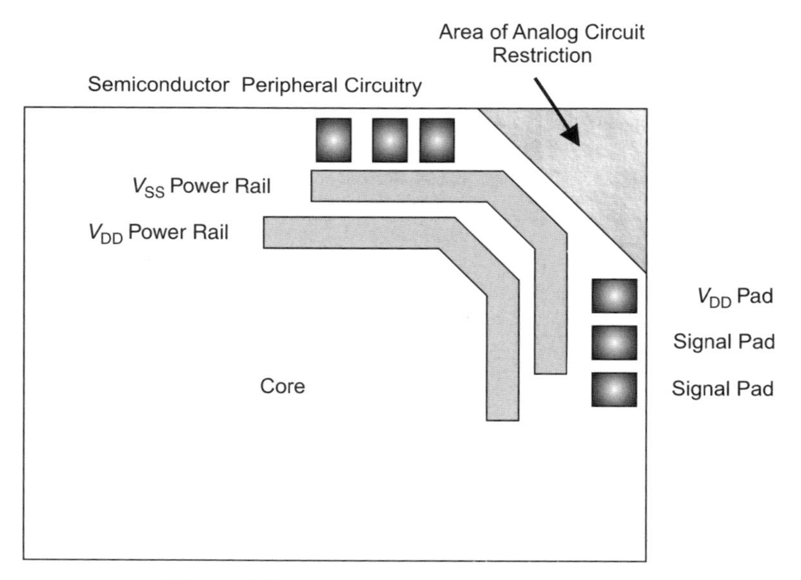

Figure 2.33 Analog circuit layout restrictions

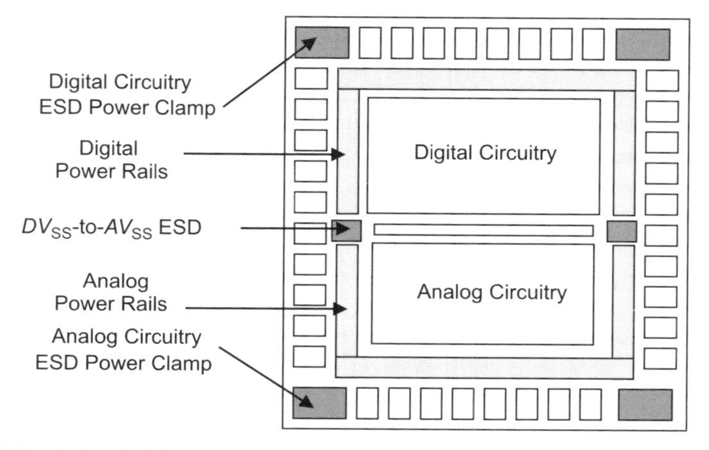

Figure 2.34 Mixed-signal digital–analog ESD co-synthesis – placement of analog circuits and ESD power clamps

domains to avoid digital noise from influencing the analog circuitry. Figure 2.34 shows the digital–analog ESD co-synthesis architecture.

2.11 MIXED-SIGNAL ARCHITECTURE – DIGITAL, ANALOG, AND RF ARCHITECTURE

In a MS architecture, the digital, analog, and RF circuitry are separated into different power domains [21–24,30–32]. Figure 2.35 shows an example of a semiconductor chip with a digital, analog, and RF power domain. To avoid ESD failures in a MS semiconductor chip, ESD protection networks are placed between the analog ground (AV_{SS}), the digital ground (V_{SS}), and the RF ground. Typical architectures contain a separate ESD power clamp in each domain. An ESD power clamp exists in the digital domain, between V_{DD} and V_{SS}, and a second ESD power clamp exists in the analog domain, between analog V_{DD} (AV_{DD}) and analog ground (AV_{SS}), and

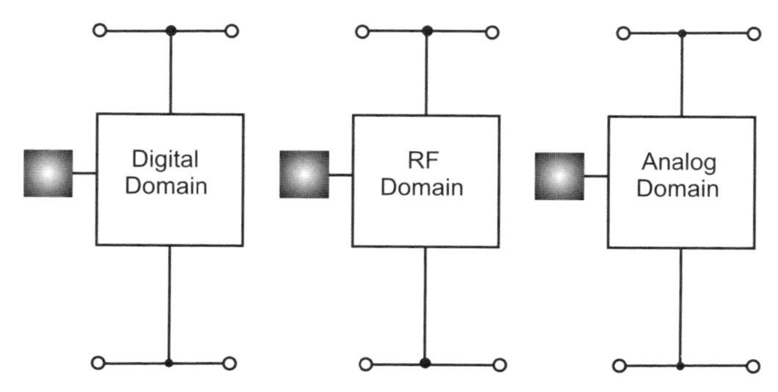

Figure 2.35 Mixed-signal architecture – digital, analog, and RF architecture

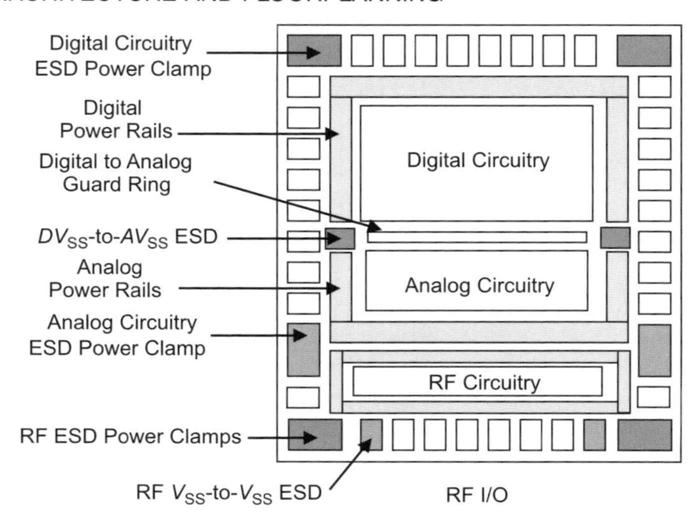

Figure 2.36 Placement and mixed-signal architecture – digital, analog, and RF architecture

a third ESD power clamp is between RF V_{DD} (RFV_{DD} or RFV_{CC}) and RF ground (RFV_{SS} or RFV_{EE}). In these mixed-signal chips, the RF application voltage is typically higher than the analog and digital application voltage.

Figure 2.36 shows an example of a floorplan for a mixed-signal chip with RF, analog, and digital circuitry [31]. To separate the analog circuitry from the digital noise, separate power rail domains exist. Additionally, a guard ring "moat" separates the two domains to produce a larger distance through the substrate region. The RF sector is separated on the lower sector of the chip floorplan. The RF circuitry is surrounded by layers of metal, forming a "faraday cage" to isolate the RF signals. The faraday cage is formed by stacking the metal layers, and passing the signals through the breaks in the faraday cage. ESD network power clamps are placed in the digital, analog, and RF domains between their power and ground rails. In addition, V_{SS} to V_{SS} ESD networks are placed to interconnect the ground rails. The V_{SS} to V_{SS} networks use series diode ESD elements, where the number of elements in series is a function of the allowed capacitive coupling between the digital, analog, and RF sectors.

2.12 SUMMARY AND CLOSING COMMENTS

In this chapter, the discussion has focused on ESD architecture and floorplan concepts. The chapter focused on "peripheral I/O" and "array I/O" architectures, and how they influence the placement of the various elements for the whole-chip design integration. The chapter also focused on native-voltage, mixed-voltage, and mixed-signal chip integration.

In Chapter 3, the discussion continues to address issues associated with full-chip ESD design synthesis. Chapter 3 focuses on the interconnects and power grid layout and design itself. The chapter will address interconnect robustness, interconnect failure, and key metrics in the whole-chip ESD design synthesis.

PROBLEMS

2.1. A chip is formed with I/O and bond pads on the periphery. Draw a semiconductor chip that has dimensions w and l. Assume the bond pads for the circuits are w_{BP} and l_{BP}, with spacing w_{sp} and l_{sp} between bond pads. Assuming the bond pads are on the perimeter, what is the maximum number of bond pads that can be placed on the semiconductor chip? Generalize an equation for the number of bond pads and the chip size relationship. Draw a picture of the chip with the bond pads and spaces. What is the area loss on the corners?

2.2. A chip is formed with I/O and bond pads in an array. Draw a semiconductor chip that has dimensions w and l. Assume the bond pads for the circuits are w_{BP} and l_{BP}, with spacing w_{sp} and l_{sp} between bond pads. Assuming the bond pads are in a mosaic array, what is the maximum number of bond pads that can be placed on the semiconductor chip? Generalize an equation for the number of bond pads and the chip size relationship. Draw a picture of the chip with the bond pads and spaces. If the ESD device for each bond pad is as large as the bond pad, what is the percentage of chip area needed for ESD protection?

2.3. In Problem 1, if the ESD device for each bond pad is 50% as large as the bond pad, what is the percentage of chip area needed for ESD protection?

2.4. Assume, in Problem 1, ESD power clamps are placed in the corners only. What is the area loss in this case?

2.5. Assume, in Problem 1, a V_{SS} power pin is added for every M bond pads, how many power pins can be placed?

2.6. Assume, in Problem 1, a V_{SS} power pin is added for every M bond pads, and a V_{DD} power pin is added for every N bond pads. What is the number of non-power I/O (e.g., signal pins) and number of bond pads for power? Assume an ESD network is placed for each bond pad, and each V_{SS} and V_{DD} pin, what is the total area utilized for ESD protection? Assume the ESD device is a percentage of the bond pad area (parameterize it as a variable p).

2.7. Assume a semiconductor chip has an analog and digital core in Problem 1. Assume there are four power rails, V_{DD}, AV_{DD}, V_{SS}, and AV_{SS}. As in Problem 1, assume a V_{SS} power pin is added for every M bond pads, and a V_{DD} power pin is added for every N bond pads. What is the number of non-power I/O (e.g., signal pins) and number of bond pads for power? Assume an ESD network is placed for each bond pad, and each V_{SS} and V_{DD} pin, what is the total area utilized for ESD protection? Assume the ESD device is a percentage of the bond pad area (parameterize it as a variable p).

2.8. Assume a semiconductor chip has an analog, digital, and RF core in Problem 1. Assume there are six power rails, RFV_{CC}, RFV_{EE}, V_{DD}, AV_{DD}, V_{SS}, and AV_{SS}. As in Problem 1, assume a V_{SS} power pin is added for every M bond pads, and a V_{DD} power pin is added for every N bond pads. What is the number of non-power I/O (e.g., signal pins) and number of bond pads for power? Assume an ESD network is placed for each bond pad, and each V_{SS} and V_{DD} pin, what is the total area utilized for ESD protection? Assume the ESD device is a percentage of the bond pad area (parameterize it as a variable p).

2.9. Assume a semiconductor chip has an analog, digital, and a smart power core in Problem 1. Assume there are six power rails, PV_{DD}, PV_{SS}, V_{DD}, AV_{DD}, V_{SS}, and AV_{SS}. As in Problem 1, assume a V_{SS} power pin is added for every M bond pads, and a V_{DD} power pin is added for every N bond pads. What is the number of non-power I/O (e.g., signal pins) and number of bond pads for power? Assume an ESD network is placed for each bond pad, and each V_{SS} and V_{DD} pin, what is the total area utilized for ESD protection? Assume the ESD device is a percentage of the bond pad area (parameterize it as a variable p). Derive the relationship where the three chip sectors are not equal areas. Define a parameter for the ratio of the areas.

2.10. As in Problem 9, smart power pins require more than one bond pad for signal pins. Assume that the smart power pins require $2\times$ pins per signal, and $4\times$ the ESD network size, how does the relationship change?

REFERENCES

1. S. Voldman. *ESD: Physics and Devices*. Chichester, UK: John Wiley and Sons, Ltd., 2004.
2. S. Voldman. *ESD: Circuits and Devices*. Chichester, UK: John Wiley and Sons, Ltd., 2005.
3. S. Voldman. *ESD: Circuits and Devices.* Beijing, China: Publishing House of Electronic Industry (PHEI); 2008.
4. S. Dabral, and T. J. Maloney. *Basic ESD and I/O Design*. Chichester, UK: John Wiley and Sons, Ltd., 1998.
5. S. Voldman. *Latchup*. Chichester, UK: John Wiley and Sons, Ltd., 2006.
6. S. Voldman. *ESD: Failure Mechanisms and Models*. Chichester, UK: John Wiley and Sons, Ltd., 2009.
7. R. Merrill and E. Issaq. ESD design methodology. *Proceedings of the Electrical Overstress/Electrostatic Discharge (EOS/ESD) Symposium*, 1993; 233–237.
8. S. Dabral, R. Aslett, and T. Maloney. Designing on-chip power supply coupling diodes for ESD protection and noise immunity, *Proceedings of the Electrical Overstress/Electrostatic Discharge (EOS/ESD) Symposium*, 1993; 239–249.
9. S. Dabral, R. Aslett, and T. Maloney, Core clamps for low voltage technologies. *Proceedings of the Electrical Overstress/Electrostatic Discharge (EOS/ESD) Symposium*, 1994; 141–149.
10. E. Adler, J. DeBrosse, S. Geissler, S. Holmes, M. Jaffe, J. Johnson, C. Koburger, J. Lasky, B. Lloyd, G. Miles, J. Nakos, W. Noble, and S. Voldman. The evolution of IBM CMOS DRAM technology. *IBM Journal of Research and Development*, Vol. 39, 1/2., Jan./March 1995; 167–188.
11. S. Voldman V. Gross, M. Hargrove, J. Never, J. Slinkman, M. O'Boyle, T. Scott, and J. Delecki, Shallow trench isolation (STI) double-diode electrostatic discharge (ESD) circuit and interaction with DRAM Circuitry, *Proceedings of the Electrical Overstress/Electrostatic Discharge (EOS/ESD) Symposium*, 1992; 277–288.
12. S. Voldman. ESD protection in a mixed voltage interface and multi-rail disconnected power grid environment in 0.5 and 0.25 μm channel length CMOS technologies, *Proceedings of the Electrical Overstress/Electrostatic Discharge (EOS/ESD) Symposium*, 1994; 125–134.
13. S. Voldman and G. Gerosa. Mixed voltage interface ESD protection circuits for advanced microprocessors in shallow trench and LOCOS isolation CMOS technology, *International Electron Device Meeting (IEDM) Technical Digest*, December 1994; 811–815.
14. S. Voldman, G. Gerosa, V. Gross, N. Dickson, S. Furkay, and J. Slinkman. Analysis of snubber clamped diode string mixed voltage interface ESD protection networks for advanced

microprocessors. *Proceedings of the Electrical Overstress/Electrostatic Discharge (EOS/ESD) Symposium*, 1995; 43–62.
15. S. Voldman. The impact of MOSFET technology evolution and scaling on electrostatic discharge protection. *Journal of Microelectronics Reliability*, 1998, **38**; 1649–1668.
16. S. Voldman. The impact of technology evolution and scaling on electrostatic discharge (ESD) protection in high-pin-count high-performance microprocessors. *Proceedings of the International Solid State Circuits Conference (ISSCC)*, Session 21, WA 21.4, February 1999; 366–367.
17. S. Voldman. CMOS-on-SOI ESD protection networks. *Proceedings of the Electrical Overstress/ Electrostatic Discharge (EOS/ESD) Symposium*, 1996; 291–301; and *Journal of Electrostatics*, 1998; **42**; 333–350.
18. S. Voldman. Electrostatic discharge (ESD) protection in silicon-on-insulator (SOI) CMOS technology with aluminum and copper interconnects in advanced microprocessor semiconductor chips. *Proceedings of the Electrical Overstress/Electrostatic Discharge (EOS/ESD) Symposium*, 1999; 105–115.
19. S. Voldman, D. Hui, L. Warriner, D. Young, R. Williams, J. Howard, V. Gross, W. Rausch, E. Leobangdung, M. Sherony, N. Rohrer, C. Akrout, F. Assaderaghi, and G. Shahidi. Electrostatic discharge protection in silicon-on-insulator technology. *Proceedings of the IEEE International Silicon on Insulator (SOI) Conference*, 1999; 68–72.
20. P. A. Juliano, and W. Anderson. ESD protection design challenges for a high pin-count Alpha microprocessor in a 0.13 μm CMOS SOI technology. *Proceedings of the Electrical Overstress/ Electrostatic Discharge (EOS/ESD) Symposium*, 2003; 59–69.
21. P. Gray, Hurst, Lewis, and Meyer. *Analysis and Design of Analog Integrated Circuits. Fifth Edition*. New York: John Wiley and Sons, Inc., 2009.
22. W.M.C. Sansen. *Analog Design Essentials*. Amsterdam: Springer, 2006.
23. A. Hastings. *The Art of Analog Layout*. Second Edition. New Jersey: Pearson Prentice Hall, 2006.
24. V. Vashchenko and A. Shibkov. *ESD Design in Analog Circuits*. New York: Springer, 2010.
25. S. K. Ghandi. *Semiconductor Power Devices*, New York: John Wiley and Sons, Inc., 1977.
26. P. Antognetti. *Power Integrated Circuits: Physics, Design, and Applications*. New York: McGraw-Hill, 1986.
27. B. J. Baliga. *High Voltage Integrated Circuits*. New York: IEEE Press, 1988.
28. B.J. Baliga. *Modern Power Devices*. New York: John Wiley and Sons, Inc. 1987.
29. V. Vashchenko, M. Ter Beek, W. Kindt, P. Hopper. ESD protection of high voltage tolerant pins in low voltage BiCMOS processes. *Proceedings of the Bipolar Circuits Technology Meeting (BCTM)*, 2004; 277–280.
30. S. Voldman. The state of the art of electrostatic discharge protection: Physics, technology, circuits, designs, simulation and scaling. *Invited Talk, Bipolar/BiCMOS Circuits and Technology Meeting Symposium*, 1998; 19–31.
31. S. Voldman. *ESD: RF Circuits and Technology*. Chichester, UK: John Wiley and Sons, Ltd., 2006.
32. R. Singh, D. Harame, and M. Oprysko, *Silicon Germanium: Technology, Modeling and Design*, New York: John Wiley and Sons, Inc., 2004.

3 ESD Power Grid Design

3.1 ESD POWER GRID

The ESD power grid and metal interconnects have a significant impact on the ESD robustness of the semiconductor chip [1–3]. In the ESD design synthesis, it is important to integrate the interconnects, vias, and power grid effectively with the bond pads, the circuitry, ESD signal pin devices, and ESD power clamp circuitry [2,3]. In this chapter, the ESD design synthesis of the interconnects and the power grid will be discussed. The discussion will address both practical and analytical examples [1–28].

3.1.1 ESD Power Grid – Key ESD Design Parameters

In the ESD design synthesis of the power grid, there are key ESD design parameters and metrics to consider. These key parameters are to be considered as part of the ESD design synthesis of the power grid:

- **Bus Width:** The bus width required to survive ESD failure for a given metal layer.
- **Bus Resistance:** The bus resistance per unit length.
- **Across ESD Bus Resistance:** The bus resistance across the length of the ESD network.
- **Critical Bus Resistance:** The worst case bus resistance allowed without failure of the signal pin (can be defined as the breakdown voltage at the signal pin divided by the ESD current through the power bus).
- **ESD Signal Pin to ESD Power Clamp Distance:** The distance between the ESD signal pin and the ESD power clamp.
- **Critical ESD Signal to ESD Power Clamp Resistance:** The worst case resistance between the ESD signal pin and the ESD power clamp prior to signal pin failure.

ESD: Design and Synthesis, First Edition. Steven H. Voldman.

- **ESD Power Clamp to ESD Power Clamp Bus Resistance:** The resistance between two adjacent ESD power clamps on a given power rail.
- **Number of ESD Power Clamps per I/O Cells:** The number of required ESD power clamps for the number of signal pins.
- **ESD Power Clamp Placement Frequency:** The spatial frequency of the placement of the ESD power clamps along the peripheral power rail.

3.1.2 ESD and the Alternative Current Path – The Role of ESD Power Grid Resistance

The semiconductor chip design can be simplified as a system with a signal path, and the power grid. The role of the signal path is to have an input signal, process the signal, and have an output signal. For area and speed, the circuits along the signal path are as small as possible to perform this function. The role of the power grid is to supply the power to the circuits to support the necessary power for this function. The circuitry is electrically connected to the power supply (e.g., V_{DD}) and the ground (e.g., V_{SS}). Figure 3.1 shows an example of a system with the signal path highlighted.

During an ESD event, the signal pins are pulsed with an ESD event where the power rails serve as reference ground potential. For ESD design synthesis, the role of the ESD circuitry is to divert the ESD current other than the signal path. Figure 3.2 shows the semiconductor chip with the alternative current path established with ESD circuitry.

The ESD circuitry establishes an alternative current path to reach the grounded reference. In order for the ESD circuitry to be effective, the alternative current path must have the characteristics of low impedance (or low resistance), as well as being of adequate robustness not to be destroyed during the ESD event below the desired ESD specification levels. The alternative current path, including the power grid, must be able to survive the peak current and energy from the ESD pulse event. The peak current of

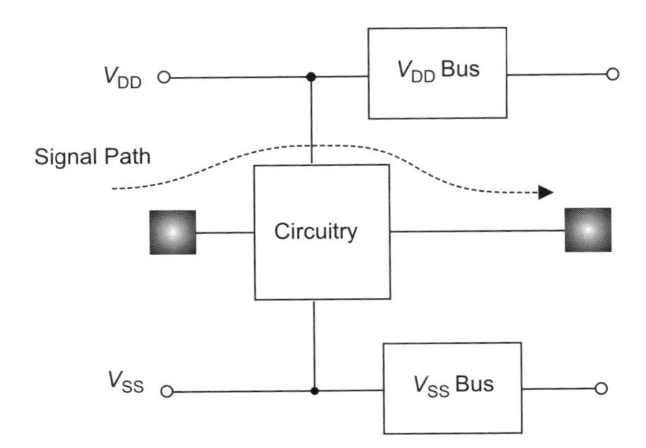

Figure 3.1 The signal path

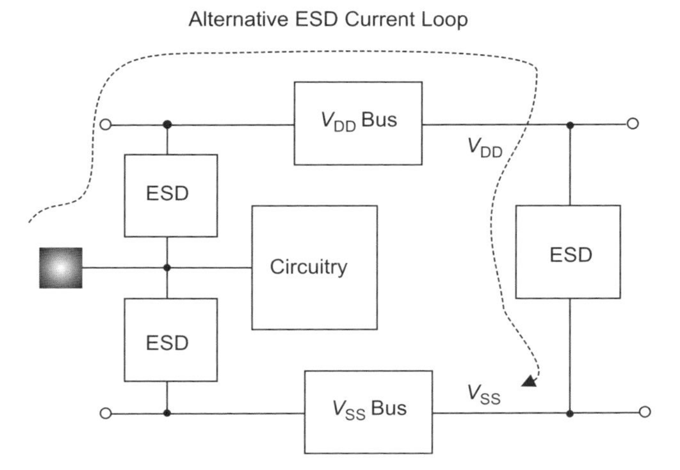

Figure 3.2 The alternative ESD current path

the ESD event is a function of the ESD specification. The ESD specifications are as follows:

- Human body model (HBM).
- Machine model (MM).
- Charged device model (CDM).
- Human metal model (HMM).
- IEC 61000-4-2.

Figure 3.3 shows the ESD Kirchoff current loop to the V_{DD} power supply rail. The figure shows the current flowing from the bond pad, the interconnect between the bond pad and the ESD

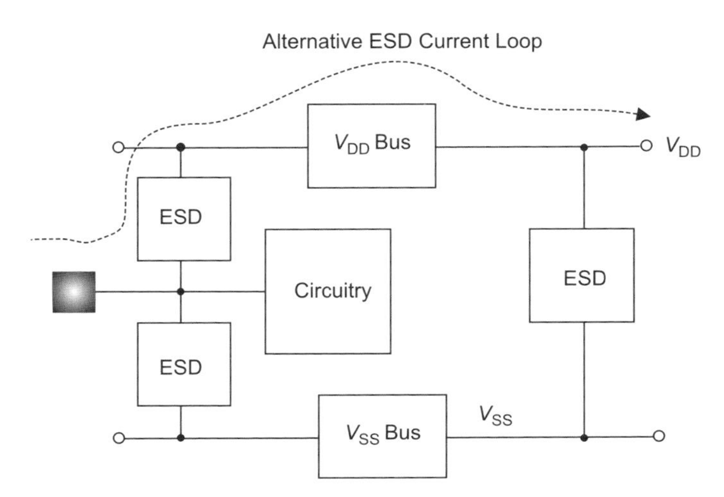

Figure 3.3 ESD Kirchoff current loop to V_{DD} power rail

signal pad device, the interconnect between the ESD device and the power grid, and through the V_{DD} power grid. Each interconnect and element in the path must have low resistance, as well as survive the ESD pulse event.

The impedance of each element must be low enough to avoid the voltage of the signal pad increasing. In order to prevent failure of the circuitry in the signal path, the bond pad voltage must be kept below the voltage-to-failure, V_f, of the circuitry. Assuming all elements are resistive in nature, according to Kirchoff's voltage law (KVL), the bond pad voltage can be represented as a function of the turn-on voltage, V_{ON}, and the IR voltage drop through the alternative current loop:

$$V_{pad} = \sum_{i}^{N} V_{ON_i} + \sum_{i}^{N} I_{ESD} R_i$$

Assuming the ESD current flows through all these elements, it can be written as

$$V_{pad} = \sum_{i}^{N} V_{ON_i} + I_{ESD} \sum_{i}^{N} R_i$$

This can be represented simply as the voltage drop through the ESD input element, the voltage drop in the power bus, and the voltage drop through the ESD power clamp element. In this case, the current flows through the ESD power clamp placed between V_{DD} and V_{SS} power rails:

$$V_{pad} = \{V_{ON} + I_{ESD} R\}_{input} + I_{ESD} R_{bus} + \{V_{ON} + I_{ESD} R\}_{clamp}$$

In the case where the grounded reference is the V_{DD} power rail, then it can be represented simply as the voltage drop through the ESD input element connected to the V_{DD} power bus, and the voltage drop in the V_{DD} power bus:

$$V_{pad} = \{V_{ON} + I_{ESD} R\}_{input} + I_{ESD}\, R_{V_{DD}}$$

Figure 3.4 shows the ESD Kirchoff current loop to the V_{SS} power supply rail. The figure shows the current flowing from the bond pad, the interconnect between the bond pad and the ESD

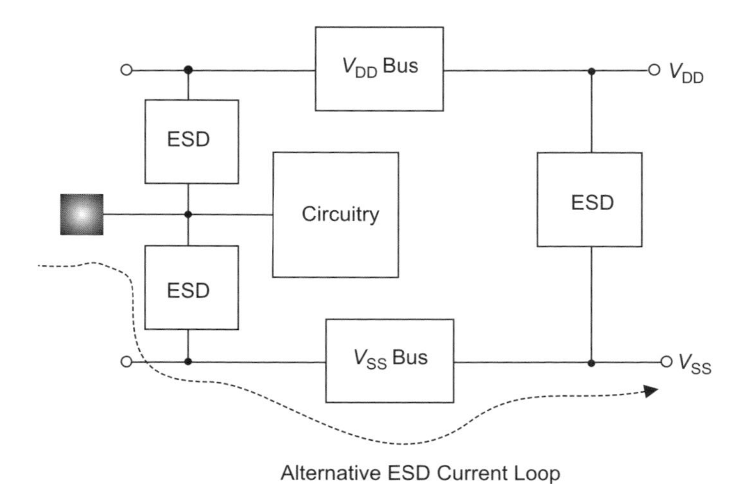

Figure 3.4 ESD Kirchoff current loop to V_{SS} rail

signal pad device, the interconnect between the ESD device and the power grid, and through the V_{SS} power grid. Each interconnect and element in the path must have low resistance, as well as survive the ESD pulse event.

In the case where the grounded reference is the V_{SS} power rail, then it can be represented simply as the voltage drop through the ESD input element which is connected to the ground power rail, and the voltage drop in the V_{SS} power bus:

$$V_{pad} = \{V_{ON} + I_{ESD}\, R\}_{input} + I_{ESD}\, R_{V_{SS}}$$

3.2 SEMICONDUCTOR CHIP IMPEDANCE

From an impedance perspective, the bond pad voltage can be represented as a function of the turn-on voltage, V_{ON}, and impedance through the alternative current loop:

$$V_{pad} = \sum_{i}^{N} V_{ON_i} + \sum_{i}^{N} I_{ESD}\, Z_i$$

Assuming the ESD current flows through all these elements, it can be written as

$$V_{pad} = \sum_{i}^{N} V_{ON_i} + I_{ESD} \sum_{i}^{N} Z_i$$

The impedance of the power grid can be represented as a lumped or distributed element. As a lumped element, the semiconductor chip can be represented as a capacitor, whose impedance is $Z_{chip} = 1/[j\omega C_{chip}]$, and the frequency is associated with the characteristic frequency of the ESD event:

$$Z_{chip} = \frac{1}{j\omega C_{chip}}$$

In a semiconductor chip with no ESD power clamps, and a single power domain, the chip capacitance scales with the chip area. In a CMOS chip, half the chip area will be a capacitor formed by the n-well to substrate metallurgical junction. In a large chip, in older technologies, there was no need for an ESD element between the V_{DD} and V_{SS} power rails.

Depending on the resistance of the power grid, the semiconductor chip impedance may be best understood as a distributed element instead of a lumped element [2]. In that case, the semiconductor chip and power bus are to be represented as a resistor–capacitor transmission line [2].

In advanced technologies, small semiconductor chips, or segmented power supply domains, ESD power clamps are introduced in each power domain. In this case, the domains can be represented as a single bus resistance element, and an ESD power clamp for the effective series elements.

3.3 INTERCONNECT FAILURE AND DYNAMIC ON-RESISTANCE

In the ESD design synthesis of interconnects, there are two primary issues. With the design synthesis of the interconnects, the failure of the interconnect is critical to the determination of the interconnect minimum requirements [1–20]. In addition, the DC resistance as well as the dynamic on-resistance is critical to the determination of the satisfactory resistances in the alternative current loop in the semiconductor chip design [1–3].

3.3.1 Interconnect Dynamic On-Resistance

Resistance is defined according to the linear relationship

$$R(T) = R_o(1 + \alpha T)$$

where $R(T)$ is the dynamic resistance at temperature T, R_o is the initial resistance, and α is the temperature coefficient of resistance (TCR). From this form, we can solve for temperature in the interconnect, as:

$$T = \frac{R(T) - R_o}{\alpha R_o} = \frac{1}{\alpha}\left(\frac{\Delta R}{R_o}\right)$$

To relate power and thermal impedance to the resistance, the differential resistance $dR(T)$ is expressed as:

$$dR = \alpha R_o dT$$

where we define temperature as the product of the heat fluence and the thermal impedance:

$$T = q\theta_{TH}$$

The total differential of temperature can be expressed as:

$$dT = q\theta_{TH} + \theta_{TH}dq$$

The heat flux is equal to the input power from conservation of energy. Substituting in power for the heat flux, the total differential resistance can be expressed as:

$$\frac{dR}{R_o} = \alpha(Pd\theta + \theta dP)$$

Since the impedance is constant, and integration of the expression from the initial to final resistance [1,11–14],

$$\frac{\Delta R}{R} = \frac{R_f - R_o}{R_o} = \alpha\theta_{TH}P$$

Hence the normalized resistance change is proportional to the product of the thermal impedance, power, and temperature coefficient of resistance. Combining Joule heating (e.g., $P = I^2R$) and resistance ($R = \rho L/A$), and substituting for the current density, J, then the expression of normalized differential resistance can be expressed as [1,12–14]:

$$\frac{\Delta R}{R} = \alpha\theta_{TH}J^2\rho LA$$

From the normalized differential resistance, we can solve for the thermal impedance as:

$$\theta_{TH} = \frac{1}{J^2} \frac{[\Delta R/R]}{\alpha \rho L A}$$

From this formulation, the thermal impedance can be extracted from the resistance change and the current density.

3.3.2 Ti/Al/Ti Interconnect Failure

Aluminum interconnects are formed on the planar dielectric film. Aluminum interconnects consist of a refractory metal film below and above the aluminum film. The refractory metal assists in the adhesion of the aluminum with the inter-level dielectric material. Typically, a standard process consists of titanium (Ti), titanium nitride (TiN) – where the TiN provides the adhesion between the dielectric and Ti film. To quantify the ESD robustness of a Ti/Al/Ti interconnect, the geometric definition is important for its evaluation [1,12].

In a Ti/Al/Ti interconnect, the interconnect cross-section is rectangular. The liner film exists on the top and the bottom of the film, but not on the sides of the interconnect. Since the interconnect is a composite film, with the refractory metal being of a high resistance value, an effective resistivity model can be used where we assume the effective resistivity is [1,13,14]:

$$\rho_{\text{eff}} = \frac{\rho_{\text{Al}}}{(1 - 2\delta/y)} = \frac{\rho_{\text{Al}}}{\aleph}$$

where the geometrical dimensions are the liner bottom, δ, and the line height, y.
Assuming a critical energy E_{crit}, at which the interconnect failure occurs, this can be related to the product of the average energy input during a pulse event. Assuming that the current is constant during the pulse, time, and the current can be removed from the average, where the average resistance is the mean between the initial and final resistance value [1]:

$$E_{\text{crit}} = \int_0^{\Delta t} P(t)dt = I^2 \int_0^{\Delta t} R(t)dt = \frac{1}{2}\left[R_o + R_f\right] I^2 \Delta t$$

We can define an expression which is the normalized change in resistance:

$$\gamma_{\text{crit}} = \frac{\left(R_f - R_o\right)}{R_o}$$

and express the critical energy according to the normalized change in resistance at failure:

$$E_{\text{crit}} = \frac{1}{2}[R_o(2 + \gamma_{\text{crit}})] I^2 \Delta t$$

Energy can be related to the heat capacity and heat of fusion [11]:

$$E = C\Delta T + m_i L_f$$

Hence, when the temperature is the critical temperature of failure, then the energy expression is the critical energy. From the resistance relationship on temperature, we can substitute for the temperature change as a function of normalized resistance change, and the temperature coefficient of resistance:

$$E_{\text{crit}} = C\left(\frac{\gamma_{\text{crit}}}{\alpha}\right) + m_i L_f$$

In the case of the composite film, the mass of each region, and specific heat of each film must be defined. In the above development, it implies that the temperature of the composite film is equal. More accurately, we can find the total mass summed over all regions:

$$E_{\text{crit}} = \sum m_i c_i \left(\frac{\gamma_{\text{crit}}}{\alpha}\right) + m_i L_f$$

From the two forms of the critical energy, the critical current to failure can be expressed as [1,12]:

$$I_{\text{crit}}^2 = \frac{\sum_i m_i c_i \left(\frac{\gamma_{\text{crit}}}{\alpha}\right) + m_i L_f}{\frac{1}{2}\Delta t R_o (2 + \gamma_{\text{crit}})}$$

Expanding this expression to address the multiple films (e.g., aluminum, cladding, and insulator materials):

$$I_{\text{crit}}^2 = \frac{\left[(m_i c_i + m_j c_j)\left(\frac{\gamma_{\text{crit}}}{\alpha}\right) + m_i L_f + m_k c_k \left(\frac{\gamma_{\text{crit}}}{\alpha}\right)\right]}{\frac{1}{2}\Delta t \left(\rho_{\text{eff}} \frac{L}{A}\right)(2 + \gamma_{\text{crit}})}$$

In this expression, the cross-sectional area of the interconnect can be brought into the numerator in order to express the equation as a function of the line width and the geometric variables associated with the films.

The physical volume of the Al region can be expressed as [1,12]:

$$V_{\text{Al}} = LWd_s\aleph$$

and

$$\aleph = \left(1 - \frac{2\delta}{d_s}\right)$$

The cladding volume can be expressed as

$$V_{\text{clad}} = LWd_s - LWd_s\aleph = LWd_s[1 - \aleph] = LW[2\delta]$$

The fully passivated thermal sheath volume can be expressed as the volume of the interconnect with the additional dimension associated with the thermal heat volume in the insulator minus the interconnect:

$$V_{\text{sheath}} = \{L + 2d_{\text{ox}}\}\{W + 2d_{\text{ox}}\}\{d_s + 2d_{\text{ox}}\} - LWd_s$$

This form assumes extension beyond the length of the interconnect. This is in the form of:

$$V_{\text{sheath}} = 4[W + d_s + L]d_{\text{ox}}^2 + 2[LW + Ld_s + 2Wd_s]d_{\text{ox}}$$

and the time-dependent oxide sheath thickness is a function of the pulse width and the heat diffusion coefficient in the oxide:

$$d_{\text{ox}}(t) = a_d\sqrt{\Delta t}$$

In the assumption that the highest-level film is unpassivated, this can be expressed as:

$$V_{\text{sheath}} = \{L + 2d_{\text{ox}}\}\{W + 2d_{\text{ox}}\}\{d_s + d_{\text{ox}}\} - LWd_s$$

As an additional assumption, for long-length structures with passivation above and below the interconnect, we can ignore the thermal component on the end of the wires and simplify the analysis to the case of:

$$V_{\text{sheath}} = \{L\}\{W + 2d_{\text{ox}}\}\{d_s + 2d_{\text{ox}}\} - LWd_s$$

Let the region be such that the indices (e.g., *ijk*) be in the order of Al, cladding, and the insulator volume, respectively. Expressing this term as a function of the volumes, heat capacity, and density:

$$I_{\text{crit}}^2 = \frac{2Wd_s\left[(\psi_i V_i C_i + \psi_j V_j C_j)\left(\frac{\gamma_{\text{crit}}}{\alpha}\right) + \psi_i V_i L_f + \psi_k V_k c_k\left(\frac{\gamma_{\text{crit}}}{\alpha}\right)\right]}{\Delta t(\rho_{\text{eff}} L)(2 + \gamma_{\text{crit}})}$$

3.3.3 Copper Interconnect Failure

In today's advanced high-performance technologies, the need for reduced resistance and capacitance interconnects has led to an evolution from Ti/Al/Ti or Al-based interconnect systems to copper (Cu)-based interconnect systems [1,13–15,17,19]. Copper interconnect systems are formed by providing troughs in the inter-level dielectric (ILD) films. The dielectric is etched using reactive ion etch (RIE) processes, followed by cladding material and Cu film deposition. The cladding, or liner material, is typically a refractory metal film. The cladding serves as a diffusion barrier, and provides adhesion to the insulator film. These materials can include TiN, WN, Ta, TaN, or TaSiN [1].

To quantify the ESD robustness of a Cu interconnect, the material and geometry are required. Copper interconnects in the damascene and dual-damascene provide both different failure levels and mechanisms compared with an aluminum-based interconnect system. Hence, a model must be established to address both the thermal physics and geometry. The liner film exists on three sidewalls, but not on the top region. Since the interconnect is a composite film, with the refractory metal being of a high resistance value, an effective resistivity model can be used where we assume the effective resistivity is [1,13,14,17]:

$$\rho_{\text{eff}} = \frac{\rho_{\text{Cu}}}{(1 - 2\Delta/x)(1 - 2\delta/y)} = \frac{\rho_{\text{Cu}}}{\aleph}$$

where the geometrical dimensions are the liner sidewall thickness, Δ, liner bottom, δ, line width, x, and line height, y. Expanding this expression to address the multiple films (e.g., copper, cladding, and insulator materials):

$$I_{\text{crit}}^2 = \frac{\left[(m_i c_i + m_j c_j)\left(\frac{\gamma_{\text{crit}}}{\alpha}\right) + m_i L_f + m_k c_k\left(\frac{\gamma_{\text{crit}}}{\alpha}\right)\right]}{\frac{1}{2}\Delta t\left(\rho_{\text{eff}}\frac{L}{A}\right)(2+\gamma_{\text{crit}})}$$

In this expression, the cross-sectional area of the interconnect can be brought into the numerator in order to express the equation as a function of the line width and the geometric variables associated with the films. The physical volume of the Cu region can be expressed as [1,13,14,17]:

$$V_{\text{Cu}} = LWd_s\aleph$$

and

$$\aleph = \left(1 - \frac{2\Delta}{W}\right)\left(1 - \frac{2\delta}{d_s}\right)$$

The cladding volume can be expressed as:

$$V_{\text{clad}} = L\left[W\left(1 - \frac{2\Delta}{W}\right)\delta + 2d_s\Delta\right]$$

The fully passivated thermal sheath volume can be expressed as [1]:

$$V_{\text{sheath}} = 4[W + d_s + L]d_{\text{ox}}^2 + 2[LW + Ld_s + 2Wd_s]d_{\text{ox}}$$

and the time-dependent oxide sheath thickness is a function of the pulse width and the heat diffusion coefficient in the oxide:

$$d_{\text{ox}}(t) = a_d\sqrt{\Delta t}$$

Let the region be such that the indices (e.g., *ijk*) be in the order of Cu, cladding, and the insulator volume, respectively.

3.3.4 Melting Temperature of Interconnect Materials

The power to failure, P_f, is proportional to the melting temperature of the material [1,4–10]. In semiconductors, the interconnect comprises a conductive material as well as a refractory metal. Table 3.1 shows a listing of materials that occur in semiconductor devices.

Table 3.1 Melting temperature of interconnect materials and the inter-level dielectric films

Material	Melting temperature (°C)
Aluminum	660
Silicon dioxide	1314
Gold	1064
Copper	1084
Titanium	1660
Tantalum	2996
Tungsten	3422

In semiconductor interconnect wiring, the metal layers comprise layers of titanium, aluminum, and titanium. The melting temperature of aluminum is 660°C. The melting temperature of titanium is 1660°C. In the failure of a Ti/Al/Ti interconnect, the aluminum melts first. This is followed by cracking of the ILD on the sides of the Ti/Al/Ti structure [3]. At a later stage, the titanium current density increases, leading to the failure of the titanium films [3].

In advanced semiconductor interconnect wiring, the metal layers comprise layers of a tantalum liner, and copper. The melting temperature of copper is 1085°C. The melting temperature of tantalum is 2996°C. In the failure of a copper interconnect, the copper melts first. This is followed by cracking of the ILD on the top of the Cu/Ta structure [3]. At a later stage, the tantalum current density increases, leading to the failure of the refractory metal film.

3.4 INTERCONNECT WIRE AND VIA GUIDELINES

In ESD design synthesis, guidelines and ground rules are needed to identify the proper wire width and via number based on the ESD objective. In the following section, guidelines for aluminum and copper interconnects are discussed [1].

3.4.1 Interconnect Wire and Via Guidelines for HBM ESD Events

For a given metal film thickness, and via size, there will be a requirement of a given interconnect width and given via number to prevent failure to a given ESD objective [1,11,13].

All interconnects in the path of the ESD current are important to avoid ESD failure. This includes the interconnect wiring layer and the vias between the interconnect layers. The interconnects of importance are as follows:

- Interconnect between the bond pad and the ESD element.
- Interconnect between the ESD element and the V_{DD} power supply rail.
- Interconnect between the ESD element and the V_{SS} power supply rail.
- Interconnect between the power rail and the V_{DD}-to-V_{SS} ESD power clamp.

Table 3.2 is an example of the signal pad to ESD clamp wire width requirements based on the wire film layer, and the HBM ESD objective. In technologies, the interconnect film thickness is a function of the wire layer level. In the table, examples of wire layers are chosen from first metal layer, M1, to last metal layer, M4. To maintain the same current density, as the metal layer becomes thinner, the metal wire width must increase accordingly. In addition, as the ESD specification is increased, the wire width must also increase.

In addition to the metal layer width, the vias between the metal layers must also sustain the ESD event [1,13–15]. ESD design rule checking and verification are integrated into the ESD design methodologies to ensure adequate via numbers. Table 3.3 is an example of the number of vias required for a given HBM ESD objective. In technologies, the physical size of the vias

Table 3.2 Interconnect wire width requirement vs. HBM ESD objective

Wire level	Wire film thickness (μm)	HBM 2 kV	HBM 4 kV	HBM 8 kV	HBM 15 kV
		Wire width (μm)	Wire width (μm)	Wire width (μm)	Wire width (μm)
M1	0.5	4	8	16	30
M2	1.0	2	4	8	15
M3	2.0	0.5	2	4	7.5
M4	4.0	0.25	1	2	3.75

Table 3.3 Via number requirement vs. HBM ESD specification

Via type	Via size (μm × μm)	HBM 2 kV Via number	HBM 4 kV Via number	HBM 8 kV Via number	HBM 15 kV Via number
Via 1–2	0.35 × 0.35	10	20	40	80
Via 2–3	0.35 × 0.35	10	20	40	80
Via 3–4	0.35 × 0.35	10	20	40	80

between metal levels remains the same. For a given ESD specification level, to maintain the same current density, the number of vias between any two metal levels will be constant.

3.4.2 Interconnect Wire and Via Guidelines for MM ESD Events

For the MM specification, the peak current is significantly higher than for the HBM. Experimental results of the ESD failure of interconnects for MM show that there is typically a 10: 1 ratio between the HBM and the MM failure levels for interconnects [13,15]. Table 3.4 shows the interconnect width requirements for MM ESD specification objectives (under the above assumption).

3.4.3 Interconnect Wire and Via Guidelines for CDM ESD Events

In the CDM specification, a semiconductor chip is charged to a voltage through a high-voltage power supply. The semiconductor chip is discharged by a grounding of a signal pin. Charge

Table 3.4 Interconnect wire width requirement vs. MM ESD objective

Wire level	Wire film thickness (μm)	MM 200 V Wire width (μm)	MM 400 V Wire width (μm)	MM 800 V Wire width (μm)
M1	0.5	4	8	16
M2	1.0	2	4	8
M3	2.0	0.5	2	4
M4	4.0	0.25	1	2

Table 3.5 Interconnect wire width requirement vs. CDM ESD objective

Wire level Ti/Al/Ti	Wire thickness (μm)	CDM 1 kV	CDM 1.5 kV	CDM 2.0 kV
		Wire width (μm)	Wire width (μm)	Wire width (μm)
M1	0.5	10	20	40
M2	1.0	5	10	20
M3	2.0	2.5	5	10
M4	4.0	1.25	2.5	5

throughout the semiconductor chip flows from the V_{SS} power rail, substrate, and V_{DD} power grid to the grounded signal pin. In the CDM event, a rise time on the order of 250 ps occurs, with a high peak current. The CDM event current flows through the power grid, and out through the signal pin wiring. As a result, the power bus (both V_{SS} and V_{DD}) and the interconnect wiring must have high enough cross-sectional area to avoid CDM failures. Table 3.5 provides a table of wire thickness and widths in order to achieve CDM specification objectives. The interconnect is assumed to be a Ti/Al/Ti interconnect structure. CDM events can achieve peak current magnitudes on the order of 10 A.

3.4.4 Interconnect Wire and Via Guidelines for HMM and IEC 61000-4-2 ESD Events

System-level events on external ports can impact semiconductor chips. As a result, system-level designers are requesting the requirement of passing the system-level IEC 61000-4-2 event. The IEC 61000-4-2 specification applies to systems that include non-contact testing. Non-contact testing has two components: the ESD current, as well as the energy associated with the E-field and H-field generated from the discharge event.

A second ESD test specification that only addresses the ESD current of the IEC 61000-4-2 specification is the HMM specification. Both these tests have the same ESD current waveform. The peak current from the IEC 61000-4-2 and HMM test can be on the order of 30 to 40 A of current. As a result, the metal width and via number for the signal pins required to pass this specification are 3× to 4× the requirement needed for CDM testing.

3.4.5 Wire and Via ESD Metrics

This section is on the wire and via ESD metrics for the interconnect and via elements. These are the values for ESD failure of the interconnect or via structure as a function of metal level for the different ESD tests (HBM, MM, and TLP).

Table 3.6 gives the values for ESD failure of the interconnect as a function of metal level for the different ESD tests (HBM, MM, and TLP). These are to be used as "rules of thumb" for wire width design for ESD robustness estimation.

Table 3.7 gives the values for ESD failure of the via structure as a function of metal level for the different ESD tests (HBM, MM, and TLP). These are to be used as "rules of thumb" for ESD robustness estimation.

Table 3.6 ESD metrics for wire width for different ESD specifications

Wire level	Wire width (μm)	ESD HBM (V/μm)	ESD MM (V/μm)	ESD TLP (A/μm)
M1	0.4	400	40	0.150
M2	0.4	400	40	0.150
M3	0.4	400	40	0.150
M3 (thick)	3.0	3000	300	1.20
M4	3.0	3000	300	1.20

Table 3.7 ESD metrics for vias for different ESD specifications

Via design layer	ESD HBM (V/via)	ESD MM (V/via)	ESD TLP (A/via)
V12	300	30	0.120
V23	300	30	0.120
V34	300	30	0.120

3.5 ESD POWER GRID RESISTANCE

ESD power grid resistance is important in the ESD design synthesis. The power grid resistance is critical to the ability of the semiconductor current to discharge the current to the power rail bond pads. As an example, assume a bond pad discharges its current to the ESD alternative current loop through the signal pin ESD device, power bus, and the ESD clamp element between V_{DD} and V_{SS}:

$$V_{pad} = \{V_{ON} + I_{ESD}\,R\}_{input} + I_{ESD}\,R_{bus} + \{V_{ON} + I_{ESD}\,R\}_{clamp}$$

Then,

$$V_{pad} - V_{ON_{input}} - V_{ON_{clamp}} = I_{ESD}\{R_{input} + R_{bus} + R_{clamp}\}$$

Assuming the ESD network resistance is zero, then we can define a worst case or critical power bus resistance as:

$$\{R_{bus}\}_{crit} = \frac{V_{crit} - V_{ON_{input}} - V_{ON_{clamp}}}{I_{ESD}}$$

The critical bus resistance, $\{R_{bus}\}_{crit}$, is the maximum the bus resistance can be prior to failure at the bond pad (see Figure 3.5). The critical voltage, V_{crit}, at the bond pad is a function of the circuitry at the signal pin. Note that in this definition, all other resistances in the alternate current loop are regarded as negligible.

3.5.1 Power Grid Design – ESD Input to Power Grid Resistance

The design of the interconnects between the ESD network and the power rail can influence the ESD robustness of the semiconductor chip. The current flow from the ESD network to the power rail is important to provide both a low-resistance path and good current distribution.

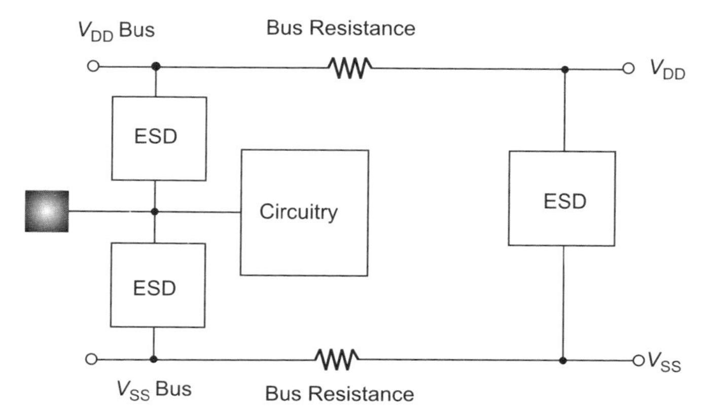

Figure 3.5 ESD power grid bus resistance

Design symmetry is important in the design to ensure good current distribution through the interconnects. Key design features of the interconnect are as follows:

- **Orientation:** Broadside orientation of the wire interconnects (e.g., the same direction as the current flow and orthogonal to the stripes).
- **Wire Width:** Adequate wire width to avoid interconnect failure, degradation, or self-heating.
- **Via Number:** Adequate via number to avoid via failures, degradation, or self-heating.
- **Resistance:** Low resistance between the ESD network and the power rail (wiring and vias).
- **Wire Design Symmetry:** Design symmetry of the wire interconnects from the ESD structure to the power grid.
- **Via Design Symmetry:** Design symmetry of the via number and placement from the ESD structure to the power grid.
- **Segmentation:** Segmentation of the wire interconnects to ensure current distribution.
- **Comb Design:** A wide comb region for the connection between the wire interconnects to the power grid for a low-resistance region across the ESD structure.

Figure 3.6 shows an example of a signal pad, an ESD network, and the wire interconnects between the ESD device and the power rail.

3.5.2 ESD Input to Power Grid Connections – Across ESD Bus Resistance

As power bus widths are scaled, the resistance along the bus as it passes through the ESD network can have an influence on the ESD signal pin network operation. In a sub-45 nm technology, in a low pin-count application, the bus resistance across the ESD network can be significant. In the case that the width of the ESD network is on the order of the spacing between an input network and its local ESD power clamp, the resistance across the ESD network can be a large percentage of the ESD bus resistance. Define the parameter "across-ESD bus

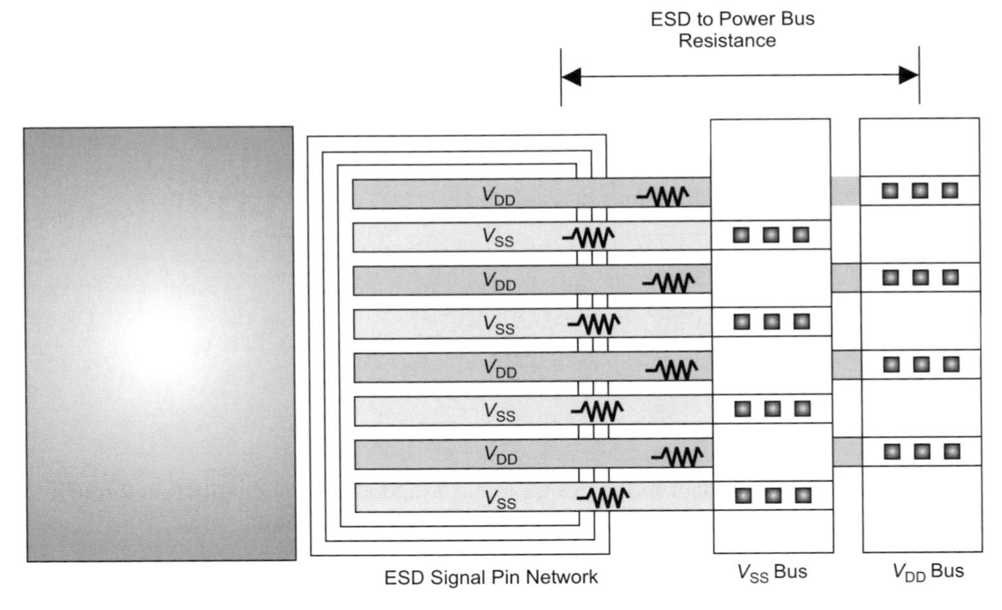

Figure 3.6 Signal pad to ESD signal pad input circuit

resistance" as the amount of bus resistance across the signal pin ESD element (Figure 3.7). In the case that the connection of the bond pad ESD to the power bus is perpendicular to the power bus, the ESD current will favor the section of the ESD input network closest to the ESD power clamp. This can lead to premature ESD failure of the signal node and via structures. In a 45 nm application, this leads to ESD failure at 150–175 V MM, and failure of the first via group which has the shortest path to the ESD power clamp. With the reduction of the "across ESD bus resistance," the ESD results increased to 250 V MM.

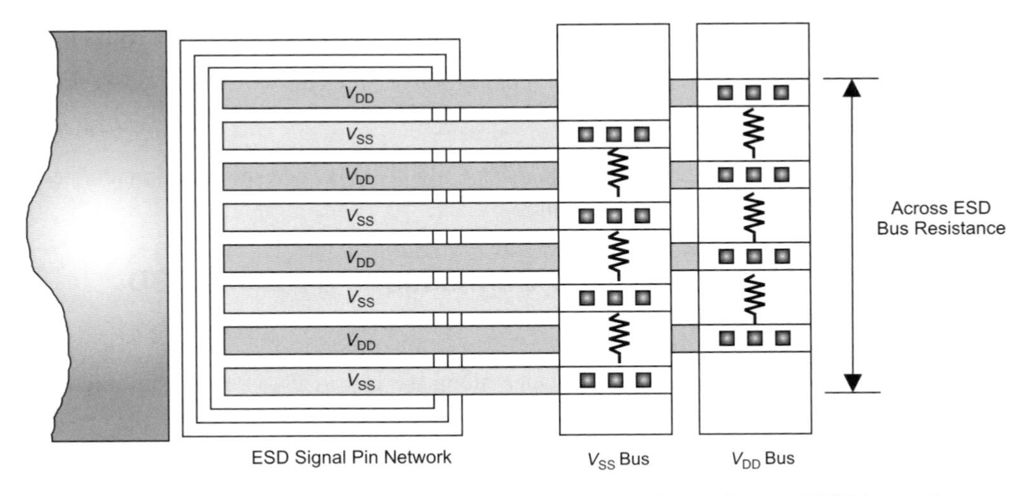

Figure 3.7 ESD input circuit to power grid resistance connections – "across ESD" bus resistance

3.5.3 Power Grid Design – ESD Power Clamp to Power Grid Resistance Evaluation

The electrical connections between the ESD power clamp and the power grid are important to ensure proper operation of the ESD power clamp element. ESD power clamps typically contain a trigger element (or network) and a clamping element. The trigger network initiates either the base (in the case of bipolar transistors) or gate (in the case of MOSFET transistors). The clamp element or clamp network provides a "sink" to discharge the ESD current. Figure 3.8 shows the electrical connections between the power bus and the ESD clamp element. Figure 3.9 highlights the resistance in the network. In the implementation, the following features are required for the electrical connection between the power bus and the ESD power clamp:

- **Wire Width and Via Number:** ESD wire width and via number must be able to support the ESD current magnitude.

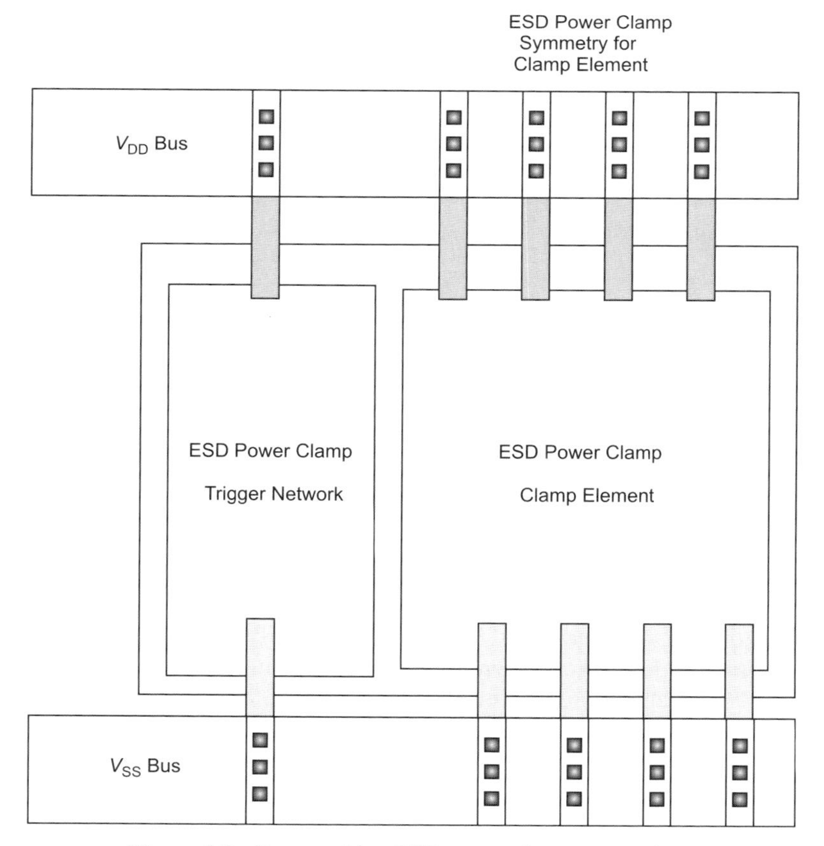

Figure 3.8 Power grid to ESD power clamp connections

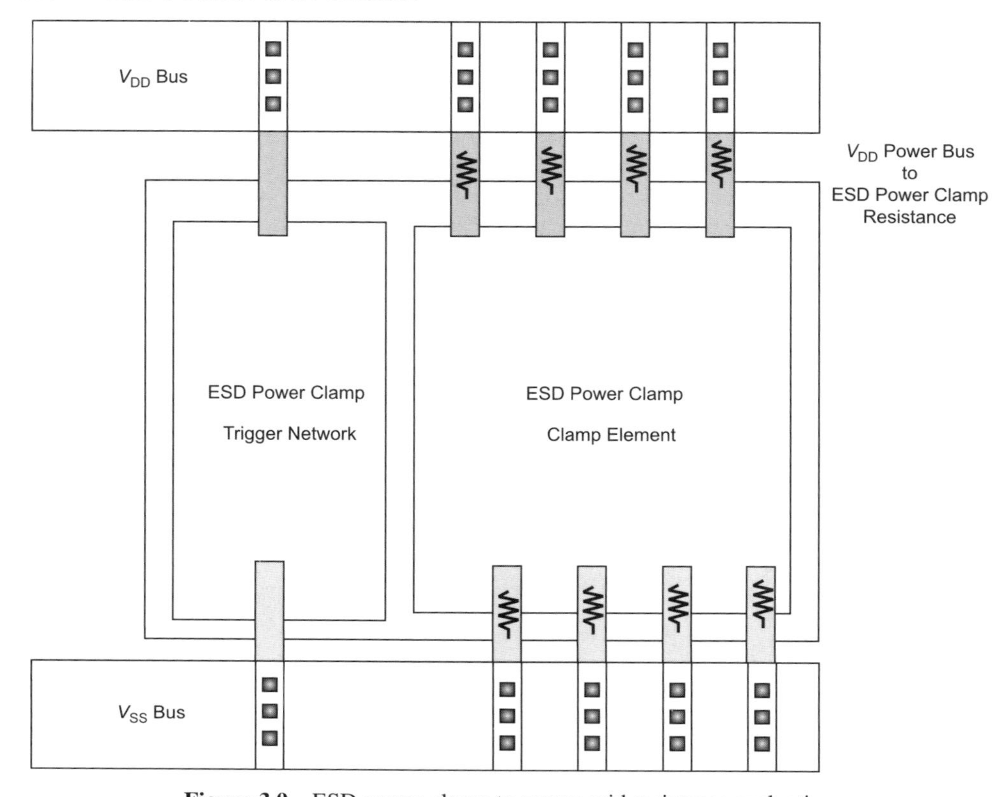

Figure 3.9 ESD power clamp to power grid resistance evaluation

- **Clamp Element Design Symmetry:** Design symmetry is required in the region of the ESD power clamping element to provide uniform current distribution in the clamp element.
- **Uniform Resistance and Symmetry:** Design symmetry to provide uniform resistance distribution.
- **Low-Resistance Connections:** Low-resistance connections in series with the ESD clamp elements.

3.5.4 Power Grid Design – Resistance Evaluation

A semiconductor chip, with peripheral I/O circuitry, typically has a peripheral power bus for the networks. Figure 3.10 shows a typical architecture with a V_{DD} power bus, and a V_{SS} power bus. In this architecture, the bus is typically of uniform width around the entire semiconductor chip. A key design parameter is the bus resistance per unit length around the semiconductor chip. A key ESD design metric is the physical resistance along the bus between any signal pin and the ESD power clamp. Figure 3.11 illustrates the resistance between a signal pin and the ESD power clamp. Note in this discussion, only a single distance in one direction is shown.

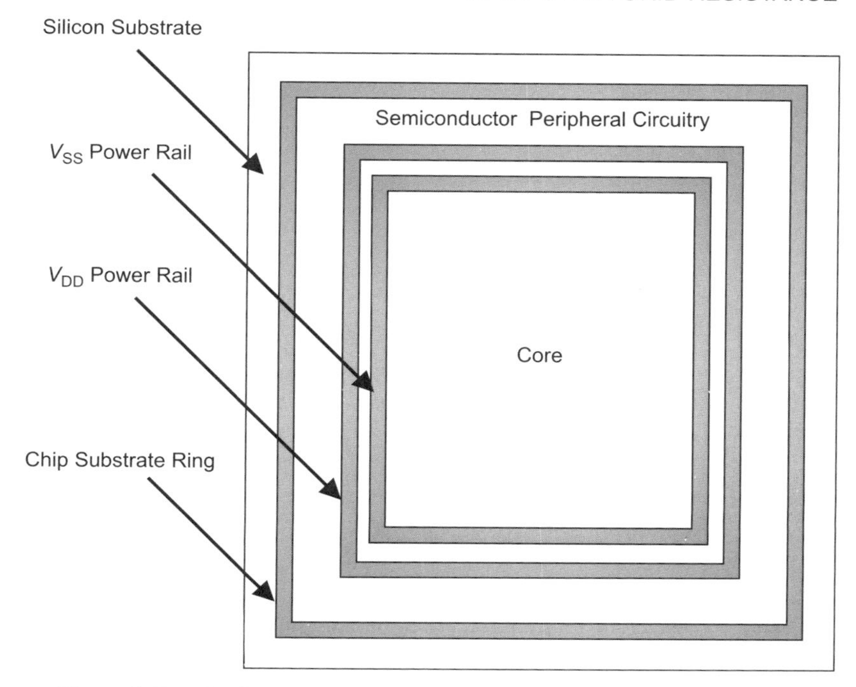

Figure 3.10 Semiconductor chip with a peripheral V_{DD} and V_{SS} power bus

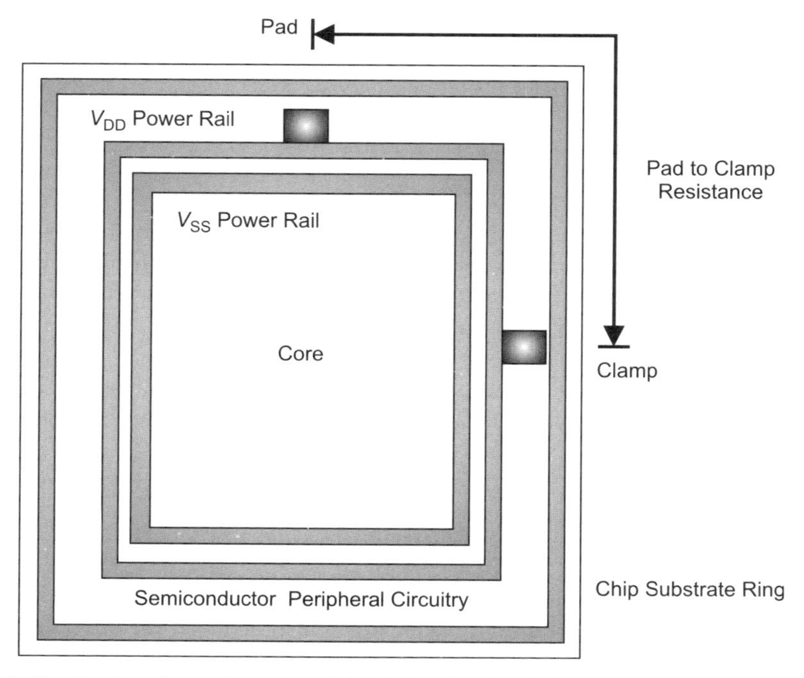

Figure 3.11 Semiconductor floorplan highlighting the signal pin to ESD power clamp distance

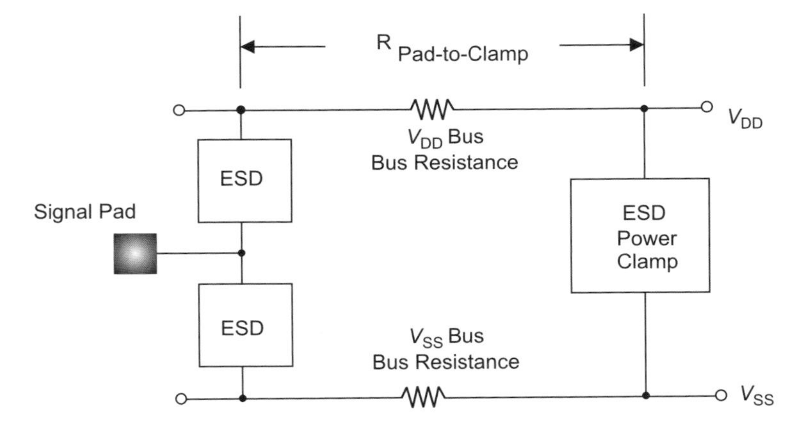

Figure 3.12 Power grid design – resistance evaluation for single current direction current path.

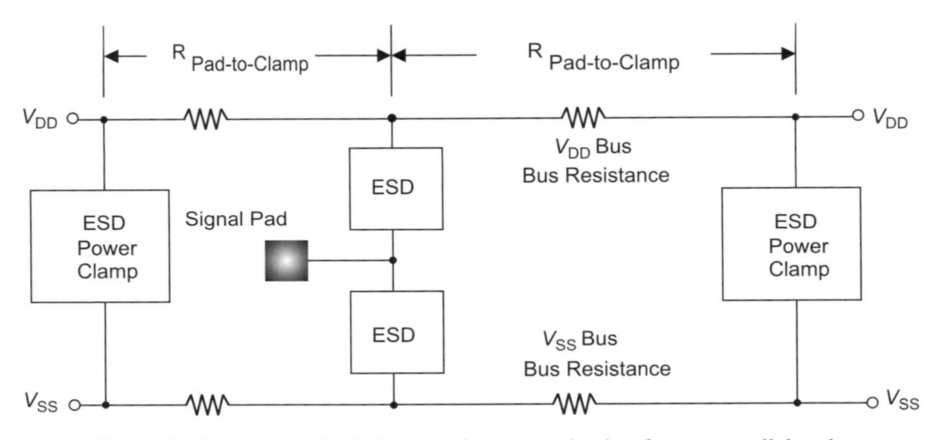

Figure 3.13 Power grid design – resistance evaluation for two parallel paths

This can be represented schematically as shown in Figure 3.12. In this schematic, the key elements are the bond pad, the ESD element, the bus resistance, and the ESD power clamp.

For completeness, in the opposite second direction, there is a second resistance in the power grid.

Figure 3.13 shows an example of the second parallel resistance to the power clamp in the other direction. The two resistance values are in parallel. In this case, the current flows in two directions to the same power clamp or to separate power clamps.

3.5.5 Power Grid Design Distribution Representation

A key parameter in ESD design is the power clamp to power clamp frequency [21]. Figure 3.14 shows a floorplan with a peripheral I/O power bus, where the ESD power clamp to ESD power clamp spacing is highlighted.

In the ESD design synthesis, the bus resistance between any signal pin to the adjacent power clamp can be plotted [21]. Figure 3.15 shows a routing resistance between

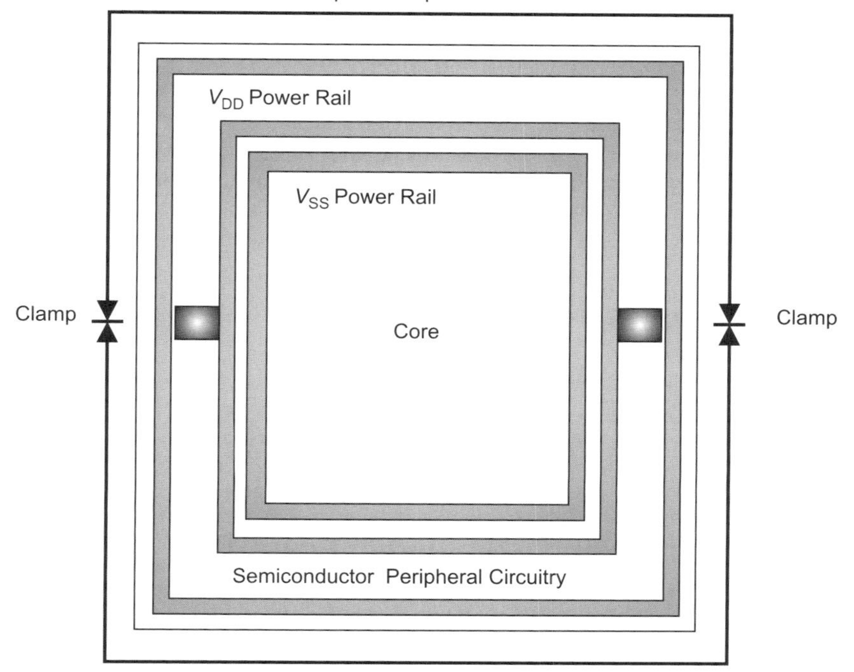

Figure 3.14 Power grid design – ESD power clamp to power clamp resistance evaluation

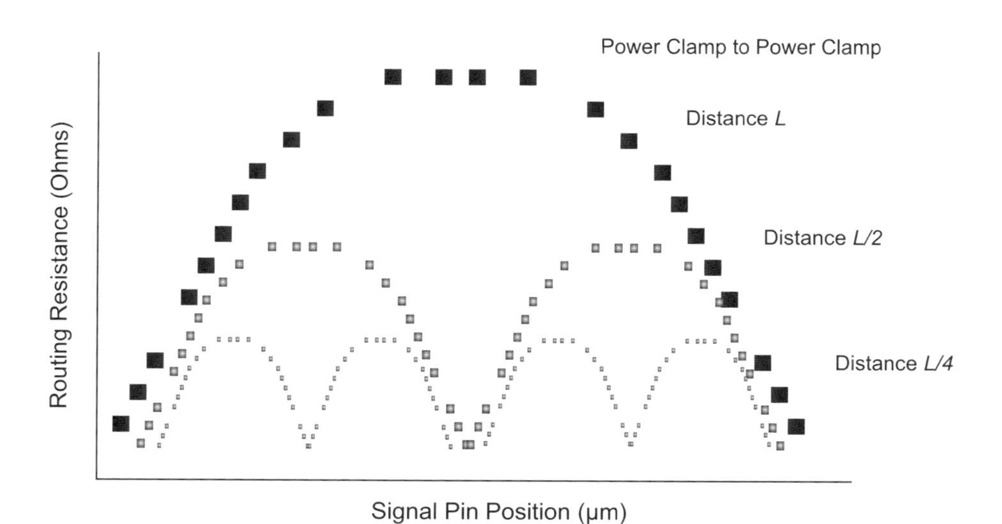

Figure 3.15 Routing resistance distribution of signal pin to ESD power clamp as a function of pin position

any signal pin and the power clamp. In the figure, the first curve is two power clamps. In this case, the worst case pin to power clamp resistance is in the center between the two power clamps. A key ESD design synthesis metric is the worst case resistance between any signal pin and the power clamp. Assume the distance between the two power clamps is L. Hence, the peak of this distribution is the worst case value at position $L/2$. Adding a third power clamp at position $L/2$ then reduces the worst case position at $L/4$ and $3L/4$. As extra power clamps are added to the design, the worst case resistance value is reduced.

3.6 POWER GRID LAYOUT DESIGN

In the power grid design, the width of the power bus is a concern for both chip area and semiconductor processing. In the following sections, a brief discussion of slotting and stacking will be given.

3.6.1 Power Grid Design – Slotting of Power Grid

In semiconductor chip design, the width of metal shapes is limited in the semiconductor process by the technology ground rules. For example, due to chemical mechanical polishing, widths are limited to avoid "dishing" of the metal shapes during the polishing process. As a result, "slots" are formed in the large metal shapes to serve as polishing stops, and lower the effect of dishing. Figure 3.16 shows an example of a power bus highlighting the slots. In the design implementation, the design rule typically states that after a given width, a slot must be formed of a specified width. For example, for every 10 μm of wire width, a slot 1 μm wide may be required to be formed. Hence, for a 20 μm wide power bus, there would be a single slot formed in the center of the power bus.

In ESD design synthesis, the power bus resistance is an ESD design rule (e.g., a resistance limit of the power bus between any signal pin and the ESD power clamp). With the placing of slots in the metal width, the effective width, and hence the resistance of the

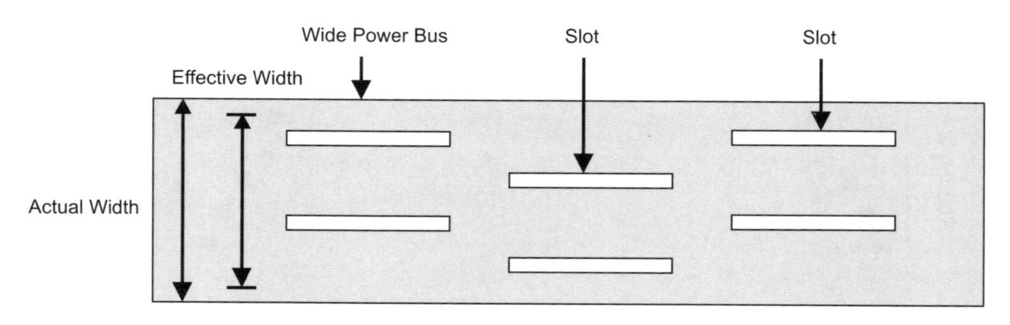

Figure 3.16 Power grid with slots

power bus, is impacted. To address this concern, the following can compensate for power bus slotting:

- **Width Compensation:** Increase the width of the power bus by the amount of width loss due to slotting.
- **Effective Width Model:** Develop a model to create an "effective width," where the effective width satisfies the ESD design ground rule for bus width.
- **Resistance and Distance Compensation:** Decrease the distance between the worst case pin to the nearest ESD power clamp.

3.6.2 Power Grid Design – Segmentation of Power Grids

In semiconductor chip design, the width of metal shapes is limited in the semiconductor process by the technology ground rules [22]. Slots are formed in the large metal shapes to serve as polishing stops, and lower the effect of dishing. Additionally, wide metal shapes can lead to significant inductance in chip design. A method to avoid "slotting" of the metal power bus is segmentation of the power grid into widths below the maximum width rule.

Figure 3.17 shows an example of a power bus highlighting a segmented power rail architecture. One of the advantages of a segmented power rail is for standard

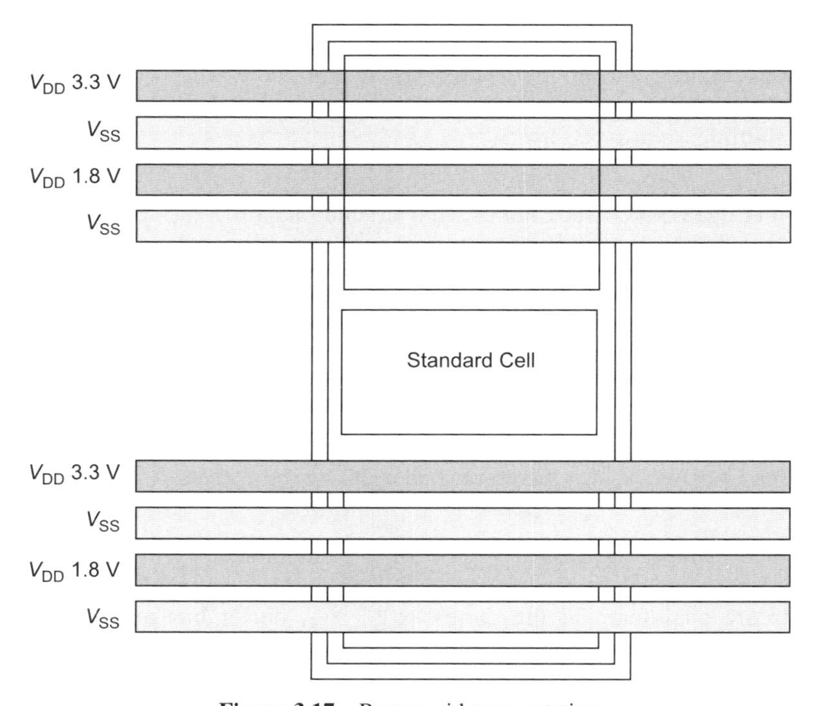

Figure 3.17 Power grid segmentation

cell implementations. In a standard cell, the peripheral cells become narrow and long. The power rails and ground rails must pass through the off-chip driver PFET and NFET, ESD network, and pre-drive circuitry. In order for the power rails to pass through the region of the peripheral rails, the power rail must be segmented into multiple rails.

An ESD concern with this concept is that adequate wire width must still be placed in the region of the ESD element for both positive and negative discharging elements. This will become more critical as technologies scale to smaller dimensions [22].

3.6.3 Power Grid Design – Chip Corners

In the ESD design synthesis, the corners of the semiconductor chip are important in the design process. The power grid design in the corners influences both the functional and ESD results by adding resistance to the power grid.

For ESD protection, it is important to have low resistance between the ESD networks at the signal pins and the ESD power clamps placed between V_{DD} and V_{SS}. Current must flow from the ESD signal pin elements, through the V_{DD} power rail, and then the V_{DD}-to-V_{SS} ESD power clamp. The bend at the corners of the power grid can lead to current crowding on the corner, increasing the resistance. When the ESD V_{DD}-to-V_{SS} power clamp is placed around the bend relative to a signal pin prior to the bend, ESD results of different signal pins can be influenced. How this manifests itself is that signal pins on the same side as the ESD V_{DD}-to-V_{SS} power clamp will be higher than signal pins around the bend (e.g., the other edge). Additionally, given there are multiple ESD power clamps on the pad ring, the series resistance of one ESD power clamp will be less than the other, preventing them from responding in parallel. To reduce these effects, there are multiple solutions:

- **Additional Edge ESD Power Clamps:** Add an additional ESD V_{DD}-to-V_{SS} power clamp on each edge of the semiconductor chip.
- **Bend ESD Power Clamps:** Place an ESD power clamp on both edges of the bend in the power grid (e.g., two ESD power clamps on the corner on each edge of the bend).
- **Bus Chamfered Corners:** Chamfer the corner of the power grid to lower the series resistance.
- **Wider Bus Corners:** Widen the power bus width on the corners.

Figure 3.18 shows the corner region of a semiconductor chip that introduces a chamfered corner to reduce the resistance in the corner region of the power rail. All power rails are chamfered at the corner (e.g., V_{DD} power bus and V_{SS} power bus). Typically, an ESD power clamp is placed in the corner region or inline with the signal pad cells.

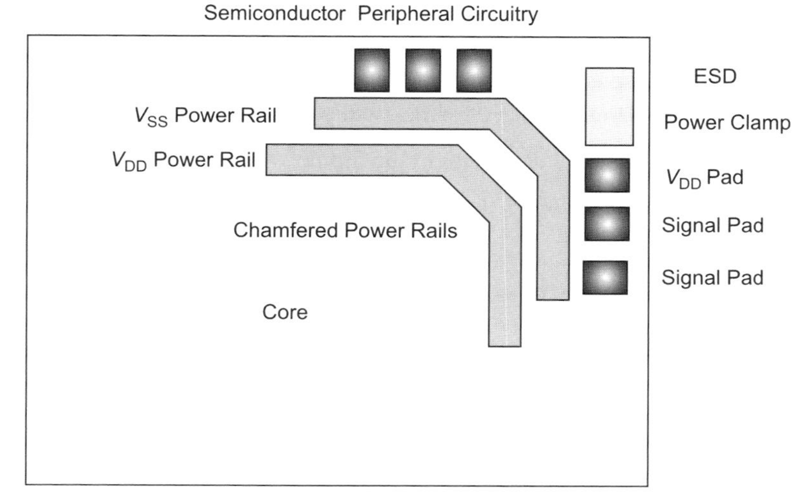

Figure 3.18 Power grid design at chip corners

3.6.4 Power Grid Design – Stacking of Metal Levels

In the ESD design synthesis of semiconductor chips, it is important to maintain low resistance between ESD signal pad networks and the power pads as well as to the ESD power clamp.

In the architecture of some applications, few metal design layers are used in the design. Additionally, the thinner metal layers are the lowest levels of metal. To maintain low resistance, and reduce chip area, it is necessary to "stack" the metal levels in the power bus.

Figure 3.19 shows an example of a "stacked metal bus" which utilizes multiple metal levels for the power or ground rail. Using multiple metal levels, a low-resistance power bus can be

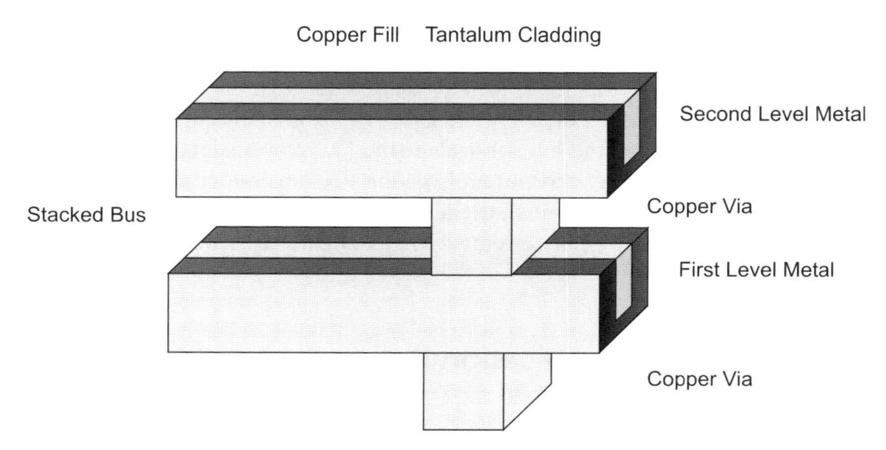

Figure 3.19 Power grid stacking of metal levels

formed without using significant area on the semiconductor chip. Design architectures where this is valuable are the following:

- **SRAM Design:** SRAM semiconductor chips with peripheral I/O.
- **DRAM Design:** DRAM semiconductors in the "spine" region.
- **Image Processing Design:** CMOS image processing chip with peripheral I/O.

3.6.5 Power Grid Design – Wiring Bays and Weaved Power Bus Designs

In chip design synthesis, some regions of the semiconductor chip area are reserved for "wiring bays" for signal lines. An ESD power bus technique is to allow "breaks" in the single or stacked buses to allow passage of the wiring bays. An ESD design synthesis practice is to allow "stitching" around the wiring bays to lower the ESD bus resistance, but allow no interference with the signal passing through the region of the power bus.

3.7 ESD SPECIFICATION POWER GRID CONSIDERATIONS

For special testing or specifications, additional power grid considerations must be addressed. In the next sections, some of the considerations for CDM, IEC 61000-4-2, and HMM specifications will be highlighted.

3.7.1 CDM Specification Power Grid and Interconnect Design Considerations

In the CDM specification, the peak current can exceed 10 A levels. CDM events can lead to failure of power buses, or signal lines, or any interconnect which prevents current flow to the grounded signal pin [22].

In bulk CMOS semiconductor chips, current can flow through the V_{DD} power grid back to the grounded signal pin; in bulk CMOS, current can flow through the entire substrate region as well. The current can flow from the V_{SS} substrate to the V_{DD} power supply through all the well structures. But in SOI technology, the current from the V_{SS} substrate to the V_{DD} power supply can only flow through the interconnects instead of the n-well regions. As a result, the power rails in SOI can fail due to the peak currents during CDM testing; whereas for the identical design in bulk CMOS, no failure occurs in the power grid.

3.7.2 HMM and IEC Specification Power Grid and Interconnect Design Considerations

For HMM and IEC 61000-4-2 specifications, the peak current can exceed 30 to 40 A. In the HMM and IEC 61000-4-2 specifications, only pins which are connected to external ports are

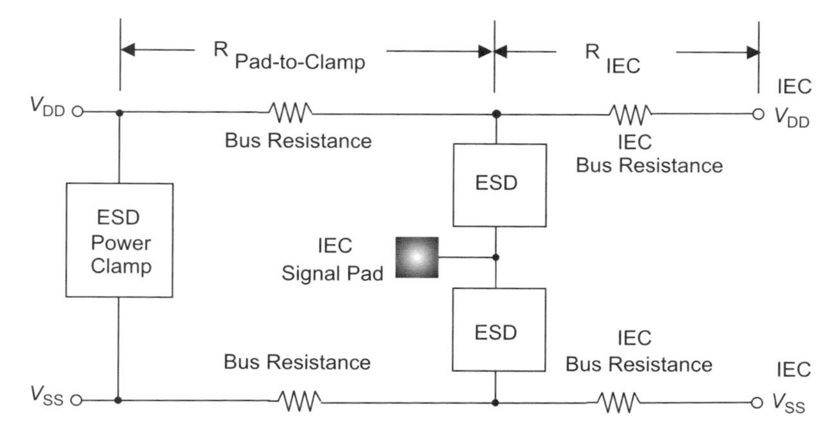

Figure 3.20 IEC bus segmentation

required to receive this high current pulse. Secondly, the large current in the substrate can influence the non-IEC tested circuitry. To protect the IEC tested pins, and avoid failure of the non-IEC pins, the IEC pins can be isolated in the power grid. This can be achieved by the following means:

- **Independent IEC power domain:** IEC signal pin, IEC V_{DD} bus, and IEC V_{SS} bus.
- **Dual width power bus:** IEC and non-IEC domains.
- **Resistance Segmentation:** Resistance separated IEC and non-IEC domains.

Figures 3.20 and 3.21 show examples of separate IEC vs. non-IEC power grids and connectivity.

3.8 POWER GRID DESIGN SYNTHESIS – ESD DESIGN RULE CHECKING METHODS

ESD design rule checking (DRC) and verification methodologies can be used to avoid ESD failures [22–28]. Examples of ESD design synthesis methods will be discussed in this section.

3.8.1 Power Grid Design Synthesis – ESD DRC Methods Using an ESD Virtual Design Level

ESD design checking and verification is performed as part of the ESD design synthesis. A simple methodology can be utilized to verify the metal layer design widths between the signal pad and the ESD, and between the signal pad and the ESD power grid.

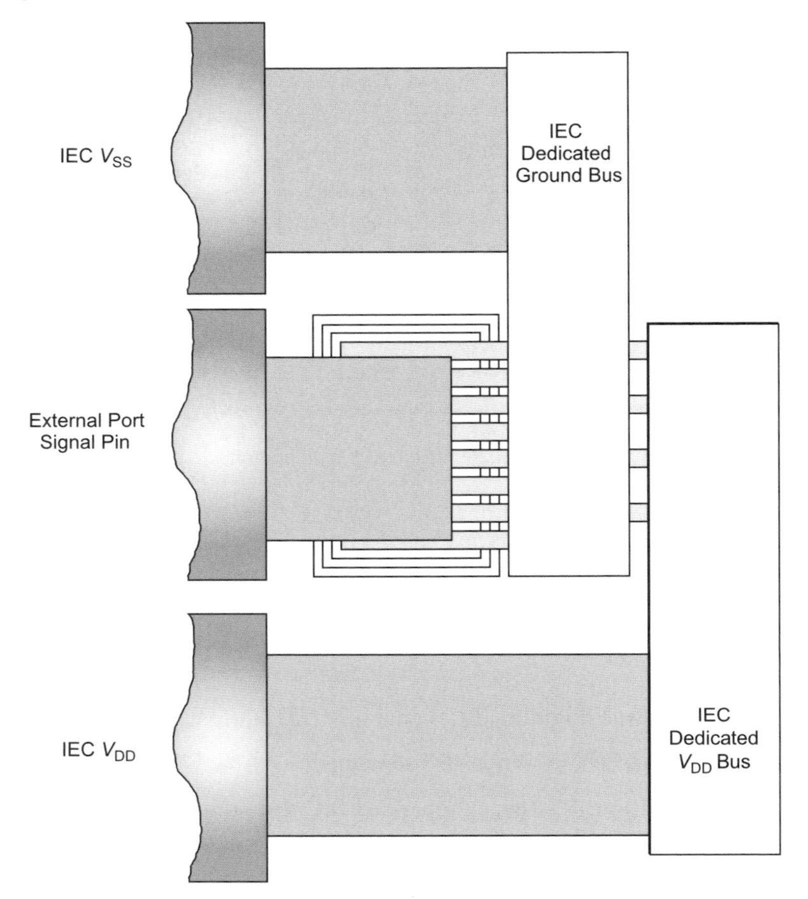

Figure 3.21 IEC bus power and ground placement

An ESD design checking and verification methodology was developed to evaluate the ESD robustness of the interconnects [23]. In this methodology, minimum wire width and maximum resistance constraints are applied to each of the chip's I/O signal pads. These constraints are propagated to the layout design and array pads wired to I/O cells located on the chip. Thus, wiring is such that interconnect wire layers and vias to the ESD protection devices are wide enough to provide adequate ESD protection level. The whole-chip semiconductor design is then verified by first identifying the chip pads, I/O cells, and ESD protection devices. All the electrical connections between these three structures are verified. In this methodology, the power rails are checked between the power rail pads and the ESD power clamps.

The methodology incorporates an "ESD dummy design layer" that identifies the existence of the ESD circuit elements [23]. The virtual ESD dummy layer is placed on the guard ring that surrounds the ESD elements. The methodology utilizes a "shrink" of the wires that are less than the desired ESD metal layer design rules. Wires between the ESD protection devices and the chip pads and I/O cells are shrunk such that unsuitable connections become "opens" (disconnected) and are found in subsequent checking. In this methodology, ESD protection on power lines on integrated chips is verified. The wire and via interconnect ESD protection level between chip pads and an ESD network, between power rails and an ESD network, and on power rails, ground rails, and between power and ground rails.

A method of checking the power grid and electrical connections of an integrated circuit chip for ESD robustness uses the following steps: (1) identify I/O pads on a chip; (2) identify ESD protection devices; (3) represent ESD protection devices as a dummy ESD shape; (4) identify a network comprising wires and vias connecting I/O pads to the ESD dummy shape; (5) determine effective wire cross-sections of the sum of any wires which are connected in parallel to the ESD dummy shape; (6) eliminate connections of vias, wires, and wire cross-sections below a minimum constraint to create "opens"; (7) check for opens to determine I/O pads not connected to the ESD dummy shape; and (8) record any of the said I/O pads not connected to said ESD dummy shape.

The formation of "opens" significantly increases the speed of the ESD design checking system from days to minutes in a large ASIC design. All small interconnect shapes are eliminated below the ESD rule dimensions. Additionally, the ESD virtual shape and the bond pads are saved. In a large ASIC design, this method provides a rapid design rule check and verification of ESD design rule conformance.

3.8.2 Power Grid Design Synthesis – ESD DRC Methods Using an ESD Interconnect Parameterized Cell

In a Cadence™ environment, different techniques can be applied to the ESD design synthesis process. Voldman, Strang, and Jordan developed an ESD methodology incorporating a hierarchical ESD parameterized cell (p-cell) in a foundry environment [24–26]. This concept can be extended by having the power bus and interconnects themselves become a p-cell [28]. In one ESD design methodology, the interconnect is converted into a parameterized cell – a p-cell known as "ESD Interconnect." In a typical environment, the wiring is not a parameterized cell but by conversion and identification of the power bus and interconnects as "ESD Interconnects," a checking and verification technique can be applied [28].

As are all circuits, high-level ESD circuits comprise a plurality of lower-level sub-circuits which, in turn, can be expressed in terms of still lower-level elements. The lowest-level ESD elements may be expressed as simple parameterized cells ("p-cells") – e.g., resistor p-cells, transistor p-cells, varactors p-cells as well as any other basic electronic component. These, in turn, can be used to express higher-level parameterized circuits. These parameterized circuits can, in turn, be connected with parameterized interconnects to ultimately form the ESD circuit. The p-cells exist in a computer aided design (CAD) environment, and are essentially a computer model of the particular element comprising all the parameters necessary for the computer to simulate that element.

In this methodology, by conversion of the interconnect wiring into a parameterized cell itself, the design system provides different means to check and verify the electrical connections. With the introduction of this concept, significant function can be provided to the ESD design synthesis. The ESD design methodology of Voldman, Strang, Collins, and Jordan provides the following concepts [28]:

- A component to verify a connection between a pad and an ESD network by verifying and checking electrical connectivity; a component to verify the width requirements to maintain ESD robustness to a minimum level.
- A component to verify that based on the ESD robustness of the ESD network, the interconnect width and via number are such as to avoid electrical interconnect failure prior to the ESD network failure.
- A component to provide for multiple lines in parallel whose cross-section can be maintained and evaluated as a set of parallel interconnects connected to a single ESD network or plurality of ESD networks.
- A component to provide for "ESD ballasting" by dividing into a plurality of lines.
- A component to provide for calculation of the ESD robustness of the interconnect based on pulse width, surrounding insulator materials (e.g., SiO_2 or low K materials), metal level, and distance from the substrate (thermal resistance based on the metal level or underlying structures).
- A component to provide for surrounding fill shapes.
- A component to provide and adjust for "cheesing" of the interconnect.

In this "ESD Interconnect" hierarchical p-cell, the flow of the method establishes the interconnection path of the pad level. Next, an ESD interconnect is verified. The method then verifies an ESD via at the level below. This method is repeated until the lowest level connects to highest metal design level of the input of the ESD network of the corresponding ESD device to that pad. The "ESD Interconnect" can be a single p-cell which contains multiple levels of metal from pad (i) to ESD p-cell (j) where the system verifies the connectivity.

In this method, the ESD interconnect p-cell and algorithm can be established, which prevents the metal level going below a given ESD width. The minimum width can be established by conversion of the metal shape into a p-cell where the metal has an algorithm with a minimum function where the width never goes below a given width defined by the minimum ESD requirement. Checking and verifying that the correct wire width and via number are never below the ESD robustness level of the circuit can be done using the information of the "inherited parameters" contained in the "translation box" formed around the electrical schematic of the hierarchical parameterized cell. The electrical schematic translation box contains the circuit type, the inherited parameters, and pin connections. The translation box may also contain functions – including, for example, ballasting, fill, and cheese. The translation box will allow the transformation of the schematic to the graphical and vice versa. From this, the ESD robustness of the circuit can be determined and stored in the circuit from electrical measurement tables of the design system, as discussed below. Also, from this, the verification that the ESD Interconnect structure p-cell is more ESD robust can be calculated from the ESD robustness wire calculations.

By way of illustration, ESD robustness of an interconnect can be calculated based on the metal level and effective metal width. For example, for a given technology file, the metal film thicknesses and materials are known. In this fashion, from the design level, and the technology file, and experimental data, a table is constructed. Also in this fashion, a "lookup table" is constructed which is based on design data and experimental results – the size of the interconnect can be judged as achieving the ESD objective. In the GUI, the input variable can choose an ESD model such as HBM. Once the GUI choice is made, the ESD HBM level can then be chosen. For example, if the GUI input is that the HBM level is to be greater than 4000 V, then all interconnects from the pad must be auto-generated to increase to a minimum width.

For the case of copper interconnects, the HBM ESD robustness level is 2× the aluminum level for the same thickness.

In the case of the MM, ESD robustness levels are 5× to 10× lower. In this fashion, in a "lookup table" which is based on design data and experimental results, the size of the interconnect can be judged as achieving the ESD objective. In the GUI, the input variable can choose an ESD model such as MM. Once the GUI choice is made, the ESD MM level can then be chosen.

In the case that multiple models, such as HBM, MM, and CDM, levels are required, the design system can be established so that the metal line width is such that all models achieve the desired levels and the minimum thickness is chosen so that all models are satisfied [28].

Analytical models, such as those developed by Wunsch–Bell or Smith–Littau, can be utilized for prediction of the critical current or power to failure. Analytical models can be used which require the heat capacity, the thermal conductivity, the melting temperature, and the pulse width of the event instead of empirical look-up tables. In this fashion, the GUI would either store the material properties and/or allow user-defined properties that are required for the analytical equation. Using the analytical models, material properties of the metal wire and the insulator properties can be used which will allow predictive capability for a given ESD event. The metal line would then be auto-generated to guarantee non-failure to that ESD event of a given current level, voltage level, or pulse width [28].

In the case of filling, the fill shapes change the effective thermal conductivity of the insulator. This can be handled in the analytical models by modifying the effective thermal conductivity of the surrounding medium. In the case of "cheesing," holes are formed in the metal. In this case, the metal width must be increased to allow for the total cross-sectional area the same.

In the case of ballasting, the metal line can be separated into a plurality of parallel wires where the wire widths are such that the total width is equal to the calculated width based on the analytical model or the lookup table result. Ballasting can be an option or a requirement of a design, or implementation.

Plurality of parallel lines and ESD ballasting needs can also be addressed by the ESD Interconnect p-cell. In the case where it is a requirement that interconnect ballasting of the ESD device is a critical need, a check can be performed where the lowest level of metal of the ESD Interconnect p-cell is divided into a plurality of interconnects which integrates with the ESD network.

If the ESD network translation box contains information that ESD ballasting is required for that specific design, the check then verifies that the ESD interconnect also contains this requirement or design failure is stated. Hence, an ESD Interconnect p-cell can have as a parameter the formation of a plurality of interconnects, and also verifies that this feature is "checked" relative to the ESD network translation box information of the ESD p-cell for the highest level of the p-cell and the lowest level of the ESD Interconnect p-cell.

To verify the presence of ESD power clamps between two power rails, the checking system provides a verification step where the labeled power pads are also connected such that the ESD Interconnect p-cell is utilized for the power grid, and an ESD power clamp. The verification and checking system will check:

- **ESD Interconnect:** The presence of the ESD Interconnect p-cell.
- **ESD Power Pad to ESD Interconnect:** The interconnection between the power pad and the ESD Interconnect p-cell.
- **ESD Power Clamp:** The presence of an ESD power clamp.
- **ESD Power Clamp Type:** The type of the ESD power clamp.
- **ESD Power Clamp Size:** The size of the ESD power clamp.

In this case, the verification and checking system will verify:

- The "connectivity" from power pad, ESD interconnect;
- ESD power clamp; and
- ESD interconnect and a power ground pad.

Between two ground power rails, or a common potential of two separated power supplies, the verification and checking of the ESD rail-to-rail device can be verified against its ESD Interconnect connection. ESD ballasting, ESD robustness, and inherited parameters can be contained in the translation box and stored for cross-comparison between the pads, the interconnect, and the ESD network [28].

3.9 SUMMARY AND CLOSING COMMENTS

In this chapter, the discussion has continued to address issues associated with full-chip ESD design synthesis. The chapter focuses on the interconnects and power grid layout and design itself, and addresses interconnect robustness, interconnect failure, and key metrics in the whole-chip ESD design synthesis. With an understanding of the inter-relationship between the power grid and the ESD networks, full-chip integration can be addressed. With our "top-down" approach, it is possible to focus on what kind of circuits are placed in the power grid. This leads us into the next chapter.

In Chapter 4, the discussion focuses on ESD power clamp networks. A key issue in ESD design synthesis is the type of ESD network used in the power grid domain. This was discussed briefly in previous chapters, but the whole of Chapter 4 will be dedicated to ESD power clamp discussions.

PROBLEMS

3.1. Assuming a peripheral pad architecture, a chip is formed with I/O and bond pads on the periphery. Draw a semiconductor chip that has dimensions w and l. Assume the bond pads for the circuits are w_{BP} and l_{BP}, with spacing w_{sp} and l_{sp} between bond pads. Assume an ESD power clamp is placed on the corners of a chip. Assuming a bus width of W_{BUS} for a single material ρ, of thickness t, derive an equation for resistance as a function of pin location on the chip, addressing only the resistance in one direction. Assume the ESD device connection to the power bus is in the center of the I/O cell (bond pad).

3.2. Assuming there are two corners with ESD power clamps, derive an equation for two resistance values from any signal pin to the power clamps (e.g., derive an equation from any signal pin position between the two power clamps). Plot the total resistance as a function of pin position. Assume the ESD device connection to the power bus is in the center of the I/O cell (bond pad position).

3.3. In a semiconductor chip, interconnect scaling theory assumes the metal film thickness scales as a MOSFET constant electric field scaling parameter, α. For each layer, the film scales thinner. Derive the relationship of Problems 3.1 and 3.2, assuming for each layer the film thickness increases as $t' = t\alpha$. Assume N layers of metal.

3.4. In semiconductor technology, interconnects are both copper and aluminum. Show the equations of Problems 3.1 and 3.2 assuming that the material type and thickness change, based on whether it is aluminum or copper. Derive the relationship with a ratio parameter of sheet resistance ρ_{Al}/ρ_{Cu} and film thickness parameters.

3.5. In semiconductor technology, with the introduction of slotting, large rectangular slots are formed in the power bus, of dimensions w_{slot} and l_{slot}. Assume slots are formed of pitch p in the width of the bus W. Derive the resistance relationship as a function of the width of the bus. *Hint*: Show the stepwise relationship. Derive the bus resistance relationship of Problems 3.1 and 3.2.

3.6. In semiconductor technology, with the introduction of copper interconnects, for polishing purposes, the wire interconnects are required to have shapes removed in the metal as a function of the width. Derive a relationship that has, when a metal bus exceeds a dimension W_{WM}, a "cheese" width W_{CHEESE} is removed (for each multiple of dimension W_{WM}). Given a bus of width w, show the relationship for width graphically, addressing the "cheesing." Derive the resistance relationship as a function of the width of the bus. *Hint*: Show the stepwise relationship. Note, assume cheesing in the length dimension follows the same relationship and cheese shapes are square.

3.7. Assume an ESD network has a width the same as the bond pad. Assume an ESD network is segmented into N parallel connection wires from the ESD network to the power rail. Assume the individual connections are of ESD wire width, W_{ESD}. Show the resistor ladder network formed between the ESD element and the power bus (e.g., include the resistance of the bus, W_{BUS}, of a single material ρ, of thickness t).

3.8. As the power bus scales in advanced technologies, the bus resistance is significant within the ESD network, leading to non-uniform ESD operation. Assume the ESD power clamp is N bond pads away. What is the percentage of the power rail resistance within the ESD network as a function of the distance of the power clamp? Show it as a function of the number of pads from the power clamp.

3.9. As in Problem 3.8, the "across ESD resistance" can be a significant part of the bus resistance. Derive a relationship for "across ESD resistance" based on the bus width parameter, the bond pad dimension, the wire interconnect width, the number of vias between the ESD and the power bus. Assume an ESD network has a width the same as the bond pad. Assume an ESD network is segmented into N parallel connection wires from the ESD network to the power rail. Assume the individual connections are of ESD wire width, W_{ESD}. Show the resistor ladder network formed between the ESD element and the power bus (e.g., include the resistance of the bus, W_{BUS}, of a single material ρ, of thickness t).

3.10. To reduce the bus resistance, stacking of metal films in parallel is done in many semiconductor chips in peripheral I/O design. Derive a relationship for the bus resistance as a function of the width, length, and MOSFET constant electric field scaling parameter, α. (e.g., the film thickness increases as $t' = t\alpha$). Assume two metal levels that are one layer from each other (e.g., M1 and M2). Assume the general case of any two layers. Assume the case of any two design layers.

REFERENCES

1. S. Voldman. *ESD: Physics and Devices*, Chichester, UK: John Wiley and Sons, Ltd, 2004.
2. S. Voldman. *ESD: Circuits and Devices*, Chichester, UK: John Wiley and Sons, Ltd, 2006.
3. S. Voldman. *ESD: Failure Mechanisms and Models*, Chichester, UK: John Wiley and Sons, Ltd, 2009.
4. D.C. Wunsch and R.R. Bell. Determination of threshold failure levels of semiconductor diodes and transistors due to pulsed voltages, *IEEE Transactions on Nuclear Science*, **NS-15**, (6), December 1968; 244–259.
5. D. Tasca. Pulse power modes in semiconductors. *IEEE Transactions on Nuclear Science*, **NS-17,** (6), 1970; 364–372.
6. J. Smith. Electrical overstress failure analysis in microcircuits. *Proceedings of the Electrical Overstress/Electrostatic Discharge (EOS/ESD) Symposium*, 1981; 41–46.
7. J. Smith and W. Littau. Prediction of thin film resistor burnout. *Proceedings of the Electrical Overstress/Electrostatic Discharge (EOS/ESD) Symposium,* 1981; 192–197.
8. D. Egelkrout. Metallization failure and thermo-mechanical shock studies, Boeing Corporation, Defense Atomic Support Agency Report 2611 (SRD), November 1971.
9. E. Kinsbron, C.M. Melliar-Smith, and A.T. English. Failure of small thin film conductors due to high current density pulses. *IEEE Transactions on Electron Devices,* **ED-26**, (1) 1979; 22–26.
10. D. Pierce. Modeling metallization burnout of integrated circuits. *Proceedings of the Electrical Overstress/Electrostatic Discharge (EOS/ESD) Symposium,* 1982; 56–61.
11. T. Maloney. Integrated circuit metal in the charged device model: Bootstrap heating, melt damage, and scaling laws. *Proceedings of the Electrical Overstress/Electrostatic Discharge (EOS/ESD) Symposium*, 1992; 129–134.

12. K. Banerjee, A. Amerasekera, and C.M. Hu. Characterization of VLSI circuit interconnect heating and failure under ESD conditions. *Proceedings of the IEEE International Reliability Physics Symposium (IRPS)*, 1996; 237–245.
13. S. Voldman. ESD robustness and scaling implications of aluminum and copper interconnects in advanced semiconductor technology. *Proceedings of the Electrical Overstress/Electrostatic Discharge (EOS/ESD) Symposium*, 1997; 316–329.
14. S. Voldman. High-current transmission line pulse characterization of aluminum and copper interconnects for advanced CMOS semiconductor technologies. *Proceedings of the IEEE International Reliability Physics Symposium (IRPS)*, 1998; 293–301.
15. S. Voldman. The impact of technology evolution and scaling on electrostatic discharge (ESD) protection on high-pin-count high-performance microprocessors. *Proceedings of the International Solid-State Circuits Conference (ISSCC)*, Session WA21, San Francisco, CA, February 15–17, 1999; 366–367.
16. K. Banerjee, A. Amerasekera, G. Dixit, and C.M. Hu. The effect of interconnect scaling and low-K dielectric on the thermal characteristics of the IC metal. *International Electron Devices Meeting (IEDM) Technical Digest*, December 1996; 65–68.
17. S. Voldman. High current characterization of dual damascene copper/SiO_2 and low-K inter-level dielectrics for advanced CMOS semiconductor technologies. *Proceedings of the IEEE International Reliability Physics Symposium (IRPS)*, 1999; 144–153.
18. K. Banerjee, D.Y. Kim, A. Amerasekera, C.M. Hu, S.S. Wong and K.E. Goodson. Microanalysis of VLSI interconnect failure modes under short-pulse stress conditions. *Proceedings of the IEEE International Reliability Physics Symposium (IRPS)*, 2000; 283–288.
19. S. Khoo, P. Y. Tan, and S. Voldman. Microanalysis and electro-migration reliability performance of transmission line pulse (TLP) stressed copper interconnects. *Journal of Microelectronics and Reliability*, 2003; **43**: 1039–1045.
20. L. Chu, W. K. Chim, K.L. Pey, and A. See. Effect of transmission line pulsing of interconnects investigated using combined low-frequency noise and resistance measurements. *Proceedings of the International Physical and Failure Analysis (IPFA) Symposium*, 2001; 97–102.
21. P. A. Juliano and W. Anderson. ESD protection design challenges for a high pin-count Alpha microprocessor in a 0.13 μm CMOS SOI technology. *Proceedings of the Electrical Overstress/Electrostatic Discharge (EOS/ESD) Symposium*, 2003; 59–69.
22. C. J. Brennan, J. Sloan, and D. Picozzi. CDM failure modes in a 130 nm ASIC technology. *Proceedings of the Electrical Overstress/Electrostatic Discharge (EOS/ESD) Symposium*, 2004; 182–186.
23. R. Bass, D. Nickel, D. Sullivan, and S. Voldman. Method of automated ESD protection level verification. U.S. Patent No. 6,086,627, July 11th 2000.
24. S. Voldman, S. Strang, and D. Jordan. An automated electrostatic discharge computer-aided design system with the incorporation of hierarchical parametrized cells in BiCMOS analog and RF technologies for mixed signal applications. *Proceedings of the Electrical Overstress/Electrostatic Discharge (EOS/ESD) Symposium*, 2002; 296–305.
25. S. Voldman, S. Strang, and D. Jordan. A design system for auto-generation of ESD circuits. *Proceedings of the International Cadence Users Group*, September 2002.
26. S. Voldman. Automated hierarchical parameterized ESD network design and checking system. U.S. Patent No. 6,704,179, March 9th 2004.
27. P. Homsiger, A. Huber, D. Korejwa, W. Livingstone, J. Panner, E. Schanzenbach, D. Stout, S. Voldman, and P. Zuchowski. Method of automated design and checking for ESD robustness. U.S. Patent No. 6,725,439, April 20th 2004.
28. D. Collins, D. Jordan, S. Strang, and S. Voldman. ESD design, verification, and checking system and method of use. U.S. Patent No. 7,134,099, November 7th, 2006.

4 ESD Power Clamps

4.1 ESD POWER CLAMPS

In this chapter, ESD power clamp networks will be explored. ESD power clamp usage began in the mid-1990s, and today is a common practice of semiconductor chip design and ESD design synthesis [1–52]. Development of ESD power clamps and the synthesis into the semiconductor chip architecture is part of the ESD design discipline and an essential component of the art of ESD design. This chapter will focus on the classification of the ESD power clamps, key design parameters, the ESD power clamp design window, trigger elements, clamp devices, and issues and problems with the ESD power clamp.

4.1.1 Classification of ESD Power Clamps

There are many different types of ESD power clamps, but conceptually they can be classified into different categories. Figure 4.1 shows a diagram of classification of ESD power clamps.

ESD power clamps must be tolerant of the power supply voltages observed in the functional semiconductor chip or system of chips [3–6]. ESD power clamps can be constructed for the native voltage power supply or mixed voltage power supplies. The ESD power clamps must be tolerant of the semiconductor chips they interface with, or the number of power rail voltages contained within a given chip.

Fundamentally, ESD power clamps contain some basic features. A first feature is the transfer of ESD current from one segment of the power grid to a second segment of the power grid. A second feature is the initiation of the ESD power clamp, commonly referred to as a "trigger" state. ESD power clamps can be as simple as one physical device or a complex circuit, or a system. In the simplest case, the trigger feature and the clamp feature can be contained within the same device. In a second classification, the trigger element is independent of the "clamp" feature (e.g., independent trigger element from the clamping feature). There is a critical concept in the ESD power clamp design synthesis in the separation of the trigger state feature from the clamp feature. The advantage of an independent trigger element is that it

ESD: Design and Synthesis, First Edition. Steven H. Voldman.

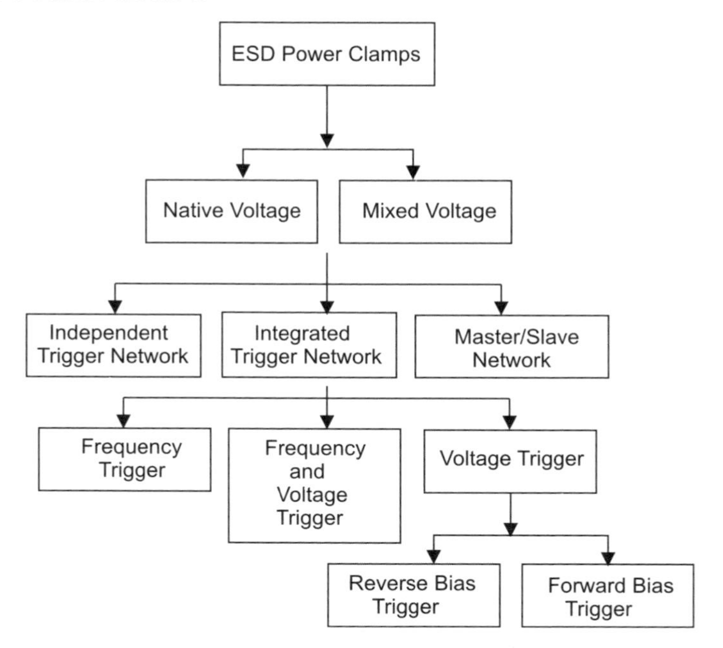

Figure 4.1 Classes of ESD power clamps

provides a second degree of freedom with the separation of the clamping feature from the trigger feature. Whereas with a single integrated element, there is a physical limitation on some devices to achieve the features desired.

In a third classification, the ESD power clamp is a system of ESD power clamps, with one trigger element for a system of clamp elements, which will be referred to as a "master/slave" architecture. A master/slave system allows integration of a single trigger element, but allows distribution of the elements in the chip system.

In the ESD power clamp "trigger feature," there are many different solutions used for ESD power clamps, but again, they can be simply stated as classifications of trigger elements.

ESD power clamps can have trigger features that respond to the ESD pulse. The response of the "trigger" network is to a given frequency or transient phenomena. This class of trigger networks will be referred to as "frequency triggering." A frequency trigger can contain elements that are frequency-dependent, such as resistors, capacitors, and inductors [3,4,6]. Frequency-triggered networks respond in the frequency domain. ESD trigger elements can also be networks that do not respond in the frequency domain. These ESD trigger networks can also be initiated by over-voltage or over-current conditions. A class of ESD trigger networks is formed by voltage-triggered elements. Voltage-triggered elements can be initiated in a forward-bias or reverse-bias state of operation [3,4].

In this classification, there are additional features that have been added to address other characteristics. Some of these features are as follows:

- Ramping of the power supplies (e.g., power-up or power-down).

- Sequencing of power supplies.
- False triggering from system events.
- ESD testing pre-charging phenomena [9,50–52].
- ESD testing "trailing pulse" phenomena [9,50–52].

4.1.2 Design Synthesis of ESD Power Clamp – Key Design Parameters

In ESD design synthesis of ESD power clamps, there are key design parameters in the decision on what type of circuit to utilize. The following are a list of key parameters in the ESD design process of ESD power clamps:

- ESD power clamp physical area.
- ESD power clamp width.
- ESD power clamp current per unit of width metric (A/μm).
- ESD power clamp "on resistance."
- ESD power clamp voltage tolerance.
- ESD power clamp latchup robustness.
- ESD power clamp false triggering immunity.
- ESD power clamp IEC 61000-4-2 responsiveness.
- ESD power clamp leakage current.
- ESD power clamp capacitance loading.
- ESD power clamp frequency response window.
- ESD power clamp trigger voltage or current.

These features and aspects of ESD power clamps will be discussed. These ESD power clamps can be made of diodes [1–10], bipolar transistors [1–10], MOSFETs [1–10], silicon-controlled rectifiers [12,13], and LDMOS transistors [10,45–49].

4.2 DESIGN SYNTHESIS OF ESD POWER CLAMPS

In the ESD power clamp, the "trigger feature" is critical to initiate the ESD power clamp. ESD power clamps can have trigger features that respond to the ESD pulse through either transient response or voltage levels. In the following sections, we focus on two major classes of "trigger networks."

4.2.1 Transient Response Frequency Trigger Element and the ESD Frequency Window

In ESD power clamps, the ESD power clamp trigger element can be a frequency-triggered network, or a transient response trigger element. Transient response trigger elements are designed to respond to the ESD events. This class of trigger networks will be referred to as "frequency triggering." Frequency-triggered networks respond in the frequency domain. The frequency trigger can contain elements that are frequency-dependent, such as resistors, capacitors, and inductors, in RC, LC, or RLC configurations. In ESD power clamps, the most widely used (and most popular) is the RC network. The RC-trigger network is also known as the "RC discriminator" network, due to it providing frequency selection in the ESD power clamp frequency domain [14,15]. By providing a separate RC-filter network, the frequency response of the trigger network will not be dependent on the inherent native frequency response of a semiconductor device, and can be "tuned" to the desired frequency. In the majority of applications, the RC-discriminator network is tuned to be responsive to the HBM and MM pulse events. One of the key advantages of frequency-triggered ESD clamps is that they are a function of the transient or rising edge, not the voltage level of the power grid.

In the frequency-trigger network, the resistor and capacitor elements can be passive or active semiconductor elements. The choice of what element to use is a function of the technology, area utilization, voltage tolerance, and device responsiveness. The resistor used for the RC network can be the following:

- Polysilicon resistor element.
- Diffused resistor element.
- "On" n-channel transistor element.
- "On" p-channel transistor element.

The capacitor element typically used for the RC network is as follows:

- MOS capacitor.
- MIM capacitor.

4.2.2 The ESD Power Clamp Frequency Design Window

Figure 4.2 shows the ESD power clamp frequency window. Figure 4.2 provides a frequency plot highlighting the typical frequency of ESD events, overlaying the typical design point for ESD power clamps. Typically, ESD power clamps are designed to respond to the HBM and MM events. ESD power clamps are not designed to respond to CDM events. In addition, the ESD power clamps are not to be initiated by the power-up and power-down of the semiconductor chip or system. The ESD power clamps are not to be initiated by system events, leading to "false triggering."

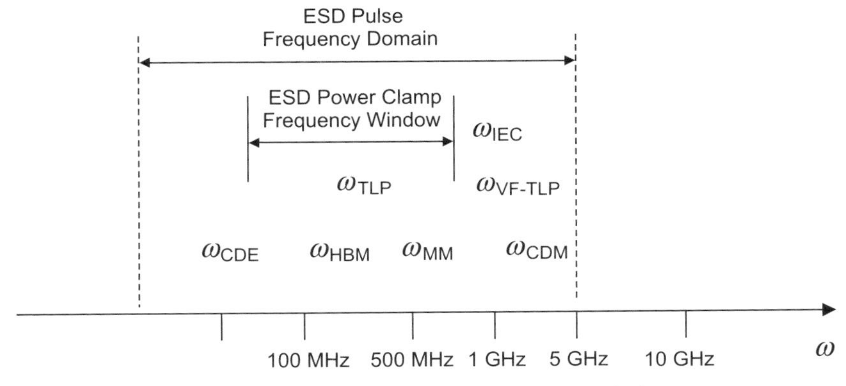

Figure 4.2 ESD power clamp frequency window

In addition, for RF applications, the ESD trigger elements are not to respond to the RF application frequency [4]. As a result, there is a defined frequency window that is acceptable for ESD power clamps and the frequency range of these networks.

4.2.3 Design Synthesis of ESD Power Clamp – Voltage Triggered ESD Trigger Elements

In ESD power clamps, the ESD power clamp trigger element can be a current- or voltage-triggered ESD network [1,3,6]. Voltage trigger elements are designed to respond to the ESD events when the voltage exceeds the trigger condition. These ESD power clamps will turn on when the voltage exceeds the trigger state. As a result, it is not dependent on the frequency of the transient event. As a result, this turns "on" the circuitry independent of whether it is an ESD or an electrical overstress (EOS) event, or any over-voltage or over-current state. These ESD power clamps are not to be initiated by the power-up and power-down of the semiconductor chip or system, except when they are in an over-voltage state.

ESD voltage-triggered elements can be either forward-bias or reverse-bias elements or circuits [3,4]. For reverse-biased trigger networks, the following are typically utilized:

- Zener breakdown diode.
- Polysilicon diode.
- CMOS LOCOS-defined metallurgical junction diode.
- CMOS shallow trench isolation (STI)-defined metallurgical junction diode.
- Bipolar transistor collector–substrate junction diode.
- Bipolar transistor base–collector junction diode.
- Bipolar transistor emitter–base junction diode.

- Bipolar transistor collector–emitter configuration.

For forward-bias trigger networks, typically a "diode string" or series cascode configured diodes are used to establish the trigger voltage [3,4,6]. For forward-biased trigger networks, the following are typically utilized:

- CMOS LOCOS-defined metallurgical junction diode.
- CMOS STI-defined metallurgical junction diode.
- Bipolar varactor (forward-bias configuration).
- Bipolar transistor base–collector junction diode.
- Bipolar transistor base–emitter junction diode.

In some applications, to achieve the desired trigger voltage, the forward-bias elements can be combined with reverse-biased elements. By using the forward-bias trigger elements in series with the reversed-bias trigger elements, higher trigger voltage states are achieved [3,4].

4.3 DESIGN SYNTHESIS OF ESD POWER CLAMP – THE ESD POWER CLAMP SHUNTING ELEMENT

For ESD protection power clamps, two basic functions are the trigger network and the "shunt" network (e.g., also referred to as the "clamp element") [14,15]. The role of the shunt element is to provide a current path in the alternative current loop to discharge the ESD current. For the effectiveness of the ESD power clamp, there are a few desired features of the ESD clamp element:

- **Low Impedance:** Provide a low-impedance path (e.g., a low "on resistance").
- **ESD Robustness:** Provide an ESD-robust solution (e.g., discharge the ESD current without failure below the desired ESD specification).
- **Scalable:** Scalable element with physical size (e.g., width, length, perimeter, or area).

As shown in Figures 4.3 and 4.4, in the ESD protection power clamps, the "shunt" network (e.g., or "clamp element") is shown. In Figure 4.3, the shunt element is a MOSFET device whereas in Figure 4.4, the shunt element is a bipolar transistor.

There are some additional desired characteristics of the ESD power clamp "shunt element." These consist of the following:

- ESD power clamp trigger condition vs. ESD power clamp shunt failure.
- ESD clamp element ESD robustness width scaling.
- ESD on-resistance.

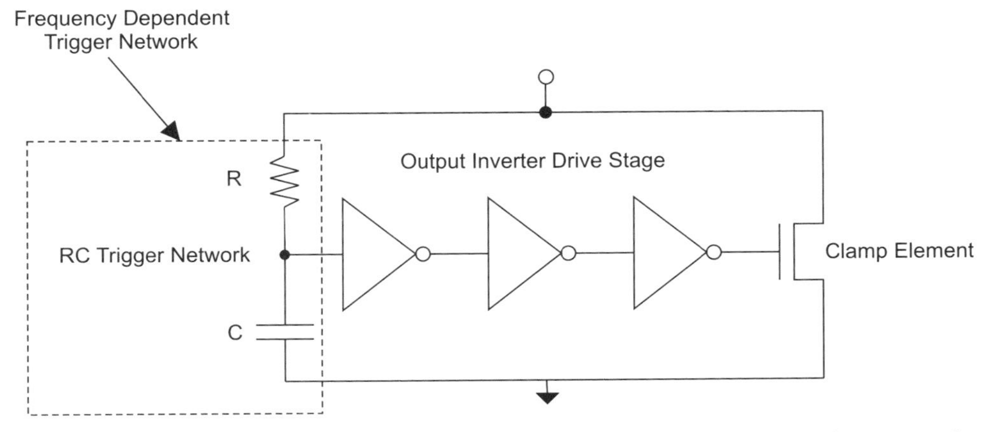

Figure 4.3 Example of frequency-triggered ESD power clamp highlighting the trigger network

4.3.1 ESD Power Clamp Trigger Condition vs. Shunt Failure

For proper operation of the ESD power clamp, the trigger network will require to initiate prior to the over-voltage or over-current of the ESD "shunt" clamp element [14,15]. In the frequency domain, if the ESD network trigger does not respond to a specific ESD event, the trigger network will not respond effectively, and the ESD "shunt clamp" will discharge according to its native breakdown event. For a MOSFET "shunt" element, the element will undergo MOSFET drain-to-source snapback. For a bipolar transistor, the bipolar element will undergo collector-to-emitter breakdown.

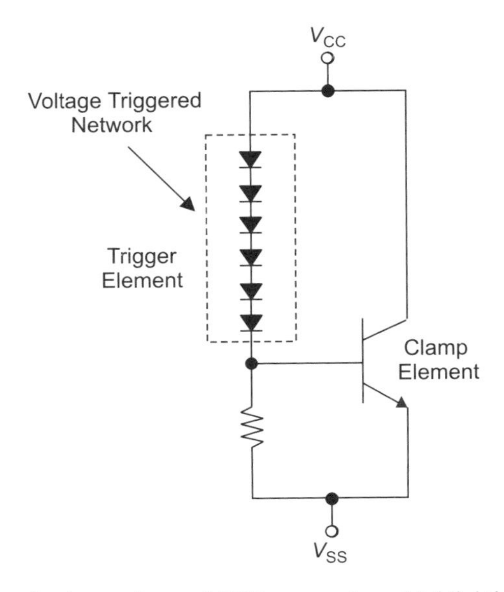

Figure 4.4 Example of voltage-triggered ESD power clamp highlighting the trigger network

4.3.2 ESD Clamp Element – Width Scaling

It is desirable to have the ESD results scale with the ESD clamp "shunt" element size. The ESD robustness will scale with the physical width given the following conditions:

- **Frequency tuning:** Proper frequency "tuning" of the trigger network (e.g., responsive to the ESD event) for MOSFET gate-driven networks or bipolar base-driven networks.
- **Drive Circuit:** Adequate current drive and current-drive distribution for bipolar base-driven networks.
- **Layout Symmetry:** Layout optimization of clamp element.
- **Ballasting:** MOSFET drain ballast (or bipolar emitter ballast) adequate to provide uniformity.
- **Power Bus Connectivity:** Electrical connection to power bus and ground rail well distributed in the ESD power clamp "clamp element" region of the circuit.

Figure 4.5 shows an example of a MOSFET power clamp width scaling for HBM pulse events [3,4]. The plot shows the improvement of the HBM ESD robustness with the increase in the power clamp width. In this study, the size of the trigger network and drive circuitry was fixed.

4.3.3 ESD Clamp Element – On-Resistance

It is desirable to have an ESD clamp on-resistance which reduces with the size of the MOSFET or bipolar clamp element. The lower the ESD clamp on-resistance, the lower the total resistance through the alternative current loop. The lower the resistance in the ESD current loop, the lower the node voltage at the bond pad node. As the impedance of the power bus and the ESD clamp element is reduced, the allowed resistance for the ESD signal pin network can be higher and achieve the same signal pin ESD robustness. Hence, lowering the

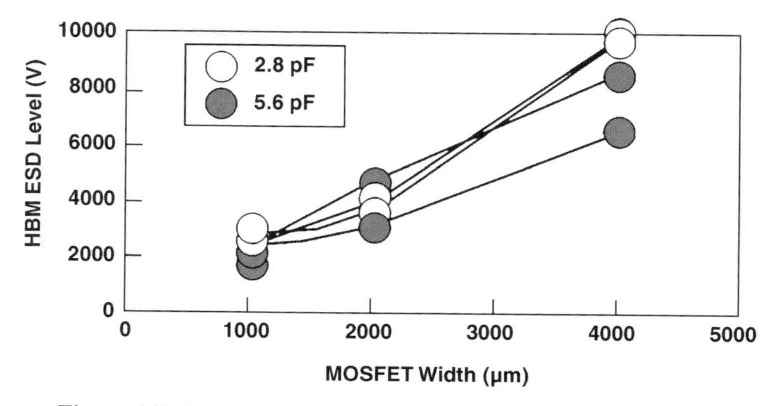

Figure 4.5 Example of ESD MOSFET power clamp width scaling

ESD power clamp resistance allows for a smaller ESD network at the signal pin (e.g., smaller network with lower capacitance).

The ESD clamp on-resistance will scale down with the clamp element device size, given that the element does not undergo current saturation effects, self-heating, or poor current distribution. Hence, if the ESD power clamp element is large enough, and self-heating is kept to a minimum, the "on-resistance" will scale with the width scaling.

4.3.4 ESD Clamp Element – Safe Operating Area

The ESD "clamp element" must remain in the SOA of the device to avoid failure of the ESD power clamp network. To avoid electrical failure of the ESD clamp element prior to achieving the ESD objective, the clamp element of the ESD power clamp must remain below a voltage absolute maximum ($V_{\mathrm{ABS\ MAX}}$) and a current absolute maximum ($I_{\mathrm{ABS\ MAX}}$) of the clamp element.

4.4 ESD POWER CLAMP ISSUES

ESD power clamps have some unique issues as a result of being placed within the power grid of a semiconductor chip. The issues will be discussed briefly, followed by examples in future sections on how to address these issues.

4.4.1 ESD Power Clamp Issues – Power-Up and Power-Down

ESD power clamps are to remain in an "off-state" when a semiconductor chip is in a power-up state, a power-down state, and in a quiescent powered state [14,15]. The different solutions to avoid initiation of the power clamps during power-up and power-down ramping are as follows:

- **Frequency Window:** Trigger networks do not respond to these frequencies.
- **Feedback Networks:** Feedback networks are placed to avoid response to power-up.
- **Enable/Disable Functions:** Logic can be integrated into the trigger network to "enable" or "disable" the ESD power clamp as desired.

4.4.2 ESD Power Clamp Issues – False Triggering

ESD power clamps can be "false triggered" as a result of pulse events from signals, over-current, over-voltage, or "spikes" during test, burn-in, or other reliability stresses [10,16,33–36]. The different solutions to avoid initiation of the power clamps during power-up and power-down are as follows:

- **Over-current Protection:** Over-current protection can be integrated to avoid the ESD power clamp outside its SOA.

- **Frequency Window:** Trigger networks do not respond to these frequencies of "spikes."
- **Feedback Networks:** Feedback networks are placed with hysteresis.
- **Enable/Disable Functions:** Logic can be integrated into the trigger network to "enable" or "disable" the ESD power clamp as desired.

4.4.3 ESD Power Clamp Issues – Pre-Charging

Pre-charging events can occur during ESD testing that can influence the ESD power clamp networks [5–52]. In the process of ESD testing, poor isolation of the test source from the DUT can lead to a pre-charging phenomenon in the semiconductor chip. After an ESD pulse is applied, a low-level current bleeds from the high-voltage source to the DUT without proper "switch" isolation. The "pre-charging" solution is as follows:

- **ESD Power Clamp Pre-charge "Bleed" Device:** A high-impedance element can be placed in parallel with the ESD power clamp to allow the bleeding of charge from the V_{DD} to the V_{SS} power rail. The "bleed device" can be a resistor. This can be placed locally to the device or non-locally to the ESD power clamp.
- **ESD Test System Modification:** Modification of the ESD stress test system by providing proper isolation.

4.4.4 ESD Power Clamp Issues – Post-Charging

A post-charging event from ESD simulators is also present that can influence the ESD test results [50–52]. After the ESD event occurs, a low-level current "tail" exists in the simulators that continues to charge the signal pins or power pins. In the process of ESD testing, poor isolation of the test source from the DUT can lead to a post-charging phenomenon in the semiconductor chip. As in the pre-charging event, the post-charging event can lead to an anomalous ESD test result.

4.5 ESD POWER CLAMP DESIGN

In this section, examples of different circuit topologies will be shown to highlight some of the previously discussed issues. Native power supply voltage and non-native ESD power clamps will be discussed.

4.5.1 Native Power Supply RC-Triggered MOSFET ESD Power Clamp

Figure 4.6 shows an example of the most commonly used ESD power clamp in the semiconductor industry, the RC-triggered MOSFET ESD power clamp. The RC-discriminator

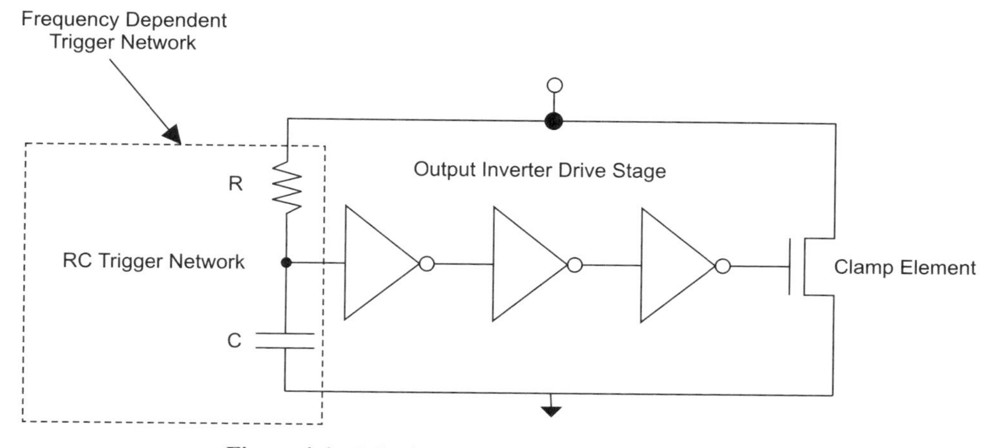

Figure 4.6 RC-triggered MOSFET ESD power clamp

network discriminates between ESD events and spurious events, or power-up and power-down if properly tuned. The RC trigger is typically "tuned" to respond to the ESD HBM and MM pulse events [2–4,14,15].

The inverter stage serves two purposes [14,15]. First, it allows for the tuning of the RC network without the loading of the first inverter gate capacitance influencing the RC "tuning." Second, it serves as a drive stage for "driving" the ESD clamp element. In recent years, to improve the responsiveness, the three inverter stages have been reduced to a single stage. The advantage of this is to improve the responsiveness. The disadvantage of the single inverter is the increase in size of the single inverter stage, and the lack of isolation between the RC-discriminator tuning and the load of the inverter stage, and output network. This network is also suitable for native voltage conditions. Given higher voltage power domains, all elements in the circuitry must be voltage-tolerant to that given power domain.

4.5.2 Non-Native Power Supply RC-Triggered MOSFET ESD Power Clamp

Figure 4.7 shows an example of an ESD power clamp for mixed-voltage semiconductor chips [3,4,6]. In many mixed-voltage or mixed-signal applications, different power clamps are required based on the voltage of the power domain. Figure 4.7 shows an RC-triggered MOSFET power clamp, where a second MOSFET is used to lower the voltage across all the elements in the lower element. In the design synthesis of this network, the "drop-down" device lowers the voltage across all elements in the ESD power clamp. Hence, it provides two roles: (1) it serves as a "level shift" of the voltage level; (2) it converts the power bus of the ESD network into a "dummy ESD power rail bus" instead of the actual power rail bus. In this case, a MOSFET is used for the "level" shifting network. Alternative ESD "level" shift elements utilized can include diode string elements.

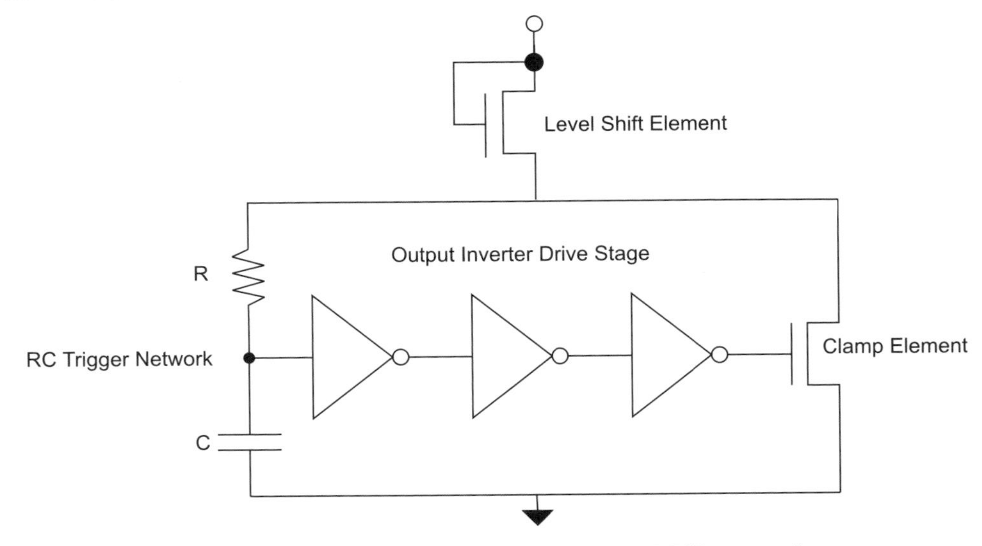

Figure 4.7 Series cascode RC-triggered MOSFET ESD power clamp

4.5.3 ESD Power Clamp Networks with Improved Inverter Stage Feedback

To provide better control of the ESD clamp element, and avoid false triggering, the "latching characteristics" of the inverter drive stage can apply well-known feedback methods. Well-known feedback techniques in CMOS logic include "half-latch" or "full-latch" circuit concepts. Figure 4.8 shows an example of an ESD power clamp with a CMOS half-latch PMOS keeper element. This provides improved control of the MOSFET output gate, which can improve intolerance to false triggering, or avoid low-level leakage of the output MOSFET [9,10,34–36].

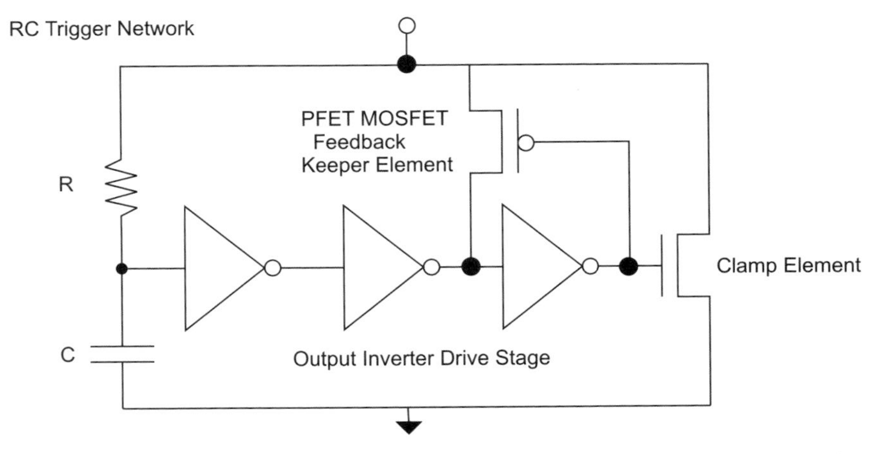

Figure 4.8 CMOS RC-trigger clamp with CMOS PFET half-latch keeper feedback

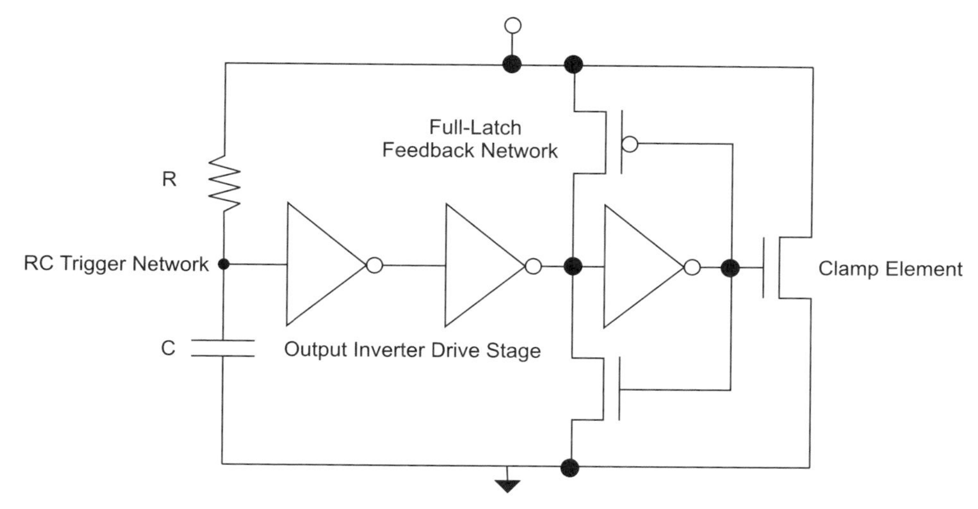

Figure 4.9 CMOS RC-trigger clamp with CMOS PFET full-latch keeper feedback

A second method to improve the "latching characteristics" of the inverter drive stage can apply well-known "full-latch" circuit concepts. Figure 4.9 shows an example of an ESD power clamp with a CMOS full-latch feedback network [9,10,34–36]. The integration of the full inverter for the feedback forms a "SRAM-like" latch between the ESD power clamp last inverter and the feedback inverter. As with the "half-latch" feedback, this provides improved control of the MOSFET output gate, which can improve intolerance to false triggering, or avoid low-level leakage of the output MOSFET.

Other techniques for improving the control of the ESD power clamp from false triggering can be applied. As the feedback is brought to the earlier stages, the size of the feedback elements can be reduced. A third method is placement of a PMOS device above the inverters. Figure 4.10 shows an example of an ESD power clamp with a PMOS element within the logic [9,10,34–36].

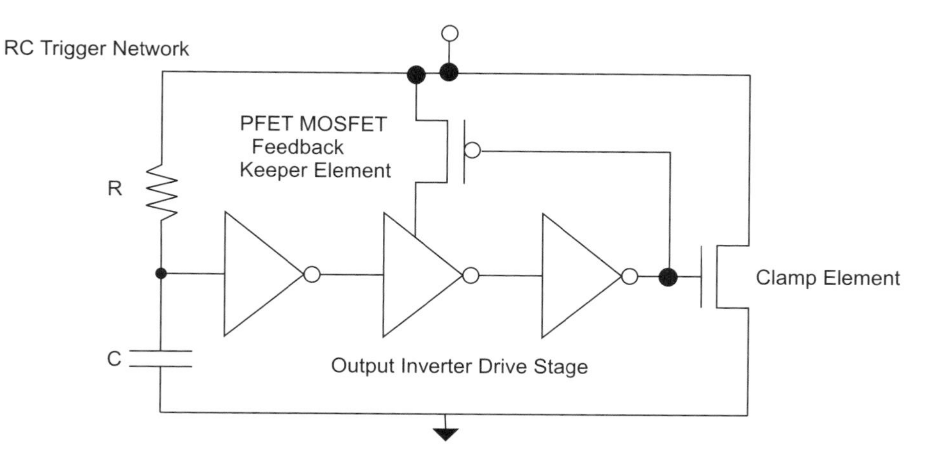

Figure 4.10 CMOS RC-trigger clamp with CMOS PFET cascade feedback

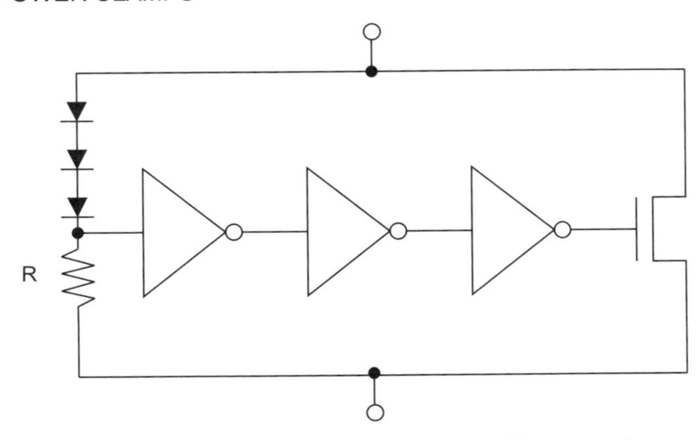

Figure 4.11 Forward-bias voltage-triggered ESD power clamps

4.5.4 ESD Power Clamp Design Synthesis – Forward Bias Triggered ESD Power Clamps

In some applications, the presence of a frequency-triggered network is undesirable – in the integration of an RC-triggered ESD MOSFET network into an RF application [4]. For example, given that the frequency response of the system, such as a cell phone, is pre-defined, it may not be advisable to place another frequency-dependent circuit in a small system (e.g., altering the frequency response of the poles and zeros in the frequency domain). As a result, some circuit design teams desire voltage-triggered networks for RF CMOS instead of frequency-triggered networks. Figure 4.11 is an example of a forward-bias voltage-triggered ESD MOSFET network. The advantage of this network responds to all over-voltage events or over-current conditions. The number of diodes is chosen to turn on prior to the MOSFET snapback voltage of the ESD clamp element. This network has a wide frequency window and is not sensitive to power-up, power-down, or false triggering events, and does provide over-current and over-voltage protection.

4.5.5 ESD Power Clamp Design Synthesis – IEC 61000-4-2 Responsive ESD Power Clamps

For applications that are required to respond to the IEC 61000-4-2 pulse event, not all circuit topologies are suitable. For the IEC 61000-4-2 event, there is a fast current pulse which is of considerable magnitude. Hence, to address the frequency response and current magnitude, many ESD power clamps are required to be modified.

During the IEC 61000-4-2 event on the chassis or ground line of a system, a negative pule occurs on the V_{SS} power rail or substrate. This can initiate the RC-triggered network from the negative pulse event. But, the elements in the RC discriminator must be responsive, or circuit failure can occur. The resistor and capacitor element choices must be responsive. Resistors, such as polysilicon resistors, may be slow to respond to fast events.

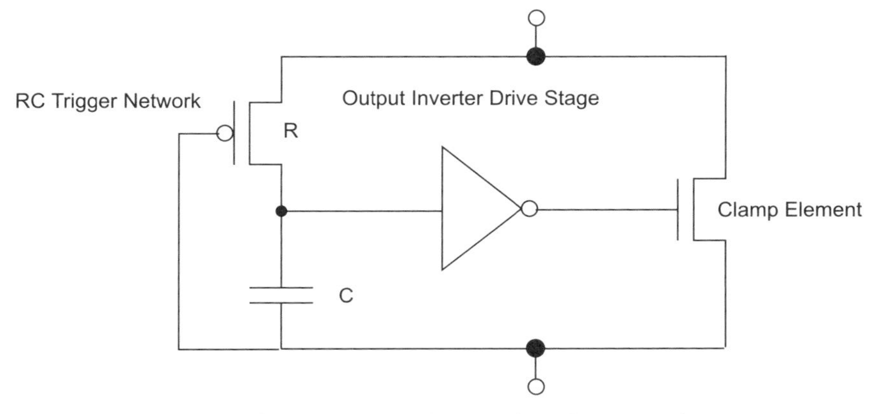

Figure 4.12 IEC 61000-4-2 responsive ESD power clamp

Figure 4.12 is an example of an IEC 61000-4-2 event-responsive ESD MOSFET network. The advantage of this network is that the p-channel MOSFET is more responsive than a polysilicon resistor element. Additionally, so that the inverter drive network is more responsive, only a single inverter stage is implemented.

4.5.6 ESD Power Clamp Design Synthesis – Pre-Charging and Post-Charging Insensitive ESD Power Clamps

ESD test systems or residual charge can influence the "state" of an RC-triggered MOSFET clamp before or after ESD stress [50–52]. With charge on the V_{DD} power rail, the voltage state of the RC-triggered MOSFET can be pre-charged, and close the MOSFET snapback voltage of the ESD clamp device. On the first discovery of this issue, it was noted by R. Ashton that products with RC-triggered power clamps which were inherently "leaky" had better ESD results than products whose V_{DD} leakage was low. It was from this that Ashton discovered the issue of an ESD test system leading to residual charge on the power grid of the semiconductor chip, influencing the pre-state of the ESD power clamp. It was noted that the charge on the V_{DD} power rail led to the MOSFET snapback of the output device prior to initiation of the RC discriminator response.

Figure 4.13 is an example of an ESD power clamp network with a "bleed" element to provide discharging of the ESD pre-charging event, or a post-charging event. Placing a high-impedance element that bleeds the charge off the power rail can avoid the ESD test system-induced operation failure of the ESD power clamp element.

4.6 ESD POWER CLAMP DESIGN SYNTHESIS – BIPOLAR ESD POWER CLAMPS

Bipolar and bipolar-CMOS (BiCMOS) technologies are used today for analog and mixed-signal semiconductor chips. In a mixed-signal semiconductor chip that utilizes bipolar

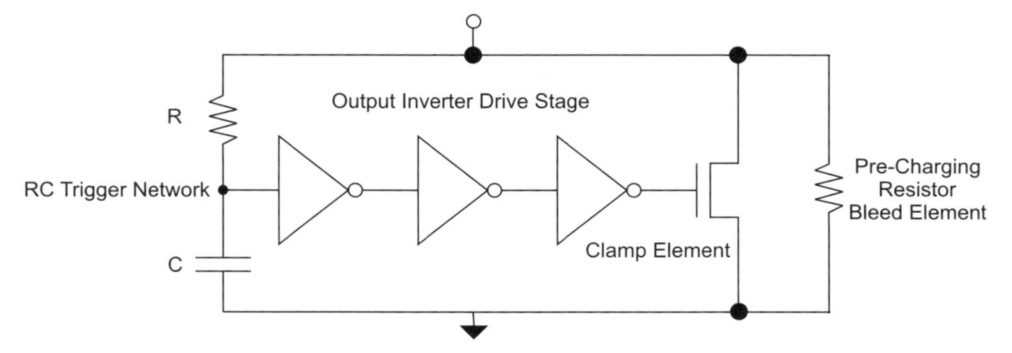

Figure 4.13 Pre-charging and post-charging insensitive ESD power clamp

transistors, analog and digital domains are separated. Bipolar transistors typically have a higher power supply voltage. Additionally, in many bipolar applications, a negative power supply voltage is also used. For these bipolar power domains, bipolar power clamps are used [3,4,10,12–14,37–44].

4.6.1 Bipolar ESD Power Clamps with Zener Breakdown Trigger Element

Figure 4.14 is an example of a bipolar ESD power clamp [38,39]. In this ESD bipolar power clamp, a single transistor is placed between the two power supplies in a collector-to-emitter configuration. The transistor is to be used to discharge the ESD current from the V_{CC} power rail to the V_{SS} ground rail. The trigger element is a Zener diode which undergoes electrical breakdown. When the voltage across the trigger element reaches the breakdown voltage of the Zener diode, the current flows through the Zener diode and into the base of the bipolar transistor. This base-driven network responds to over-voltage conditions in the semiconductor chip. Since it is a voltage-triggered network, it has a wide frequency window of operation; the frequency response is limited to the frequency response of the Zener diode, and its bipolar transistor.

4.6.2 Bipolar ESD Power Clamps with Bipolar Transistor BV_{CEO} Breakdown Trigger Element

Figure 4.15 is an example of a BV_{CEO} voltage-triggered bipolar ESD power clamp. In this ESD bipolar power clamp, a single transistor is placed between the two power supplies in a collector-to-emitter configuration [3,4,40–42]. A second transistor is used as the trigger element and is also placed in a common emitter (C-E) configuration. The clamp transistor element is a high-breakdown (HB) transistor, and is to be used to discharge the ESD current from the V_{CC} power rail to the V_{SS} ground rail. The trigger element is a low breakdown (LB) voltage (e.g., BV_{CEO}) npn transistor which undergoes electrical breakdown. When the voltage across the trigger element reaches the BV_{CEO} breakdown voltage, the current flows through the trigger element

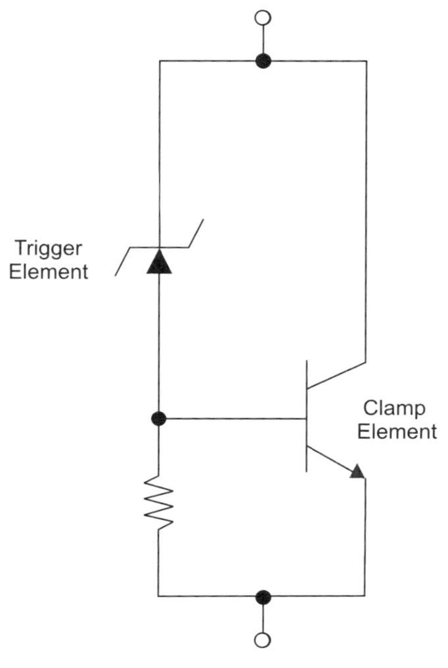

Figure 4.14 Reverse-breakdown Zener-triggered bipolar ESD power clamps

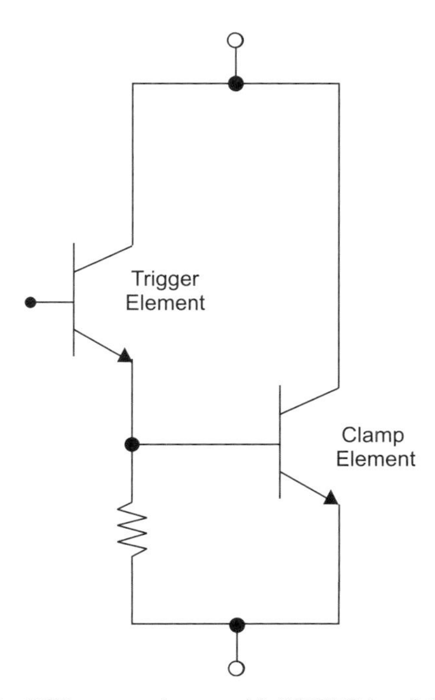

Figure 4.15 Bipolar ESD power clamps with BVCEO breakdown trigger element

and into the base of the bipolar transistor. This base-driven network responds to over-voltage conditions in the semiconductor chip. Since it is a voltage-triggered network, it has a wide frequency window of operation; the frequency response is limited to the frequency response of the two transistors.

4.6.3 Bipolar ESD Power Clamps with BV_{CEO} Bipolar Transistor Trigger and Variable Trigger Diode String Network

One of the limitations of breakdown-triggered networks is that the breakdown voltage is the native voltage of the element. Hence, it is an advantage to provide a variable trigger voltage for different power supply conditions [3,4,41]. Figure 4.16 is an example of a bipolar ESD power clamp with a variable trigger condition. In this ESD bipolar power clamp, a single transistor is placed between the two power supplies in a collector-to-emitter configuration. A series of diode elements (in forward-bias state) and a second transistor is placed in a C-E configuration. The clamp transistor element is a high-breakdown transistor, and is to be used to discharge the ESD current from the V_{CC} power rail to the V_{SS} ground rail. The trigger network is a forward-biased diode string and a low-breakdown voltage (e.g., BV_{CEO}) npn transistor. When the voltage across the trigger network reaches the sum of the forward diode voltage of the diode string and the BV_{CEO} breakdown voltage of the npn transistor, the current flows through the trigger element and into the base of the bipolar transistor. This base-driven network responds to over-voltage

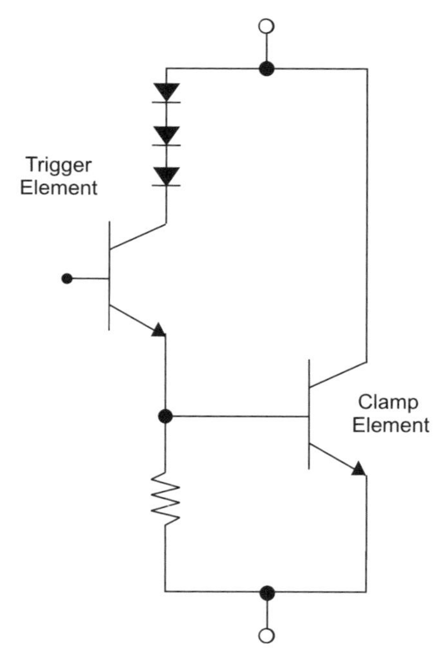

Figure 4.16 Bipolar ESD power clamp with BVCEO breakdown trigger element and diode string elements

conditions in the semiconductor chip. Since it is a voltage-triggered network, it has a wide frequency window of operation; the frequency response is limited to the frequency response of the two transistors.

4.6.4 Bipolar ESD Power Clamps with Frequency Trigger Elements

One of the limitations of breakdown-triggered networks is that the power supply voltage must reach the breakdown voltage prior to initiating the ESD power clamp. Frequency-triggered networks can be established using bipolar devices. Figure 4.17 is an example of a bipolar ESD power clamp with a capacitively coupled trigger network [43,44].

In this ESD bipolar power clamp, two transistors are placed in a Darlington configuration, with a capacitor element as part of the trigger network. The capacitor is placed between the power supply and the capacitor of the first transistor stage in the Darlington network. A second transistor is placed in a Darlington C-E configuration. A diode, used to drop voltage across the single transistor, is placed between the two power supplies in a collector-to-emitter configuration. This base-driven network responds to frequency conditions in the semiconductor chip. Since it is a frequency-triggered network, it has a smaller ESD power clamp frequency window of operation; the frequency response is defined by the capacitor element.

4.7 MASTER/SLAVE ESD POWER CLAMP SYSTEMS

ESD power clamps can be lumped or distributed through a semiconductor chip. In the case of a distributed system, a design synthesis concept is to provide a single trigger element for many

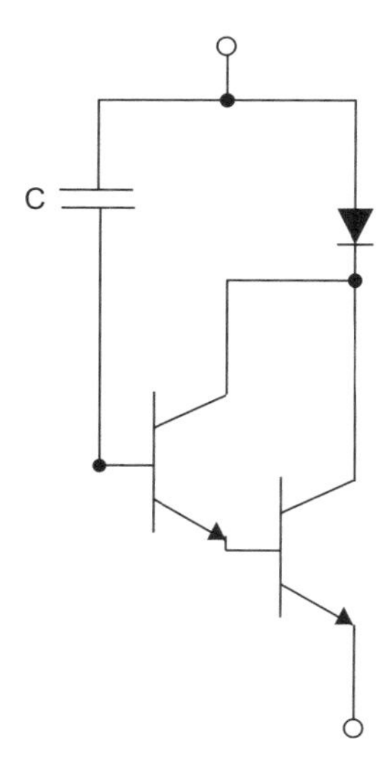

Figure 4.17 Capacitive-triggered bipolar ESD power clamp

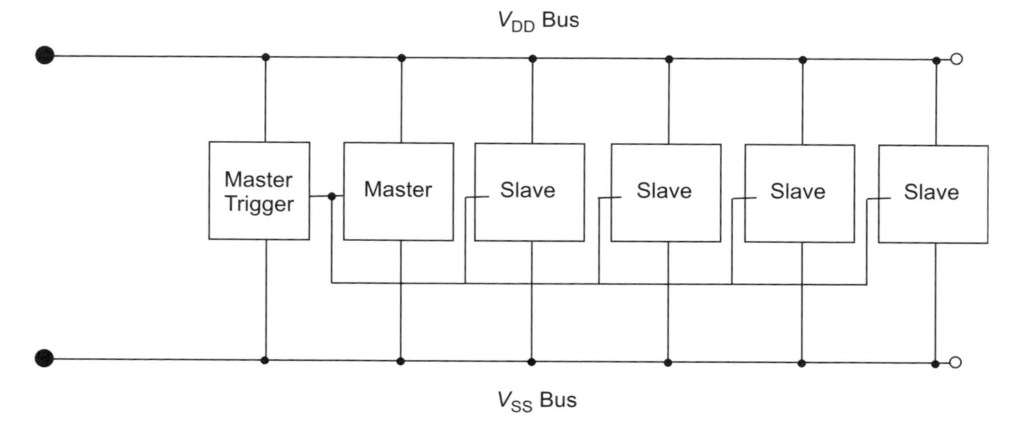

Figure 4.18 ESD power clamp master/slave system

ESD "clamp" elements [10]. There are two advantages of this concept. First, a single trigger will initiate all elements in parallel, instead of independent triggers which may vary across a semiconductor chip. Second, there is a saving of semiconductor chip area. Figure 4.18 shows a high-level diagram of the master/slave ESD system for a full-chip ESD design implementation. In this implementation, only one trigger element exists.

Figure 4.19 shows a master/slave ESD system for a full-chip ESD design implementation using an RC-triggered MOSFET network. In this implementation, only one RC-trigger element and one drive circuit is used. In this fashion, all MOSFET gate connections can be triggered simultaneously, and the trigger network area is saved around the semiconductor chip. For this network system to be effective, the electrical connectivity between the MOSFET gate drive network and all the ESD clamp elements must be provided with low resistance bussing around the semiconductor chip.

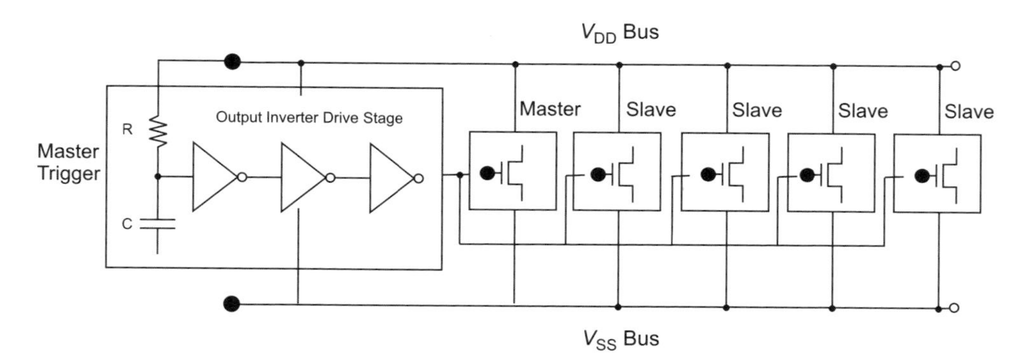

Figure 4.19 ESD power clamp master/slave system

4.8 SUMMARY AND CLOSING COMMENTS

In this chapter, the discussion focuses on ESD power clamp networks. A key issue in ESD design synthesis is the type of ESD network used in the power grid domain. This was discussed briefly in previous chapters, and this entire chapter is dedicated to ESD power clamp discussions, circuit topology, and issues. There is a vast amount of publications and literature in this area.

In Chapter 5, the discussion focuses on ESD networks on the signal pads. In this case, the entire chapter will be dedicated to ESD signal pin device layout and design.

PROBLEMS

4.1. What are the advantages and disadvantages of reverse-bias triggered ESD power clamps? Are they scalable? How does one make a reverse-bias triggered ESD network scalable with power supply and technology scaling?

4.2. What are the advantages and disadvantages of forward-bias triggered ESD power clamps? Are they scalable? How does one make a reverse-bias triggered ESD network scalable?

4.3. A bipolar transistor open-base collector-to-emitter breakdown voltage (BV_{CEO}) is inversely proportional to the unity current gain cutoff frequency (according to the well-known Johnson limit relationship). How can this be synthesized and utilized for ESD protection in ESD power clamps? Show an example of this. Show how this can be used in technology scaling for bipolar technology.

4.4. Show an example of an RC-triggered ESD power clamp containing three inverter stages: resistor, capacitor, and NFET output clamp. What are the concerns? Discuss false-triggering concerns, power-up, power-down, latchup, and electrical overstress issues. What elements in this network can form a parasitic pnpn and form a latchup issue? How does this network respond to increased temperature?

4.5. Show an example of an RC-triggered ESD power clamp containing three inverter stages: resistor, capacitor, and PFET output clamp. What are the concerns? Discuss false-triggering concerns, power-up, power-down, latchup, and electrical overstress issues. What elements in this network can form a parasitic pnpn and form a latchup issue? How does this network respond to increased temperature?

4.6. In a mixed-voltage application, with two MOSFET transistors there are two choices in the construction of RC-triggered clamps: (a) use of only the higher-voltage MOSFET; or (b) cascode of the low-voltage output. What are the tradeoffs, performance, over-voltage, and area considerations between the two implementations?

4.7. In a bipolar power application, the signal is required to swing from a positive to a negative voltage (e.g., from V_{CC} to V_{EE}, where $V_{CC} = +5$ V and $V_{EE} = -5$ V). Show multiple configurations for ESD networks in a three-rail architecture with V_{CC}, V_{SS}, and V_{EE} power rails.

4.8. A 5.0 V ESD MOSFET RC-triggered power clamp is re-mapped from a 3.3 V MOSFET power clamp. What adjustments are made in the re-mapping from an over-voltage case? In the MOSFET output transistor size, on-resistance, channel length, and area? Show the terms in the MOSFET current model that are influenced.

4.9. In the operation of the circuit, show the resistances in the alternative current path assuming a turn-on voltage and on-resistance for the ESD input network, power bus, and ESD power clamp. Discuss the area and placement tradeoffs between the power bus, input device, and ESD power clamp elements. What are the competing factors?

4.10. As technologies scale, the ESD input device is required to scale to smaller dimensions. Assuming the on-resistance of the ESD device increases as $R' = R\alpha$ (where α is the constant electric field scaling theory), how should the ESD power bus resistance and ESD power clamp resistance scale accordingly to maintain a constant ESD protection level (e.g., constant ESD scaling theory)?

4.11. As technologies scale, assume the voltage scaling on the input node is as $V' = V/\alpha$, where ${V_{BR}}' = V_{BR}/\alpha$ (where α is the constant electric field scaling theory); also, the ESD input device is required to scale to smaller dimensions. Assume the on-resistance of the ESD device increases as $R' = R/\alpha$. How should the ESD power bus resistance and ESD power clamp resistance scale accordingly to maintain a constant ESD protection level (e.g., constant ESD scaling theory)? Given the bus width is also scaled, as $W' = W/\alpha$, show that the placement requirement is modified to maintain constant ESD results. (*Hint*: derive the placement relationship of the ESD input device to power clamp based on α scaling theory.)

REFERENCES

1. A. Amerasekera and C. Duvvury. *ESD in Silicon Integrated Circuits*. Chichester, UK: John Wiley and Sons, Ltd, 1995.
2. S. Voldman. *ESD: Physics and Devices*. Chichester, UK: John Wiley and Sons, Ltd, 2004.
3. S. Voldman. *ESD: Circuits and Devices*. Chichester, UK: John Wiley and Sons, Ltd, 2005.
4. S. Voldman. *ESD: RF Technology, and Circuits*. Chichester, UK: John Wiley and Sons, Ltd, 2006.
5. S. Voldman. *ESD: Failure Mechanisms and Models*. Chichester, UK: John Wiley and Sons, Ltd, 2009.
6. S. Dabral and T. J. Maloney. *Basic ESD and I/O Design*. New York: John Wiley and Sons, Inc., 1998.
7. R. Troutman. *Latchup in CMOS Technology: The Problem and the Cure*. New York: Kluwer Publications, 1986.
8. S. Voldman. *Latchup*. Chichester, UK: John Wiley and Sons, Ltd, 2009.
9. M. D. Ker and S.F. Hsu. *Transient-Induced Latchup in CMOS Integrated Circuits*. Singapore: John Wiley and Sons (Asia) Pte Ltd, 2009.
10. V. Vashchenko and A. Shibkov. *ESD Design for Analog Circuits*. New York: Springer, 2010.
11. N. Maene, J. Vandenbroeck, and L. Van dem Bempt. On chip electrostatic discharge protections for inputs, outputs, and supplies of CMOS circuits. *Proceedings of the Electrical Overstress/Electrostatic Discharge (EOS/ESD) Symposium*, 1992; 228–233.
12. G. D. Croft. Dual rail ESD protection using complementary SCRs. *Proceedings of the Electrical Overstress/Electrostatic Discharge (EOS/ESD) Symposium*, 1992; 243–249.

13. G. D. Croft. ESD protection using SCR clamping. U.S. Patent No. 5,574,618, November 12, 1996.
14. W. Mack and R. Meyer. New ESD protection schemes for BiCMOS processes with application to cellular radio designs. *Proceedings of the IEEE International Symposium on Circuits and Systems*, 1992.
15. R. Merrill and E. Issaq. ESD design methodology. *Proceedings of the Electrical Overstress/ Electrostatic Discharge (EOS/ESD) Symposium*, 1993; 233–238.
16. S. Dabral, R. Aslett, and T. Maloney. Designing on-chip power supply coupling diodes for ESD protection and noise immunity. *Proceedings of the Electrical Overstress/Electrostatic Discharge (EOS/ESD) Symposium*, 1994; 239–249.
17. G. D. Croft. ESD protection using a variable voltage supply clamp. *Proceedings of the Electrical Overstress/Electrostatic Discharge (EOS/ESD) Symposium*, 1994; 135–140.
18. D. Krakauer, K. Mistry, and H. Partovi. Circuit interaction during electrostatic discharge. *Proceedings of the Electrical Overstress/Electrostatic Discharge (EOS/ESD) Symposium*, 1994; 113–119.
19. N. Tanden. ESD trigger circuit. *Proceedings of the Electrical Overstress/Electrostatic Discharge (EOS/ESD) Symposium*, 1994; 120–124.
20. S. Dabral, R. Aslett, and T. Maloney. Core clamps for low voltage technologies. *Proceedings of the Electrical Overstress/Electrostatic Discharge (EOS/ESD) Symposium*, 1994; 141–149.
21. S. Voldman. ESD protection in a mixed voltage interface and multi-rail disconnected power grid environment in 0.5- and 0.25-μm channel length CMOS technologies. *Proceedings of the Electrical Overstress/Electrostatic Discharge (EOS/ESD) Symposium*, 1994; 125–134.
22. T. Maloney. Novel clamp circuits for IC power supply protection. *Proceedings of the Electrical Overstress/Electrostatic Discharge (EOS/ESD) Symposium*, 1995; 1–12.
23. E. Worley, R. Gupta, B. Jones, R. Kjar, C. Nguyen, and M. Tennyson. Sub-micron chip ESD protection schemes which avoid avalanche junctions. *Proceedings of the Electrical Overstress/ Electrostatic Discharge (EOS/ESD) Symposium*, 1995; 13–20.
24. G. D. Croft. Transient supply clamp with a variable RC time constant. *Proceedings of the Electrical Overstress/Electrostatic Discharge (EOS/ESD) Symposium*, 1996; 276–279.
25. J. Smith. A substrate triggered lateral bipolar circuit for high voltage tolerant ESD protection applications. *Proceedings of the Electrical Overstress/Electrostatic Discharge (EOS/ESD) Symposium*, 1998; 63–71.
26. W. Anderson, J. Monanaro, and N. Howorth. Cross-referenced ESD protection supplies. *Proceedings of the Electrical Overstress/Electrostatic Discharge (EOS/ESD) Symposium*, 1998; 86–96.
27. M.D. Ker. Whole chip ESD protection design with efficient V_{DD}-to-V_{SS} clamp circuit for submicron CMOS VLSI. *IEEE Transactions on Electron Devices*. **ED-46** (1), 1999; 173–183.
28. T. Maloney. Stacked PMOS clamps for high voltage power supply protection. *Proceedings of the Electrical Overstress/Electrostatic Discharge (EOS/ESD) Symposium*, 1999; 70–77.
29. S. Poon and T. Maloney. New considerations for MOSFET power clamps. *Proceedings of the Electrical Overstress/Electrostatic Discharge (EOS/ESD) Symposium*, 2002; 1–5.
30. T. Maloney, S. Poon, and L. Clark. Methods for designing low-leakage power supply clamps. *Proceedings of the Electrical Overstress/Electrostatic Discharge (EOS/ESD) Symposium*, 2003; 27–43.
31. J. Smith and G. Boselli. A MOSFET power clamp with feedback enhanced triggering for ESD protection in advanced CMOS technologies. *Proceedings of the Electrical Overstress/Electrostatic Discharge (EOS/ESD) Symposium*, 2003; 8–16.
32. M. Stockinger, J. Miller, M. Khazhinsky, C. Torres, J. Weldon, B. Preble, M. Bayer, M. Akers, and V. Kamat. Boosted and distributed rail clamp networks for ESD protection in advanced CMOS technologies. *Proceedings of the Electrical Overstress/Electrostatic Discharge (EOS/ESD) Symposium*, 2003; 17–26.
33. M. Stockinger and J. Miller. Advanced ESD rail clamp network design for high voltage CMOS applications. *Proceedings of the Electrical Overstress/Electrostatic Discharge (EOS/ESD) Symposium*, 2003; 280–288.

34. P. Tong, W. Chen, and R. Jiang. Active ESD shunt with transistor feedback to reduce latchup susceptibility or false triggering. *Proceedings of the IEEE International Symposium on the Physical and Failure Analysis of Integrated Circuits (IPFA)*, 2004; 89–92.
35. M.D. Ker and S.F. Hsu. Component-level measurement for transient-induced latchup in CMOS ICs under system-level ESD considerations. *IEEE Transactions on Device and Materials Reliability*, **6** (3), 2006; 461–472.
36. C.C. Yen and M.D. Ker. Failure of on-chip power-rail ESD clamp circuits during system-level ESD test. *Proceedings of the International Reliability Physics Symposium (IRPS)*, 2007; 598–599.
37. J. Z. Chen, X.Y. Zhang, A. Amerasekera, and T. Vrotsos. Design and layout of a high ESD performance NPN structure for submicron BiCMOS/bipolar circuits. *Proceedings of the International Reliability Physics Symposium (IRPS)*, 1996; 227–232.
38. S. Joshi, P. Juliano, E. Rosenbaum, G. Katz, and S. M. Kang. ESD protection for BiCMOS circuits. *Proceedings of the Bipolar Circuit Technology Meeting (BCTM)*, 2000; 218–221.
39. S. Joshi, R. Ida, P. Givelin, and E. Rosenbaum. An analysis of bipolar breakdown and its application to the design of ESD protection circuits. *Proceedings of the International Reliability Physics Symposium (IRPS)*, 2001; 240–245.
40. S. Voldman, A. Botula, and D. Hui. Silicon germanium heterojunction bipolar transistor ESD power clamps and the Johnson limit. *Proceedings of the Electrical Overstress/Electrostatic Discharge (EOS/ESD) Symposium*, 2001; 326–336.
41. S. Voldman. Variable trigger voltage ESD power clamps for mixed voltage applications using a 120 GHz/100 GHz (f_T/f_{MAX}) silicon germanium heterojunction bipolar transistor with carbon incorporation. *Proceedings of the Electrical Overstress/Electrostatic Discharge (EOS/ESD) Symposium*, 2002; 52–61.
42. S. Voldman and E. Gebreselasie. Low-voltage diode-configured SiGe:C HBT triggered ESD power clamps using a raised extrinsic base 200/285 GHz (f_T/f_{MAX}) SiGe:C HBT device. *Proceedings of the Electrical Overstress/Electrostatic Discharge (EOS/ESD) Symposium*, 2004; 57–66.
43. Y. Ma and G.P. Li. A novel on-chip ESD protection circuit for GaAs HBT RF power amplifiers. *Proceedings of the Electrical Overstress/Electrostatic Discharge (EOS/ESD) Symposium*, 2002; 83–91; and *Journal of Electrostatics*, **59**, October 2003; 211–227.
44. Y. Ma G.P. Li. InGaP/GaAs HBT DC-20 GHz distributed amplifier with compact ESD protection circuits. *Proceedings of the Electrical Overstress/Electrostatic Discharge (EOS/ESD) Symposium*, 2004; 50–54.
45. C. Duvvury. F. Carvajal, C. Jones, and M. Smayling. Device integration for ESD robustness of high voltage power MOSFETs. *International Electron Device Meeting (IEDM) Technical Digest*, 1994; 407–410.
46. M. Mergens, W. Wilkening, S. Mettler, *et al.* Analysis of lateral DMOS power devices under ESD stress conditions. *IEEE Transactions on Electron Devices*, **ED-47**, 2000; 2128–2137.
47. V. De Heyn, G. Groeseneken, B. Keppens, *et al.* Design and analysis of new protection structures for smart power technology with controlled trigger and holding voltage. *Proceedings of the International Reliability Physics Symposium (IRPS)*, 2001; 253–258.
48. G. Bertrand, C. Delage, M. Bafluer, *et al.* Analysis and compact modeling of a vertical grounded base npn transistor used as ESD protection in a smart power technology. *IEEE Journal of Solid State Circuits*, **36**, 2001; 1373–1381.
49. J.H. Lee, J.R. Shih, C.S. Tang, *et al.* Novel ESD protection structure with embedded SCR LDMOS for smart power technology. *Proceedings of the International Reliability Physics Symposium (IRPS)*, 2002; 156–161.
50. C. Duvvury, R. Steinhoff, G. Boselli, V. Reddy, H. Kunz, S. Marum, and R. Cline. Gate oxide failures due to anomalous stress from HBM ESD testers. *Proceedings of the Electrical Overstress/Electrostatic Discharge (EOS/ESD) Symposium*, 2004; 132–140.

51. T. Meuse, R. Barrett, D. Bennett, M. Hopkins, J. Leiserson, and L. Ting. Formation and suppression of a newly discovered secondary EOS event in HBM test systems. *Proceedings of the Electrical Overstress/Electrostatic Discharge (EOS/ESD) Symposium*, 2004; 141–145.
52. R. A. Ashton, B. E. Weir, G. Weiss, and T. Meuse. Voltages before and after HBM stress and their effect on dynamically triggered power supply clamps. *Proceedings of the Electrical Overstress/ Electrostatic Discharge (EOS/ESD) Symposium*, 2004; 153–159.

5 ESD Signal Pin Networks Design and Synthesis

5.1 ESD SIGNAL PIN STRUCTURES

In this chapter, ESD signal pin networks will be explored. Today, it is common practice to have ESD networks on all signal pins and they form a critical part of the ESD design synthesis [1–39]. Development of ESD signal pin networks and synthesis into the semiconductor chip architecture is part of the ESD design discipline and an essential component of the art of ESD design. Due to the wealth of material on this subject, the focus will instead be on the integration, design synthesis, key parameters, and layout, as opposed to discussing all forms of ESD input circuitry. First, the focus of the chapter will address integration of the ESD signal pin with the bond pad. This will be followed by discussion of MOSFET layout design [1–3,13–16], diode layout [1–3,8–21], SCR design [3,6,13–16], as well as passive elements (resistors and inductors) [5,6]. Examples will be addressed from CMOS digital [1–21], analog [6], and RF applications [4,5,27–38].

Today, it is common practice to have ESD networks on all signal pins. Typically, the only cases that do not require ESD networks are "self-protecting networks." ESD signal pin structures are a critical part of the ESD design synthesis. In the following sections, active and passive elements used in ESD protection circuits will be discussed. In the discussion, the characteristics, design practices, and physical layout will be highlighted. As this subject is covered significantly in other ESD textbooks, this text will focus on design layout and ESD design synthesis. Orientation, arrangement, and rearrangement play a role in the ESD design synthesis process. A focus of this chapter is the placement of the ESD signal pad elements with respect to the bond pad and power bus. In different design architectures, this is critical to the floorplan integration of the bond pad, the ESD network, the power bus, and the I/O circuitry.

ESD: Design and Synthesis, First Edition. Steven H. Voldman.

5.1.1 Classification of ESD Signal Pin Networks

ESD signal networks can be defined in different classes of ESD networks. Figure 5.1 shows classifications of ESD networks. A possible list of features or classes is as follows:

- **Bond Pad Condition:** ESD network relative to bond pad position (ESD under bond pad vs. not under bond pad).
- **Rail Conditions:** Single-rail, dual-rail, or triple-rail electrical connectivity.
- **Power Supply Voltage Level:** Native power supply versus mixed-voltage interface ESD signal pin networks.

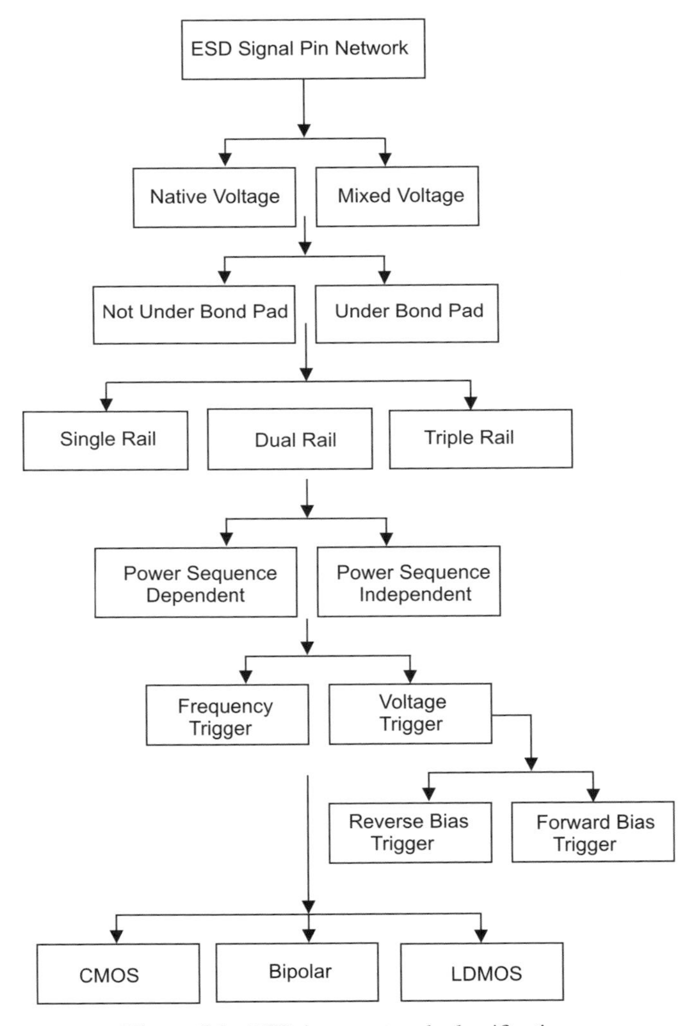

Figure 5.1 ESD input network classifications

- **Power Supply Sequencing Condition:** Power supply sequence-dependent or sequence-independent.
- **Trigger Integration:** Integrated vs. non-integrated trigger element.
- **Trigger Type:** Frequency-triggered vs. voltage-triggered.
- **Technology Type:** CMOS, bipolar, BiCMOS, or BCD-type devices.
- **Device Type:** MOSFETs, diodes, SCRs, LDMOS transistors, resistors, capacitors, or inductors.

ESD signal pins can be defined as under a bond pad [5,22–26,37,38], or not under a bond pad. The relative orientation of the ESD network with the bond pad will be discussed in depth. A second classification is the number of electrical power rails the ESD network is connected to between the signal pin and the power rails. A third classification is whether the network is for native voltage power supply, or for mixed-voltage interfaces [1,18–20]. ESD signal pin networks can be defined according to whether they are sequence-independent or sequence-dependent applications. In the ESD signal pin network, the ESD trigger element can be integrated with the shunt element or separate (as was true in the ESD power clamp networks). The "turn-on" of the clamp can be frequency-initiated or voltage-triggered. Lastly, these ESD signal pin networks can be constructed from CMOS, SOI, bipolar, or LDMOS technologies, and the corresponding elements contained within these technologies.

5.1.2 ESD Design Synthesis of ESD Signal Devices – Key Design Parameters

In ESD design synthesis of ESD signal pins, there are key design parameters in the decision on what type of circuit to utilize (Figure 5.2). The following is a list of key parameters in the ESD design process of ESD networks:

- ESD signal pin physical area.
- ESD signal pin width.
- ESD signal pin current per unit of width metric (A/μm).
- ESD signal pin "on resistance."
- ESD signal pin voltage tolerance.
- ESD signal pin leakage current.
- ESD signal pin capacitance loading.
- ESD signal pin frequency response.
- ESD signal pin trigger voltage or current.
- ESD signal pin latchup robustness.

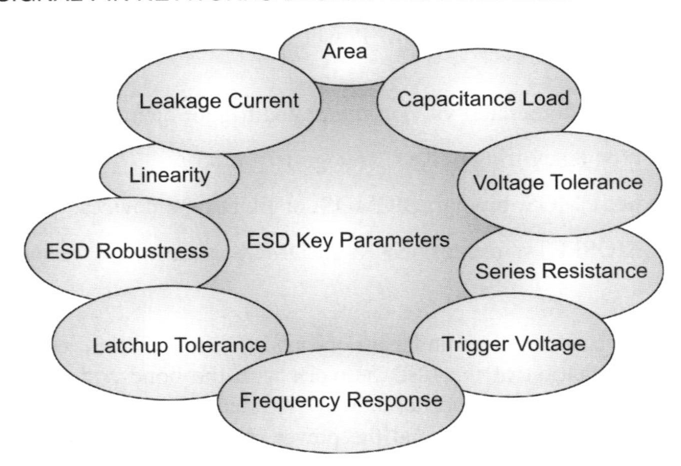

Figure 5.2 ESD input network key design parameters

5.2 ESD INPUT STRUCTURES – ESD AND BOND PADS LAYOUT

ESD and bond pad layout integration is a key part of any ESD design synthesis. In this section a few examples of the arrangement and rearrangement of ESD networks and bond pads are discussed.

5.2.1 ESD and Bond Pad Layout and Synthesis

In design synthesis of a semiconductor chip, the integration of the bond pads and the ESD network is important for effective implementation of the ESD network. Figure 5.3 shows an

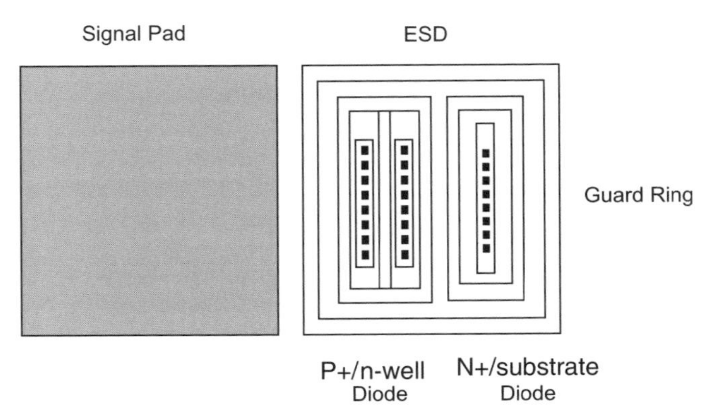

Figure 5.3 ESD input network and bond pad

example of a bond pad with adjacent ESD network. The advantages of the spatial placement of the ESD network adjacent to the bond pad are as follows:

- **Wide Interconnect:** Allowance of maximum wide interconnect wire width from the bond pad to the ESD element at either lowest metal level (e.g., M1 wire layer) or highest metal level (e.g., M-last wire layer) without obstruction of power or core wiring.
- **Low Resistance:** Provide lowest resistance between the bond pad and the ESD network.
- **Broadside Orientation:** Allowance of broad-side wiring from the bond pad to the ESD element.

5.2.2 ESD Structures Between Bond Pads

In design synthesis of a semiconductor chip, the placement of the bond pad with the ESD network can influence the utilization of the chip area. In the integration process, the placement of the ESD network in the same physical dimension with the I/O network extends the length of the standard cell and chip size. The extension of the ESD element toward the chip interior extends the width of the peripheral area used in the design. At the same time, the area between the bond pads is not always suitable for circuitry. To avoid extension of the standard cell, and to utilize wasted area between the bond pads, the ESD network can be placed between the bond pads.

Figure 5.4 shows an example of a bond pad with adjacent ESD network between the bond pads. The advantages of the spatial placement of the ESD network between the bond pads are as follows:

- **Unused Area:** Utilizes wasted or unused area between bond pads.
- **Reduced Standard Cell Height:** Reduces the "height" of a standard cell peripheral circuit.
- **Reduced Bond Pad Width:** Reduces the required width of the peripheral bond pad ring.
- **Wide Interconnect:** Allowance of maximum wide interconnect wire width from the bond pad to the ESD element at either lowest metal level (e.g., M1 wire layer) or highest metal level (e.g., M-last wire layer) without obstruction of power or core wiring.
- **Low Resistance Path:** Provide lowest resistance between the bond pad and the ESD network.
- **Broadside Orientation:** Allowance of broad-side wiring from the bond pad to the ESD element.
- **Dual Broadside Orientation:** Allows for wiring of the ESD elements on both sides (negative and positive ESD solutions can be separated, allowing for double the effective wiring for the different test conditions).
- **Spatial Separation:** Spatially separates ESD element from the I/O cells, allowing for reduced guard rings required.
- **Spatial Separation:** Spatial separation of the ESD element from the I/O cell reduces risk of latchup and injection phenomena [4,7].

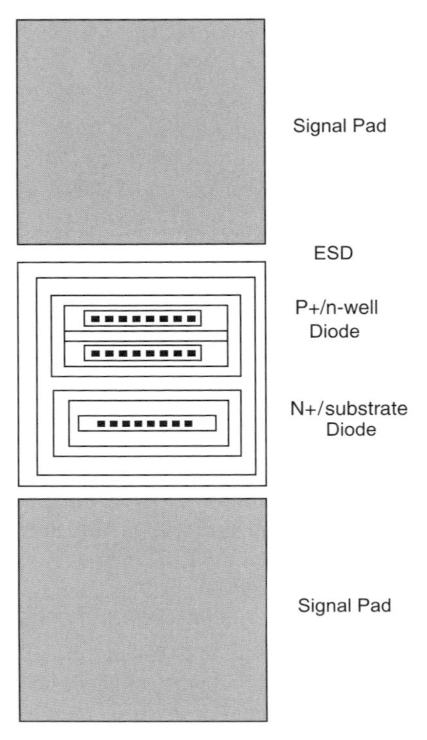

Figure 5.4 ESD network between bond pads

5.2.3 Split I/O and Bond Pad

In design synthesis of a semiconductor chip, the integration of the bond pads and the I/O circuit can be arranged in such a fashion that the I/O circuit is separated for layout advantages. Figure 5.5 shows an example of a bond pad where the I/O network is spatially separated on both sides of the bond pad structure. The advantages of the spatially separated network are as follows:

- **I/O Length:** Length of the I/O network is reduced.
- **Latchup Avoidance by Spatial Separation:** NFET pull-downs and PFET pull-ups are separated to avoid excessive guard rings and latchup concerns between the NFET and PFET elements [4,7].
- **Power Rail Separation:** Power rails can be separated on both sides of the bond pad.
- **NFET Compatibility:** NFET-based ESD networks can be placed adjacent to NFET pull-down.

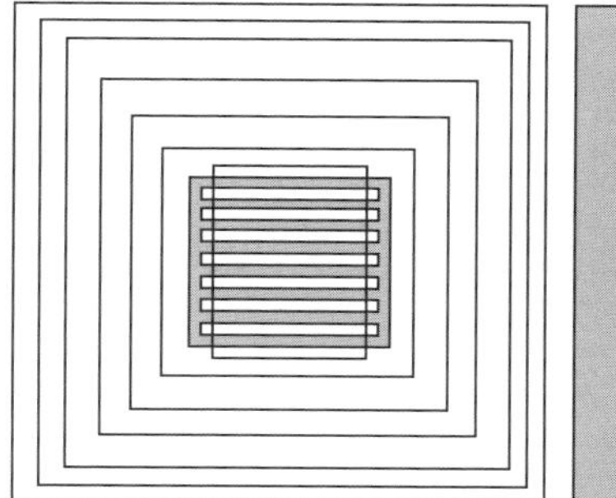

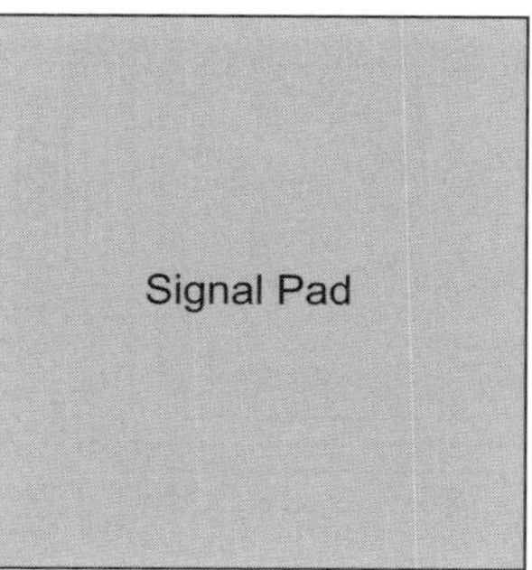

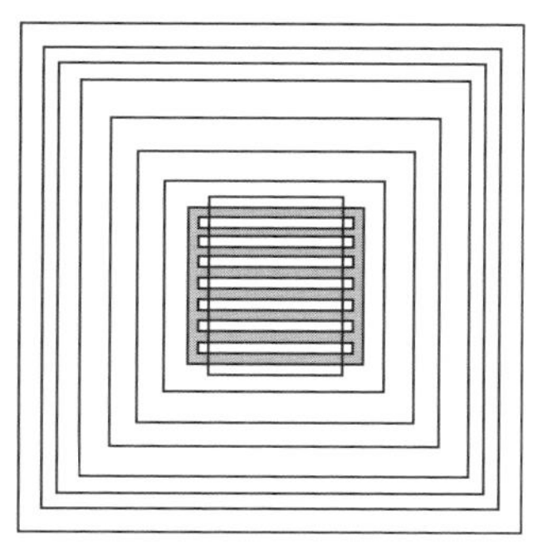

Figure 5.5 Spatially separated I/O network adjacent to bond pad

- **PFET Compatibility:** PFET-based ESD networks can be placed adjacent to PFET pull-up.
- **Between Bond Pad Area:** PFET pull-up and NFET pull-down can be placed between the bond pads.

5.2.4 Split ESD Adjacent to Bond Pad

In design synthesis of a semiconductor chip, ESD networks provide both ESD protection solutions for positive and negative pulse events. As a result, there are at times at least two separate networks or devices to address these events. In many cases, these consist of p-type elements for positive ESD events and n-type elements for negative ESD events. In the integration process, the placement of the ESD network in the same physical dimension with the I/O network extends the length of the standard cell and chip size. The extension of the ESD element toward the chip interior extends the width of the peripheral area used in the design. One solution is the placement of the ESD network adjacent to the bond pad, but to separate or "split" the spatial orientation of the ESD element on either side of the bond pad.

Figure 5.6 shows an example of a bond pad with adjacent ESD network between the bond pads. The advantages of the spatial splitting of the ESD network are as follows:

- **Interconnects:** Allowance of maximum wide interconnect wire width from the bond pad to the ESD element at either lowest metal level (e.g., M1 wire layer) or highest metal level (e.g., M-last wire layer) without obstruction of power or core wiring.
- **Resistance:** Provide lowest resistance between the bond pad and the ESD network.
- **Broadside Wiring:** Allowance of broad-side wiring from the bond pad to the ESD element.
- **Bi-directional Wiring:** Allows for wiring of the ESD elements on both sides (negative and positive ESD solutions can be separated, allowing for double the effective wiring for the different test conditions).

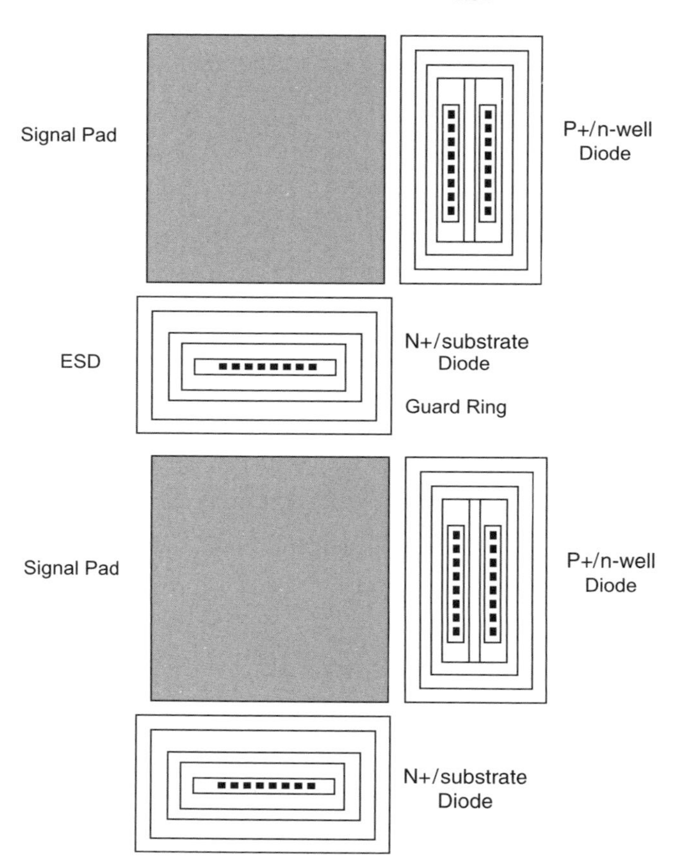

Figure 5.6 ESD network split adjacent to bond pads

- **Spatial Separation:** Spatially separates ESD elements within the ESD network allowing for reduced guard rings required (or none at all).
- **Spatial Separation:** Spatial separation of the ESD elements within the ESD network to avoid latchup within the ESD network.
- **Power and Ground Bus Separation:** Allows separation of ground and power buses to two sides of the bond pad in the architecture of the peripheral pad ring, or interior "spine."
- **I/O Height Reduction:** Reduces the "height" of a standard cell peripheral circuit (if placed between bond pads).
- **Bond Pad Ring Width Reduction:** Reduces the required width of the peripheral bond pad ring (if placed between bond pads).

Figure 5.7 shows an example of a split dual-diode network where the "up" diode is separated from the "down" diode element.

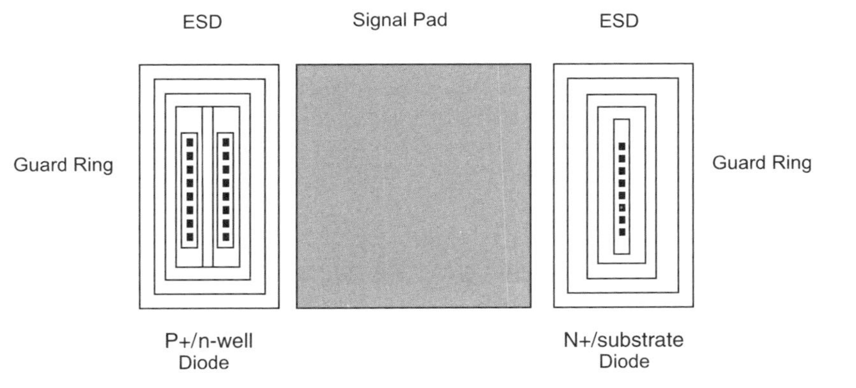

Figure 5.7 ESD network split adjacent to bond pads

5.2.5 ESD Structures Partially Under Bond Pads

One of the problems of placement of active devices under bond pads, is that not all semiconductor devices are qualified to be placed under a bond pad [4,5,22–26]. For some technologies, only structures without MOSFET gates can be placed under a bond pad. For RF applications, the bond pad structure influences the RF model s-parameters and is not allowed. As a result, the integration of ESD circuit networks that contain both approved and non-approved devices can be only partially placed under the bond pad area.

Figure 5.8 shows an example of an ESD network partially under a bond pad. The advantages of the spatial placement of the ESD network adjacent to the bond pad are as follows:

- **Bond Pad Ring Width:** Reduction of bond pad ring width.
- **Chip Size:** Reduction of total semiconductor chip size.
- **Bond Pad Area Utilization:** Allowance of full area under a bond pad for ESD utilization.

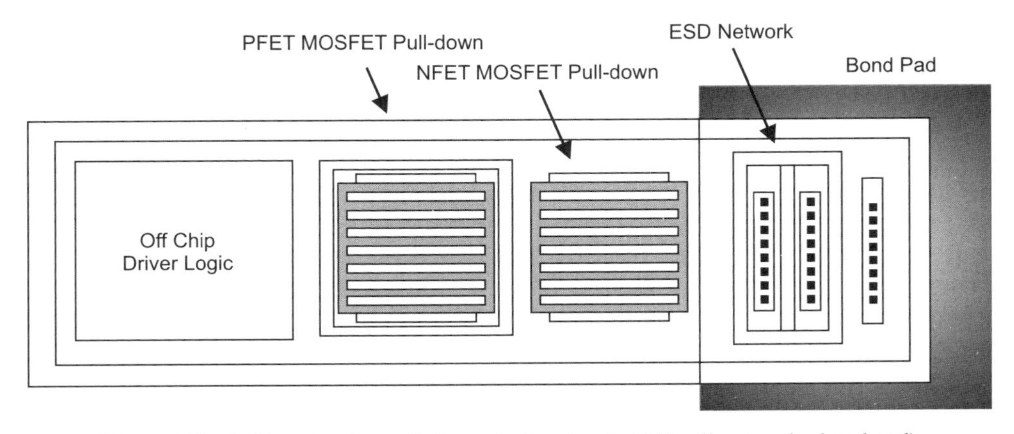

Figure 5.8 ESD networks partially under bond pads (I/O cell not under bond pad)

- **Wire Interconnection:** Allowance of maximum wide interconnect wire width from the bond pad to the ESD element at either lowest metal level (e.g., M1 wire layer) or highest metal level (e.g., M-last wire layer) without obstruction of power or core wiring.
- **Series Resistance:** Provide lowest resistance between the bond pad and the ESD network.

5.2.6 ESD Structures Under and Between the Bond Pads

In ESD design synthesis, it is possible to utilize all the area in the bond pad region. Secondly, in ESD input circuit design there are primary stage and secondary stage ESD networks. Third, there is at times a need for ESD networks between differential pair circuits. Additionally, there are implementations which are uncertain, and a "spare" ESD element, experiment, or back-up design is needed. One solution to this is the placement of one ESD design under the bond pad, and a second one between the bond pads.

Figure 5.9 shows an example of an ESD network under a bond pad, and a second element of the network between the bond pads. The advantages of the spatial placement of the ESD networks are as follows:

- **Chip Size:** Reduction of total semiconductor chip size.
- **Area Utilization:** Allowance of full area under a bond pad for ESD utilization.
- **Secondary Stages:** Secondary stage design between bond pads.
- **Differential Pair:** Differential pair ESD network between two bond pads.
- **Spare ESD:** Spare experimental backup design.

5.2.7 ESD Circuits and RF Bond Pad Integration

For RF applications, the wire bond pad can influence the RF circuit performance [5,26]. Both the inductance and the capacitance of the bond pad influence the impedance of the RF network. The wire bond itself introduces a series inductance. The bond pad introduces an inductive shunt component and a capacitor shunt to the chip substrate. The inductance of the RF bond pad can be reduced by the introduction of thick inter-level dielectric (ILD) films, conductive field shields, and insulator regions in the chip substrate. With the construction of inductors, thick ILD films are introduced to provide high-quality factor (Q) inductors. Using semiconductor interconnect metal films, slotted field plates are placed under RF bond pads to reduce eddy currents in the chip substrate [37,38]. Additionally, trench structures (e.g., shallow trench isolation, trench isolation (TI), and deep trench (DT)) are formed in a mesh to prevent the flow of substrate eddy currents.

For the RF network, the capacitance of the RF bond pad introduces a shunt capacitance term. With the introduction of an RF ESD network, such as an ESD diode network, or a grounded gate n-channel MOSFET, the capacitance of the RF bond pad and the ESD network is an additive

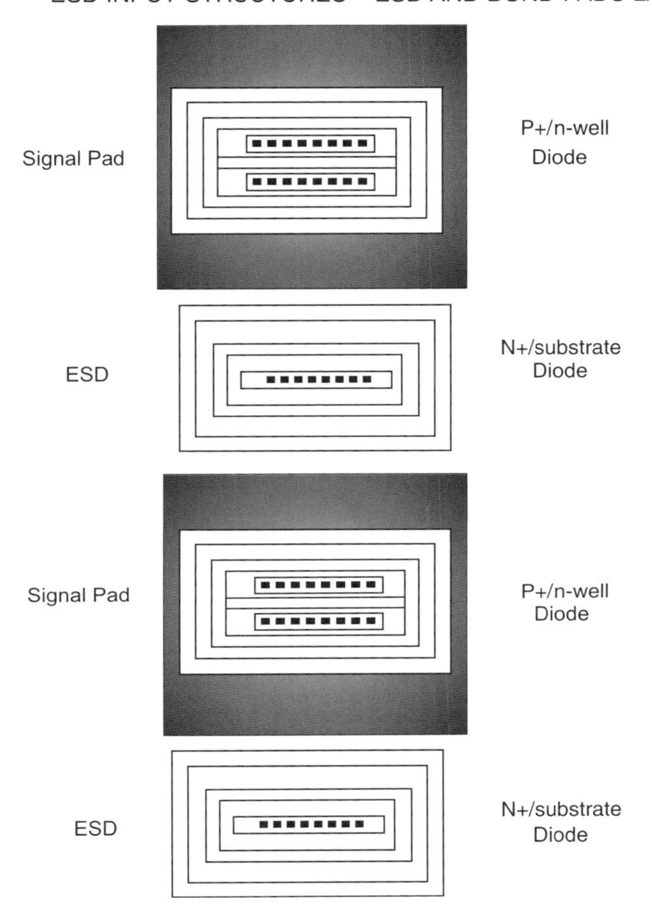

Figure 5.9 ESD networks under bond pad and additional ESD network element between the bond pads

shunt capacitance term to the RF circuit. The capacitance load of the components can be expressed as:

$$C_{\text{load}} = C_{\text{wire bond}} + C_{\text{bond pad}} + C_{\text{ESD}} + C_{\text{ckt}}$$

Since the capacitance loads of the bond pad and the ESD network are additive, in order to reduce the total capacitance load on the RF network, it is advantageous to reduce the capacitance of the bond pad structure. A solution to reduce the bond pad structure without additional semiconductor processing cost is to use a smaller bond pad for RF pins compared with the other analog and digital signal pins. Hence it is common to introduce octagonal pads by forming an octagon shape within the square shape of the standard pad used in a given technology.

Figure 5.10 shows an example of an octagonal pad structure [5]. The octagon is effectively formed by the removal of the corners of the square pad structure of the standard analog and digital pad structures. Note that in some applications, all the pads can be designed identically to the same shape – whether RF pins, analog pins, or digital pins. The removal of the corners of the

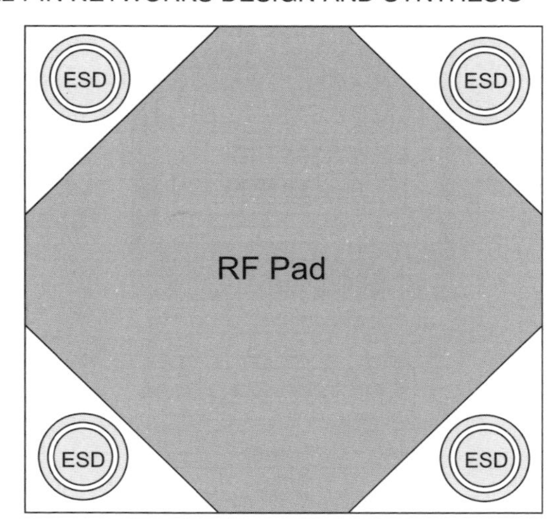

Figure 5.10 RF octagonal pad structure with circular ESD elements

square pad introduces a reduction in the pad capacitance load; this capacitance reduction allows for either higher RF performance or a larger acceptable load capacitance for the RF ESD element.

With the removal of the corners of the square pad structure, in RF applications, this additional area can be utilized for additional circuit function, or ESD networks. Figure 5.10 shows an example of utilization of circular ESD structures on the corners of the octagonal pad structure. Figure 5.11 shows an example of utilization of octagonal ESD structures on the corners of the octagonal pad structure. Note that the octagonal and circular ESD designs are designed so as to fit within the normal square pad foot print; this imposes a limit on the area and size of the ESD elements (e.g., the ESD device diameter). In the case of mixed-signal

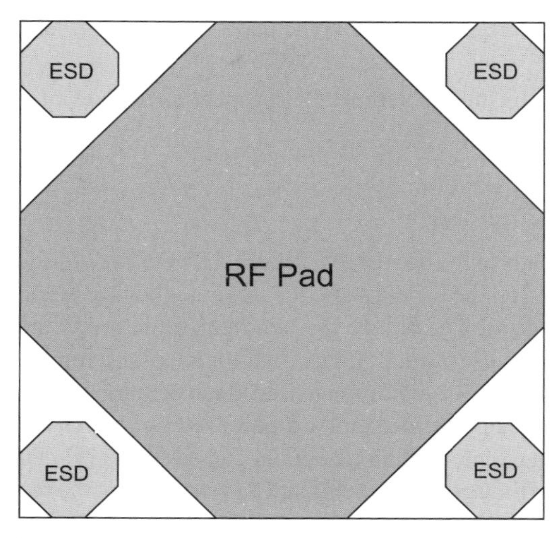

Figure 5.11 RF octagonal pad structure with octagonal ESD elements

applications with analog, digital, and RF pins all using octagonal pad structures, different ESD design practices are established. The implementations can be as follows:

- **Identical Elements:** Each element is an independent ESD diode element where they are all of identical design. In this case, the anode and cathode elements can be interchanged and provide 4× variations in the load capacitance and ESD structure size.
- **Non-identical ESD Elements:** Two of the four corner elements are independent ESD diode elements used for diodes between the input signal and the power supply (e.g., "up" diodes), and two are diodes used between the input signal and the ground rail (e.g., "down" diodes). In this fashion, a 2× variation in the load capacitance and ESD structure size is possible.

A common mixed-signal and RF ESD design practice is to utilize the identical structures for analog and RF pins, but for analog pins and digital circuits, the area utilized is double, and for RF pins the load capacitance is reduced by not utilizing all the four corners of the given signal pin [5]. An RF ESD design synthesis practice is as follows:

- **Octagonal Pads and Pad Corner ESD:** Octagonal pads are used for RF pins, where octagonal or circular ESD structures are placed on the corner regions.
- **Pad Corner ESD:** Identical or non-identical ESD elements are placed on the corners of the octagonal pad structure.
- **Selective Usage of Pad Corner ESD:** For analog and digital pins, ESD diode networks are formed using all four of the ESD diode elements on the corners of the octagonal pad structures, when octagonal pads are used on a given design for all signal and power pins.
- **Pad Corner RF ESD:** For RF pins, only one or two of the ESD diode elements are used on the corners of the octagonal pad structure (e.g., one-quarter or one-half of the capacitance load).

5.2.8 RF ESD Signal Pad Structures Under Bond Pads

RF technology applications extend from small single transistor chips (e.g., GaAs power amplifiers) to system-on-chip applications. In the wireless marketplace, the small die size allows for low-cost components. In many of the small die size chips, a few active and passive elements are connected. In these RF applications, the number of circuits on a chip is such that the total bond pad area is a significant percentage of the total chip area. To save space, RF ESD structures can be placed under bond pads [5,26,37,38]. The following concerns exist for placement of structures under bond pads in RF applications:

- Changes in the semiconductor device characteristics from the mechanical strain.
- Failure of the semiconductor component (e.g., gate dielectric failure, dislocation).
- Metal deformation and ILD cracking.
- Mechanical failure of the bond pad (e.g., separation from the semiconductor component).
- Change in the RF characteristics and models.

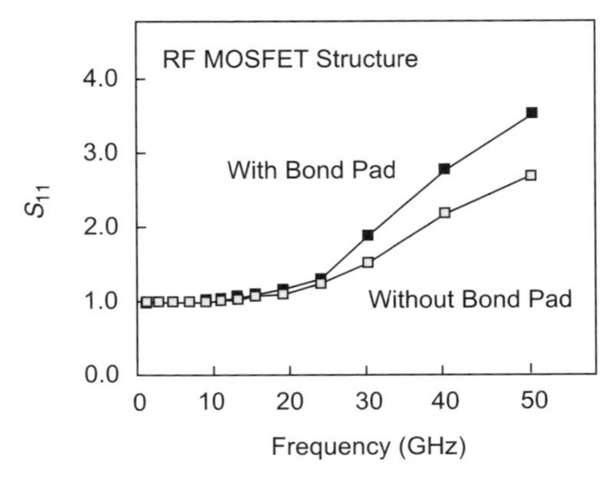

Figure 5.12 RF MOSFET scattering parameter S_{11} with and without bond pad

The placement of RF components under wire bond pads can influence the RF model. Figures 5.12 and 5.13 show the evaluation of the s-parameters as a function of frequency from 0 to 50 GHz. The parameters S_{11} and S_{22} both show that as the frequency increases, the difference between the structure with wire bond pad separates from the structure without wire bond pad. For the case of S_{11}, at low frequency S_{11} is near a unity value. As the frequency increases, S_{11} also increases. At approximately 20 GHz, the case of with and without bond pad separates, and below 20 GHz, there is little observed difference. Note that for both s-parameters, the separation between the two cases occurs after 30 GHz [5,26].

One of the issues with integrating the RF ESD network with the bond pad is that for RF applications, a deep trench mesh exists under the bond pad. A method to integrate circuit components under bond pads includes establishing a trench border on a circuit element and synthesizing a set of trench mesh edges of a trench mesh to be coincident with the trench border on the circuit element. The method further includes eliminating a trench mesh contained within the trench border of the trench circuit element [39].

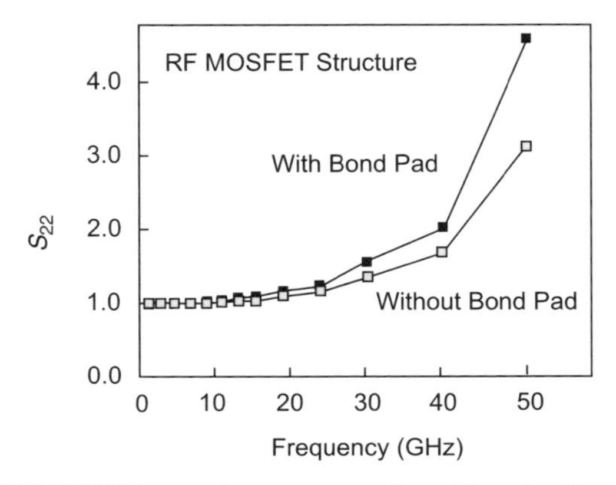

Figure 5.13 RF MOSFET scattering parameter S_{22} with and without wire bond pad

On the issue of RF ESD structures under pads, an RF ESD design synthesis is as follows [5]:

- **Area Reduction:** Significant area reduction in semiconductor chip size for RF components can be achieved in low pin count, pad-limited chip design by placement of RF ESD elements under pads (e.g., RF input ESD networks, and RF ESD power clamps).
- **Thick Insulator Films:** RF CMOS and RF BiCMOS technologies have lower risks of insulator mechanical failure because of the thicker ILD films utilized for high-quality factor inductor elements and passives (compared to CMOS technology).
- **RF ESD Interconnect Capacitance Load:** RF ESD structures under bond pads must address the tradeoff between the mechanical cracking issues (which prefer metal connections close to the silicon surface) and capacitance loading issues of the interconnects (which prefer metal connections far from the silicon surface).
- **RF ESD and RF Models:** Placement of RF ESD structures under wire bond pads must evaluate the DC, RF, and ESD measurements with and without the bond pad. Placement of RF ESD structures under wire bond pads must evaluate the change in the RF circuit model to evaluate the influence on the ESD elements, and total ESD circuit.

5.3 ESD DESIGN SYNTHESIS AND LAYOUT OF MOSFETs

ESD design synthesis of MOSFET devices is a key part of the ESD design discipline [1–5,12–16]. The physical layout of the MOSFET devices, in both the ESD networks and the output circuits (e.g., off-chip drivers) has considerable focus. For ESD networks, and for the off-chip driver circuitry, almost every feature of the MOSFET design is optimized for ESD robustness. In this section, the key parameters for ESD design of MOSFETs will be discussed.

5.3.1 MOSFET Key Design Parameters

In CMOS technology, the most common failure mechanism is the n-channel MOSFET transistors. As a result, a very large focus on the semiconductor process, layout, and design of almost every feature is defined. Figure 5.14 shows an example of a CMOS MOSFET cross-section. For ESD optimization, a commonly used process feature utilized is the ESD implant and the silicide blocking mask. In this ESD design practice, the key MOSFET parameters are as follows [12–16]:

- MOSFET channel length.
- MOSFET channel width.
- MOSFET drain contact to MOSFET gate space.
- MOSFET source contact to MOSFET gate space.
- Substrate pickup placement.
- Substrate pickup orientation.

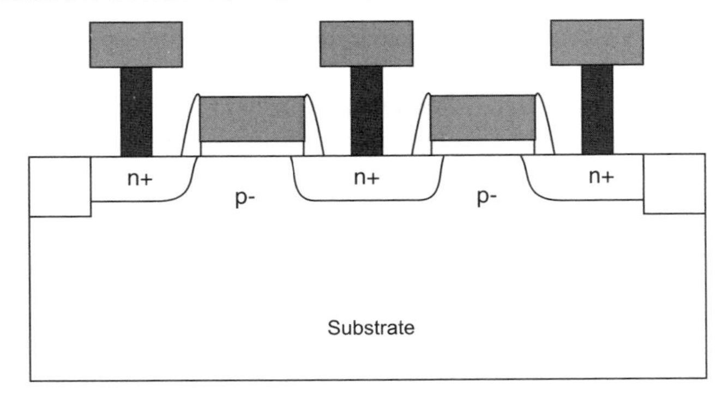

Figure 5.14 MOSFET layout cross-section

- Drain contact-to-contact spacing.
- Source contact-to-contact spacing.
- Drain last contact-to-drain edge spacing.
- Drain silicide block mask width.
- Source silicide block mask width.
- Silicide block mask shape and geometry.
- Silicide block mask-to-MOSFET gate space or overlap.
- ESD implant placement.

MOSFET channel length: The MOSFET channel length used is a function of the technology generation. The smaller the MOSFET channel length, the higher the MOSFET current drive and the lower the MOSFET snapback voltage.

MOSFET width: The MOSFET channel width used for ESD protection devices and self-protecting networks is a function of the ESD specification objective, ESD product result to the ESD specification (e.g., ESD product margin), and the ESD metric of ESD result for a given MOSFET width (e.g., kV/μm). ESD results do not scale with MOSFET width, without the optimization of the MOSFET layout. Hence, the objective of the ESD designer and technologist is to maximize the ESD MOSFET width metric to achieve the highest result, and scalability.

MOSFET drain contact-to-MOSFET gate space: The MOSFET drain contact to the MOSFET gate edge influences the MOSFET series resistance and drain ballasting. In some implementations, the silicide block mask is placed between the MOSFET drain contact and the MOSFET gate. The ESD design synthesis choice is a tradeoff between area, capacitance, series resistance, and ESD results.

MOSFET source contact-to-MOSFET gate space: The MOSFET source contact to the MOSFET gate edge influences the MOSFET series resistance and source ballasting. In some implementations, the silicide block mask is placed between the MOSFET source contact and the MOSFET gate. The ESD design synthesis choice is a tradeoff between area, capacitance, series resistance and ESD results.

Substrate pickup placement: The placement of the substrate "pickup" (e.g., p+ contact) can be an abutted or a non-abutted structure. The substrate pickup can be a linear stripe, and can be placed either as a single stripe, or placed between each MOSFET finger. The ESD design synthesis choice is a tradeoff between area, capacitance loading, and ESD results. The best ESD results will be achieved given the placement of a single "pickup" stripe for each MOSFET finger.

Substrate pickup orientation: The placement of the substrate "pickup" (e.g., p+ contact) orientation can influence the ESD robustness of the MOSFET. The substrate pickup can be a linear stripe, either parallel or orthogonal orientation, or a ring. The typical design choice is the placement parallel to the MOSFET finger or a complete ring.

Drain contact-to-contact spacing: The MOSFET contact-to-contact spacing is typically chosen as a minimum space to maximize the MOSFET current drive, and lowest MOSFET resistance. This parameter is not typically modified in the design layout.

Source contact-to-contact spacing: The MOSFET contact-to-contact spacing is typically chosen as a minimum space to maximize the MOSFET current drive, and lowest MOSFET resistance. This parameter is not typically modified in the design layout.

Drain last contact-to-drain edge spacing: In ESD design, there are at times failure mechanisms associated with the edge of a MOSFET drain laterally. As a result, the spacing between the MOSFET drain edge and the last contact is increased to avoid failure mechanisms. The tradeoff is the introduction of asymmetry of the number of contacts between the drain and the source, MOSFET current drive, and ESD.

Drain silicide block mask width: The MOSFET drain silicide mask is placed between the MOSFET drain contacts and the MOSFET gate. Depending on the ground rules, the space of the silicide mask relative to the contact and to the gate region varies. The MOSFET silicide mask influences the MOSFET series resistance and drain ballasting. In some implementations, the silicide block mask is placed between the MOSFET drain contact and the MOSFET gate; in other designs, the silicide block mask can overlay the MOSFET gate, and even over the entire MOSFET gate. These choices are dependent on overlay, alignment, MOSFET gate sheet resistance, and silicide issues. The ESD design synthesis choice is a tradeoff between area, capacitance, series resistance, and ESD results.

Source silicide block mask width: As in the prior section, the silicide mask can be placed on the MOSFET source region.

Silicide block mask shape and geometry: The MOSFET source/drain silicide block layout can be a rectangle or ring geometry. As a rectangular shape it is placed between the MOSFET contact and gate structures. When used over the MOSFET gate completely, it will extend as a single rectangle from MOSFET drain contacts to the MOSFET source contacts. As a ring structure, the MOSFET silicide block mask forms a "ring" around the MOSFET drain contact region. This achieves separating the end drain contact from the MOSFET drain edge, and achieves ballasting.

Silicide block mask to MOSFET gate space or overlap: The MOSFET silicide block mask is required either to be separated from the polysilicon MOSFET gate, or over the MOSFET gate. This layout issue is highly dependent on the technology generation, rules, and manufacturing control.

ESD implant placement: In some technologies, there is an additional ESD implant which modifies the implants in the MOSFET device. The ESD implants are placed over the MOSFET drain to improve the ESD robustness of the transistor. The layout placement and rules associated with the ESD implant are dependent on the technology (Table 5.1).

Table 5.1 MOSFET ESD layout key parameters

Parameter	Typical dimension	Note
Channel length	Minimum	
Channel width	Technology and ESD protection level dependent	200–700 μm
Contact-to-gate space (drain)	1–5 μm	
Contact-to-gate space (source)	1–3 μm	
Contact-to-contact	Minimum	
Contact-to-drain edge	2× to 10×	
Silicide block width	1–5 μm	
Silicide block mask to gate	Technology dependent	Non-overlap or overlapped
Silicide block mask to contacts	Technology dependent	
Silicide block shape	Rectangle or ring	
ESD implant	Technology dependent	

5.3.2 Single MOSFET with Silicide Block Masks

In Figure 5.14, a single MOSFET cross-section is shown in a single or dual-well CMOS technology [13–16]. As discussed in the previous section, the design is optimized to provide uniform current distribution through the entire layout structure for both ESD and performance objectives. In multi-finger implementations, the axis of symmetry and the thermal axis are in the center of the MOSFET drain structure.

Figure 5.15 shows a typical MOSFET layout including a source and drain silicide block mask [13–16]. In this example, a rectangular shape is used for the MOSFET silicide block mask. Additionally, it does not overlap the MOSFET gate region. The axis of symmetry is the center line through the MOSFET drain structure. To provide the maximum uniformity in current, design symmetry for the entire structure – including the p+ source pickup – is critical.

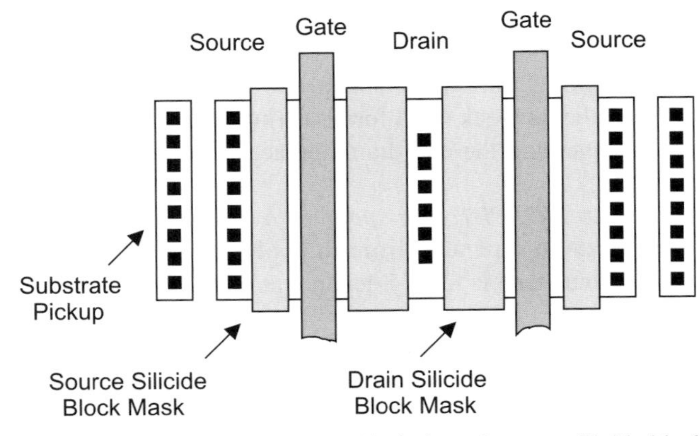

Figure 5.15 MOSFET layout with drain and source silicide block

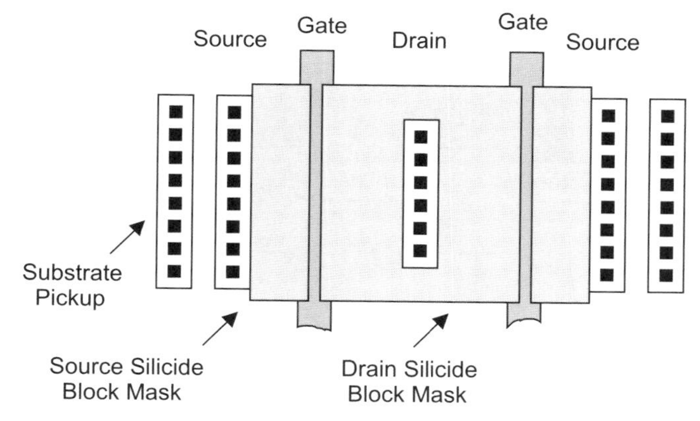

Figure 5.16 MOSFET layout with enclosed drain silicide block

Figure 5.16 shows a typical MOSFET layout including a drain silicide block mask using a ring configuration on the MOSFET drain [13–16]. In this example, the MOSFET silicide block mask forms an "enclosed drain" region. In this example, the MOSFET silicide block mask overlaps the polysilicon gate structure. The axis of symmetry is through the center of the MOSFET drain.

5.3.3 Series Cascode MOSFET

Series cascode MOSFET structure are used in self-protecting off-chip drivers, and ESD networks where a higher voltage tolerance is required. In both mixed-voltage, mixed-signal, high-voltage, and smart power applications, cascoded MOSFETs are utilized for ESD protection [1,4]. There are two styles of layout for a series cascode MOSFET:

- Separate physical layout for each MOSFET structure in separated physical regions.
- Integrated physical layout in the same physical region.

In the first case, the separated MOSFETs can be designed in the prior section, where they are electrically connected in a cascode manner with metal layers. In the second case, for density and performance reasons, the two transistors can be formed locally together. Figure 5.17 shows a cross-section of a series cascode MOSFET. The unique key feature for the series cascode network is as follows:

- MOSFET gate-to-gate spacing.

In this layout, a parasitic npn is formed between the top transistor MOSFET drain and the lower transistor MOSFET source. As a result, the ESD response of this structure does not respond like two separate transistors, but is influenced by the third lateral parasitic npn transistor as well. This effect was first discovered to have a lower MOSFET snapback voltage and a function of the gate-to-gate spacing [1,2].

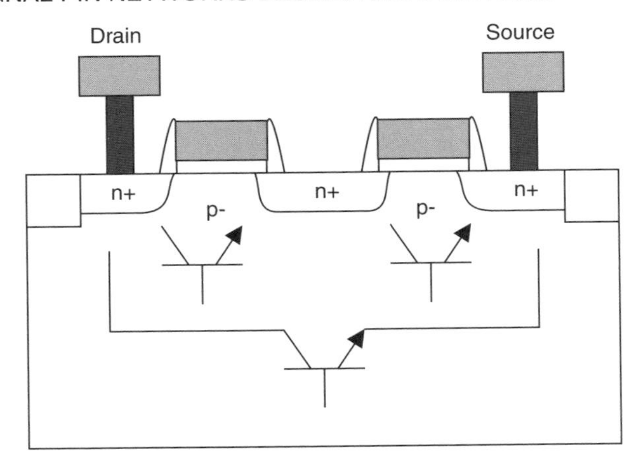

Figure 5.17 MOSFET cross-section of series cascode MOSFETs

5.3.4 Triple-well MOSFETs

In the ESD design synthesis and layout of a triple-well MOSFET, consideration must be made of the epitaxial region and the vertical parasitic bipolar transistor [7]. A parasitic npn is formed between the MOSFET source and drain junctions, the epitaxial region, and the isolating well region (Figure 5.18). As a result, the CMOS latchup response is influenced by the vertical parasitic npn transistor as well [7]. Critical ESD and latchup design parameters are as follows:

- Isolated p-epitaxial region sheet resistance.
- Triple-well, buried-layer vertical resistance.
- Triple-well, lateral resistance.

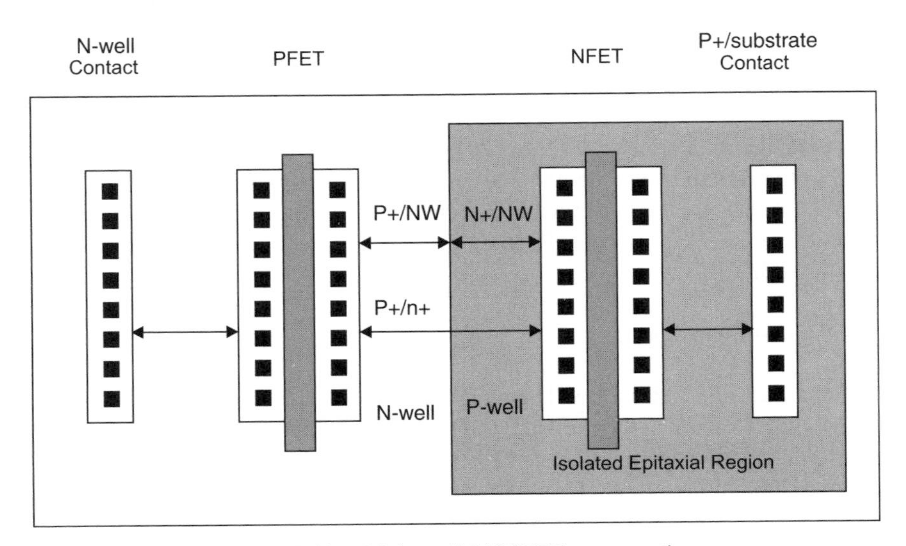

Figure 5.18 Triple-well MOSFET cross-section

Placement of the p+ isolated epitaxial "pickup" can strongly influence the ESD robustness of a multi-finger MOSFET and vertical npn "turn-on." Additionally, the electrical connection to the triple-well buried layer and its corresponding vertical and lateral resistance.

5.4 ESD DESIGN SYNTHESIS AND LAYOUT OF DIODES

In CMOS technology, diodes are used for ESD protection circuits [1,2,13–16]. Diodes have significant value for CMOS digital, analog, and RF applications. In the following section, key design parameters, layout, and design synthesis will be discussed.

5.4.1 ESD Diode Key Design Parameters

In CMOS technology, diodes are used for ESD protection circuits [1,2,13–16]. The effectiveness of a semiconductor diode for an ESD protection circuit is a strong function of the diode semiconductor process, layout, and design. Figure 5.19 shows an example of a CMOS p + /n-well diode. In this ESD design practice, the key diode parameters are as follows:

- Diode anode width.
- Diode cathode width.

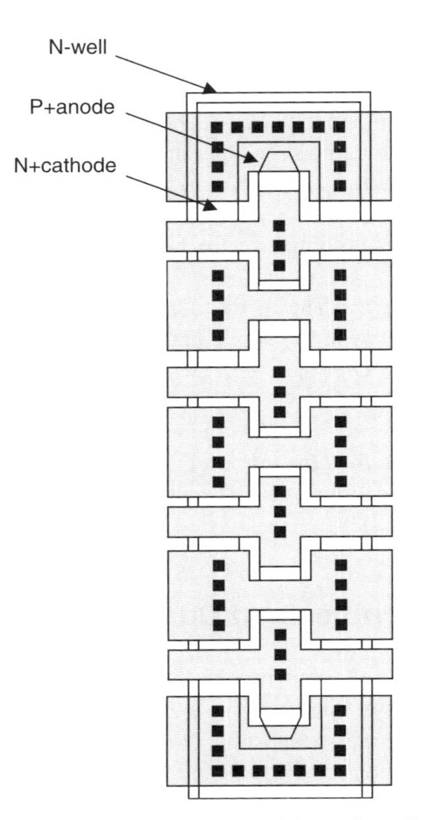

Figure 5.19 ESD input network with p + /n-well diode

Table 5.2 Diode ESD layout key parameters

Parameter	Typical dimension	Note
Anode width	Minimum	Technology dependent
Anode total perimeter	Technology and ESD protection level dependent	100–800 μm
Diode anode-to-cathode spacing	Minimum	
Diode contact-to-diffusion edge	2× to 3× minimum	
Diode contact-to-contact spacing	Minimum	
Diode anode end contact to diffusion edge	2× to 3× Minimum	

- Diode perimeter.
- Diode anode-to-cathode space.
- Diode contact-to-diffusion edge spacing.
- Diode contact-to-contact spacing.
- Diode anode end contact to diffusion edge.

Diode anode width: The anode width can be minimum to reduce the total capacitance associated with the area. The center of the anode in the physical layout is the axis of symmetry. ESD results can increase with anode width (technology-dependent).

Diode cathode width: The cathode width can be of the same scale as the anode width. The cathode should surround the anode symmetrically, where the anode center is the axis of symmetry.

Diode perimeter: ESD results improve for high perimeter, or maximized perimeter-to-area ratio.

Diode anode-to-cathode space: The anode-to-cathode spacing should be reduced to provide the lowest resistance path between the anode and the cathode.

Diode contact-to-diffusion edge spacing: The anode contact-to-diffusion edge space can be minimum.

Diode contact-to-contact spacing: The contact-to-contact space can be minimum.

Diode anode end contact to diffusion edge: At the end of the anode, to avoid three-dimensional effects and device failure, the last contact can be increased to weaken any potential failure mechanisms (Table 5.2).

5.4.2 ESD Design Synthesis of Dual-Diode Networks

In CMOS technology, typically, there are two diodes placed on a signal pin node [1,2,4,5,13–16]. One is for the positive discharge events (e.g., p + /n-well diode) and the second is for negative discharge events (e.g., n+ or n-well to substrate diode). As discussed previously, these elements can be spatially separated, or integrated into a common region. In the case where the n-well diode is placed adjacent to the p + /n-well diode, the key parameters in Figure 5.20 are important for

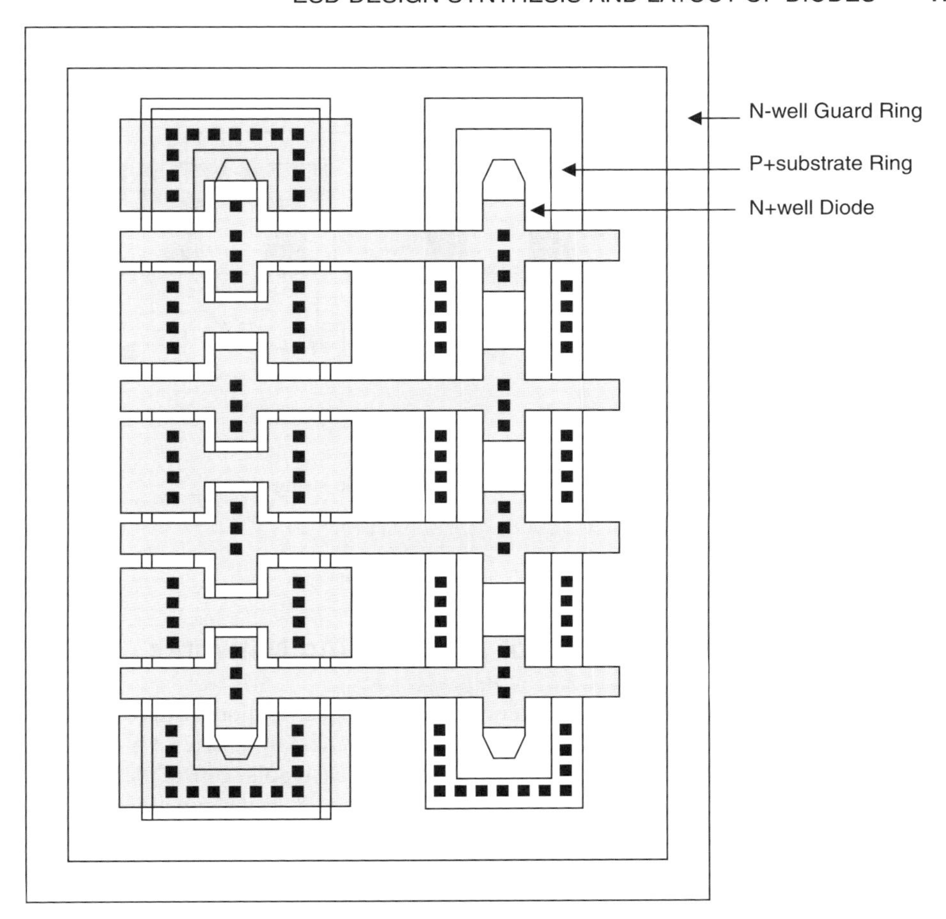

Figure 5.20 ESD input network with n-well diode

proper ESD device operation. In this ESD design practice, the key diode parameters are as follows (Table 5.3):

- N-well diode cathode width.
- N-well to adjacent n-well (p + /n-well tub).

Table 5.3 N-well diode ESD layout key parameters

Parameter	Typical dimension	Note
Cathode width	Minimum	
Cathode total perimeter	Technology and ESD protection level dependent	50–200 μm
N-well to n-well space	2× to 2.5× minimum	Lateral NPN
Diode contact-to-diffusion edge	Minimum	
Diode contact-to-contact spacing	Minimum	

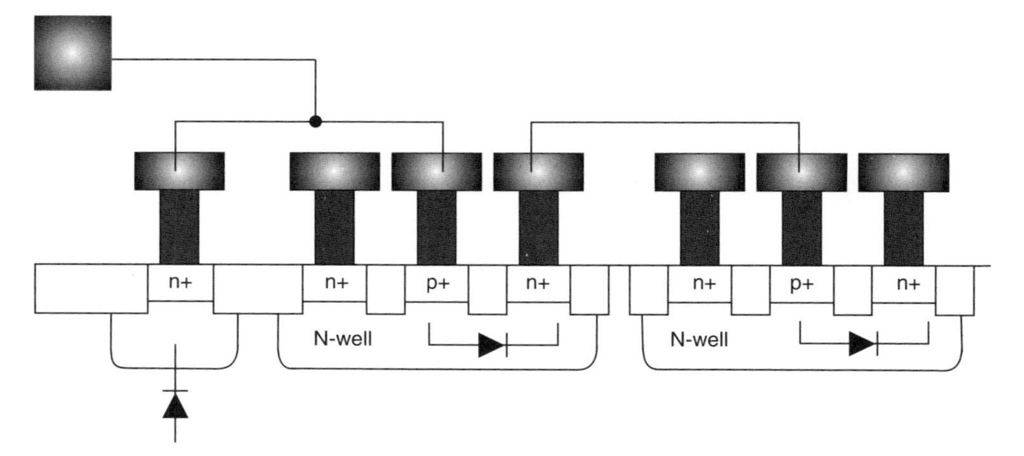

Figure 5.21 Mixed-interface ESD diode string circuit

- N-well to guard ring spacing.
- N-well diode contact-to-contact spacing.

5.4.3 ESD Design Synthesis of Diode String Networks

In CMOS technology, for mixed-voltage interface networks, series diodes are commonly used for ESD protection circuits [1,2,4,18–20,22]. The effectiveness of a series p + /n-well diode string is a function of the total series resistance, and the vertical bipolar gain of the pnp elements within the diode string. The ESD series diode resistance is a strong function of the diode semiconductor process n-well sheet resistance, layout, and diode perimeter. Figure 5.21 shows an example of a CMOS p+/n-well diode series string ESD network. In this ESD design practice, the key diode parameters are as follows (Table 5.4):

- Diode anode width.
- Diode perimeter.
- Diode perimeter-to-area ratio.

Table 5.4 Diode string ESD layout key parameters

Parameter	Typical dimension	Note
Anode width	Minimum	
Anode total perimeter (per diode)	Technology and ESD protection level dependent	100–400 μm
Diode anode-to-cathode spacing	Minimum	Non-critical
Diode contact-to-diffusion edge	2× to 3× minimum	Non-critical
Diode contact-to-contact spacing	Minimum	Non-critical
Diode anode end contact to diffusion edge	2× to 3× minimum	
Perimeter-to-area ratio	Maximum	Optimize
N-well to n-well tub spacing	Minimum	Lateral NPN

- Diode anode-to-cathode space.
- Diode contact-to-diffusion edge spacing.
- Diode contact-to-contact spacing.
- Diode anode end contact to diffusion edge.
- N-well-to-n-well spacing.

As in the prior developments, the same features are used. In this implementation, it is key to maximize the perimeter-to-area ratio, and minimize all non-critical dimensions to provide a low-area design. In this design, the space between adjacent diodes can be minimum n-well-to-n-well spacing.

5.4.4 ESD Design Synthesis of Back-to-Back Diode String

In CMOS technology, for RF applications, back-to-back series diodes are commonly used for ESD protection circuits due to the low signal swing on both RF inputs and RF outputs (Figure 5.22) [5]. For RF applications, where the ESD objectives may be lower, the size of these structures may be significantly smaller than used in CMOS digital circuitry. The ESD series diode resistance is a strong function of the diode semiconductor process n-well sheet resistance, layout, and diode perimeter. For RF applications, the more diodes in series, the

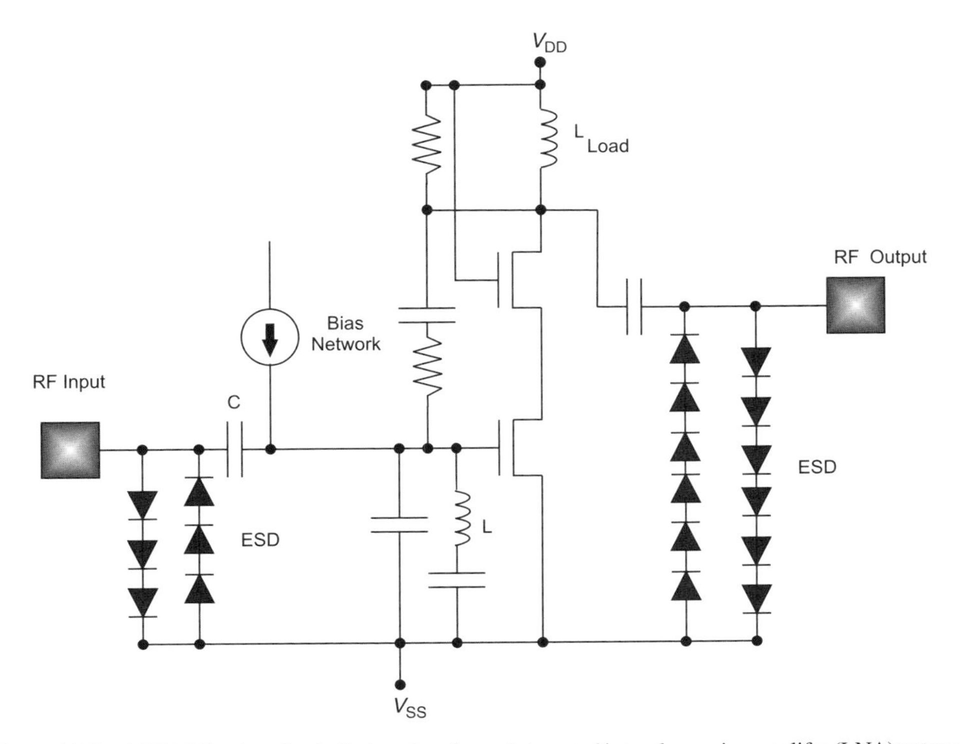

Figure 5.22 ESD RF back-to-back diode string shown integrated into a low noise amplifer (LNA) network

lower the capacitance (e.g., due to the formation of series capacitors). These networks are used between the RF(IN) to ground (V_{SS}) for the RF(IN) pins. They are also used for RF (OUT) to ground (V_{SS}).

5.4.5 ESD Design Synthesis for Differential Pair

In CMOS technology, for digital, analog, and RF applications, a common failure is the ESD failure between two differential pair receiver networks. For differential pair signals, the signal swing between the differential pair is small. There are two ESD design synthesis methods to establish ESD protection between differential pairs:

- ESD signal pin symmetric back-to-back diode strings.
- Cross-coupled ESD diode pair with parasitic element between the differential pair.

Figure 5.23 shows an example of an ESD differential pair network circuit schematic for a differential pair. In this case, dual diode networks are used between each input signal, IN(+) and IN(−), and the corresponding V_{DD} and V_{SS} power rails. An additional symmetric ESD network is placed between IN(+) and IN(−) to provide ESD protection between the differential pair.

Figure 5.24 shows an example of an ESD differential pair network circuit schematic for a differential pair where no additional ESD network is placed between each input signal, IN(+) and IN(−), but a parasitic device from the dual-diode network is utilized. In this case, the previously discussed dual-diode ESD network is wired so that the "up diode" of signal IN(+) is placed adjacent to the "down diode" of IN(−); and the "up diode" of signal IN(−) is placed adjacent to the "down diode" of IN(−). A parasitic pnpn element can be formed between IN(+) and IN(−). The parasitic pnpn SCR is formed by the p + /n-well diode in the n-well (anode), the n-well, the p-substrate, and the n-well diode (cathode). Through layout design, the differential pair SCR is formed from the conventional ESD network, with a metal wiring change. The effectiveness of this ESD network for ESD protection between differential pair IN(+) and IN(−) will depend on the technology process, physical dimensions, and design parameters. In the next section, the pnpn SCR will be discussed. Table 5.5 shows a brief summary of some of the critical dimensions for the pnpn parasitic. The critical dimensions are optimized to the desired gain and trigger conditions.

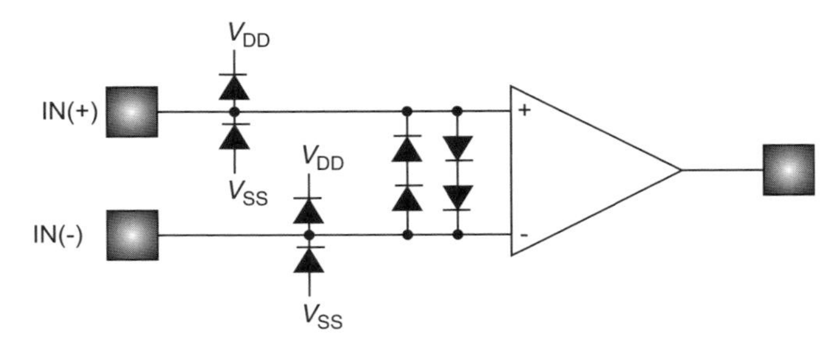

Figure 5.23 ESD input network for differential pair receivers

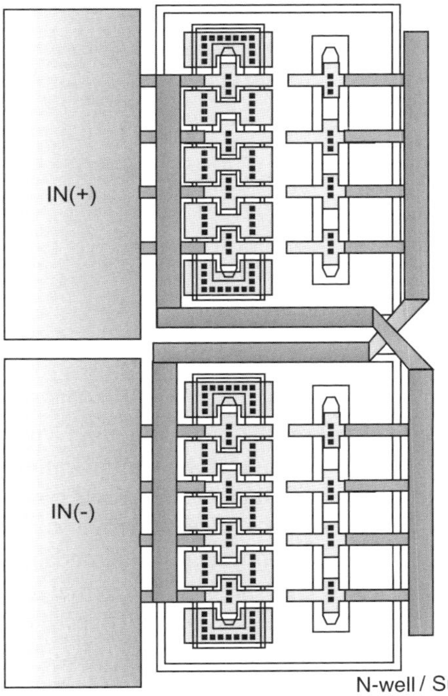

P+/n-well Diode
Parasitic PNPN
N-well / Substrate Diode
IN(+)
IN(-)
N-well / Substrate Diode
Parasitic PNPN
P+/n-well Diode

Figure 5.24 ESD input network for differential pair receivers layout

Table 5.5 Differential pair parasitic PNPN ESD layout key parameters

Parameter	Typical dimension	Note
Anode width	1–2 μm	ESD protection level dependent
Anode finger width	25–100 μm	ESD protection level dependent
Anode to n-well edge spacing	Technology dependent	Optimized
Cathode width	25–100 μm	Same as anode width
Cathode (n-well) to n-well tub spacing	2–5 μm	Optimized

5.5 ESD DESIGN SYNTHESIS OF SCRs

Silicon controlled rectifiers, also known as pnpn devices, are used for ESD signal pins and ESD power clamp circuits [3]. There are many different types of silicon controlled rectifier networks used today in semiconductors, from low-voltage CMOS to high-voltage applications. The different classes of SCRs are as follows:

- Integrated native trigger device.
- Independent trigger device.
- Voltage-triggered SCR.
- Frequency-triggered.
- Uni-directional SCR.
- Bi-directional SCR.
- High-voltage, medium-, and low-voltage SCR.

Integrated native trigger device: In many SCR ESD circuits, the trigger condition is initiated by a semiconductor element contained within the pnpn structure. The trigger network can be initiated by the native regenerative feedback of the pnpn structure. The triggering means to initiate the regenerative feedback is typically the avalanche multiplication of the n-well region within the pnpn element. In the native triggered pnpn, the triggering condition is a function of the technology and the pnpn layout choices. In some integrated trigger, such as the low-voltage trigger SCR (LVTSCR) [3], the trigger element is physically integrated into the SCR structure.

Independent trigger device: In many SCR ESD circuits, an external trigger circuit is used to initiate the regenerative feedback current for the silicon controlled rectifier or to switch the state of the SCR from the high-voltage/low-current state, to the low-voltage/high-current state. The triggering circuit can be a forward-bias diode string circuit, RC-trigger network, grounded gate MOSFET (also known as GGNMOS), or other circuits. The advantage of the independent trigger network is that the switching of the SCR can be initiated without the dependence on the pnpn layout and the semiconductor process. From a design layout perspective, the trigger element can be placed spatially separate from the pnpn circuit. Examples include:

- DTSCR (diode-triggered SCR).
- GGNMOS-triggered SCR.

Voltage-triggered SCR: Voltage-triggered SCR circuits are initiated by either forward-bias or reverse-bias breakdown of an element within the native device, or an independent trigger circuit.

Frequency-triggered: Frequency-triggered SCR circuits are initiated by RC discriminator or $C\,\mathrm{d}V/\mathrm{d}t$ transient response. The advantage of this circuit is that it is not dependent on the voltage state of the signal pin and responds to the "rising edge" or transient condition. It is also not dependent on the voltage breakdown of a region in the semiconductor process technology.

Uni-directional SCR: SCR circuits can be uni-directional and only respond to a single polarity between the anode and the cathode of the pnpn.

Bi-directional SCR: SCR circuits can be bi-directional and only respond to a two polarity between the anode and the cathode of the pnpn. This bi-directionality can be utilized for differential signal pins, or circuits with positive and negative signal swing.

High-voltage, medium-, and low-voltage SCR: SCR circuits can be applied from high-voltage power applications to low-voltage CMOS networks. In smart power technologies, the trigger voltages must exceed the signal pin voltages, and must provide isolation voltages to prevent electrical breakdown. In smart power technologies, BCD technologies, the SCR circuits utilize the high-voltage implants, isolation, and other features available to achieve these objectives.

5.5.1 ESD Design Synthesis of Uni-directional SCRs

Figure 5.25 shows an example of a pnpn structure that is uni-directional [13–16]. Table 5.6 shows a listing of key design parameters. In this design, the trigger state is typically the breakdown voltage of the n-well, and is typically utilized for higher voltage states. For high-voltage applications, additional well implants are introduced that can lead to breakdown voltage at significantly higher voltages.

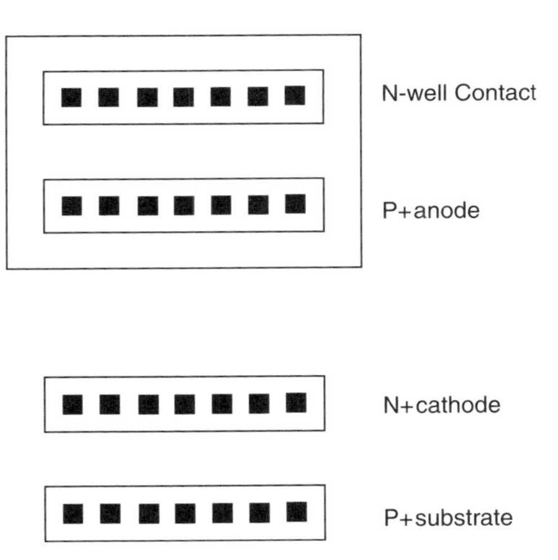

Figure 5.25 Uni-directional SCR for high-voltage signal pins

Table 5.6 PNPN ESD layout key parameters

Parameter	Typical dimension	Note
Anode width	1–2 μm	ESD protection level dependent
Anode finger height	25–100 μm	ESD protection level dependent
Anode to n-well edge spacing	Trigger voltage dependent	Optimized
N-well tub pickup to anode spacing	Trigger voltage dependent	
Cathode width (n+)	25–100 μm	Same as anode width
Cathode (n+) to n-well tub spacing	2–5 μm	Optimized
Substrate pickup to cathode space	Trigger voltage dependent	Optimized

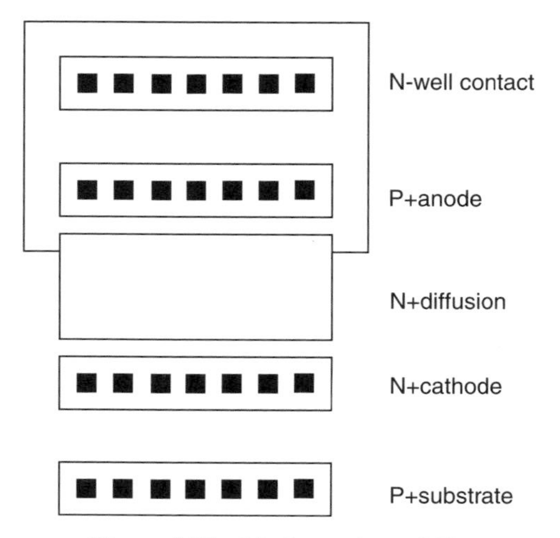

Figure 5.26 Medium-trigger SCR

Figure 5.26 shows an example of a pnpn structure that is uni-directional using a medium-voltage trigger element. In this implementation, the n+ diffusion extends outside the n-well region, to initiate triggering of the SCR at the n+ diffusion breakdown voltage. Table 5.7 shows a listing of key design parameters. In this design, the trigger state is typically the breakdown voltage of the n-well, and is typically utilized for medium-voltage states.

Figure 5.27 shows an example of a low-voltage trigger SCR (LVTSCR) pnpn structure that is uni-directional using a scalable low-voltage trigger element [3]. In this implementation, the n+ diffusion extends outside the n-well region, and forms a MOSFET between the n+ region and the cathode. This is used to initiate triggering of the SCR at the MOSFET snapback voltage. Table 5.8 shows a listing of key design parameters. In this design, the trigger state is the MOSFET snapback voltage, and is typically utilized for low-voltage signal pins.

5.5.2 ESD Design Synthesis of Bi-directional SCRs

Bi-directional SCR circuits can be bi-directional and only respond to a two polarity between the anode and the cathode of the pnpn. Bi-directional SCRs are also known as "triac" devices. This

Table 5.7 PNPN medium voltage SCR ESD layout key parameters

Parameter	Typical dimension	Note
Anode width	1–2 μm	ESD protection level dependent
Anode finger height	25–100 μm	ESD protection level dependent
Anode to n-well edge spacing	Trigger voltage dependent	Optimized
N+ trigger to n-well edge	Trigger voltage dependent	Optimized
N-well tub pickup to anode spacing	Trigger voltage dependent	
Cathode width (n+)	25–100 μm	Same as anode width
Cathode (n+) to n+ trigger	2–5 μm	Optimized
Substrate pickup to cathode space	Trigger voltage dependent	Optimized

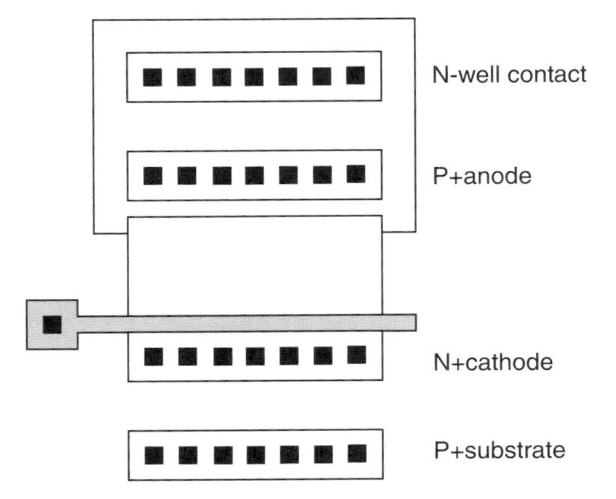

Figure 5.27 Low-voltage trigger SCR (LVTSCR)

Table 5.8 LVTSCR PNPN ESD layout key parameters

Parameter	Typical dimension	Note
Anode width	1–2 μm	ESD protection level dependent
Anode finger height	25–100 μm	ESD protection level dependent
Anode to n-well edge spacing	Trigger voltage dependent	Optimized
N+ trigger to n-well edge	Trigger voltage dependent	Optimized
N-well tub pickup to anode spacing	Trigger voltage dependent	
Cathode width (n+)	25–100 μm	Same as anode width
MOSFET channel length (*L*)	Trigger voltage dependent	Optimized
Substrate pickup to cathode space	Trigger voltage dependent	Optimized

bi-directionality can be utilized for differential signal pins, or circuits with positive and negative signal swing. Figure 5.28 shows an example of a bi-directional SCR element. This design is symmetrical around the center point to provide a symmetrical switching characteristic in the positive and negative polarity.

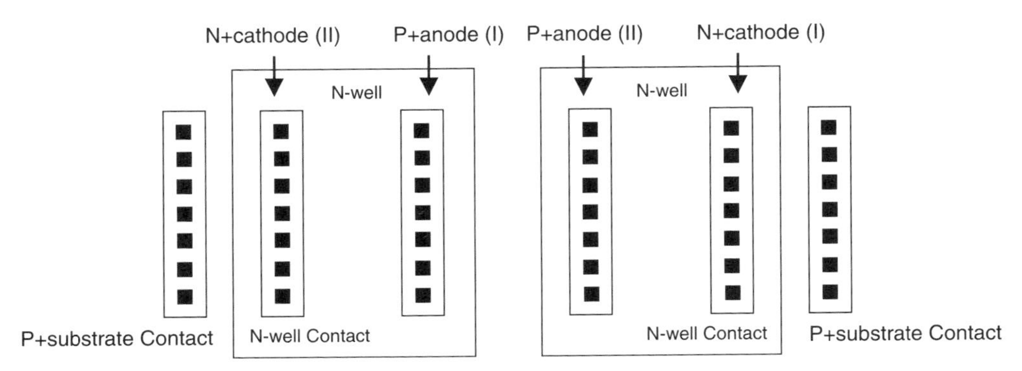

Figure 5.28 Bi-directional SCR structure

5.5.3 ESD Design Synthesis of SCRs – External Trigger Element

In silicon controlled rectifiers, external trigger networks can be used to initiate the switching of an SCR ESD network. The following is a list of SCR ESD networks with external trigger elements:

- CMOS diode-triggered SCR.
- External GGNMOS-triggered SCR.
- External RC-triggered SCR.
- High holding voltage SCR.

5.6 ESD DESIGN SYNTHESIS AND LAYOUT OF RESISTORS

Resistor element design is an important part of the ESD design synthesis of both ESD networks and functional circuits. Resistor elements are used in ESD signal pin circuits and ESD power clamps. Resistor elements are also used for input and output circuits for both functional and ESD requirements [1–4].

Resistor elements are used for input circuitry. For input receiver networks, resistors are used in series between the bond pad and the inverter circuit. In the network, a typical configuration is bond pad, a dual-diode network, a series resistor, and a secondary stage dual-diode network. Figure 5.29 shows an example of an ESD network with a resistor element.

5.6.1 Polysilicon Resistor Design Layout

In the case of a non-diffused element, the resistor can be a polysilicon film, or metal level (e.g., M1 metal layer or M0 tungsten metal layer) [1–6,13–16]. Figure 5.30 shows an example of a

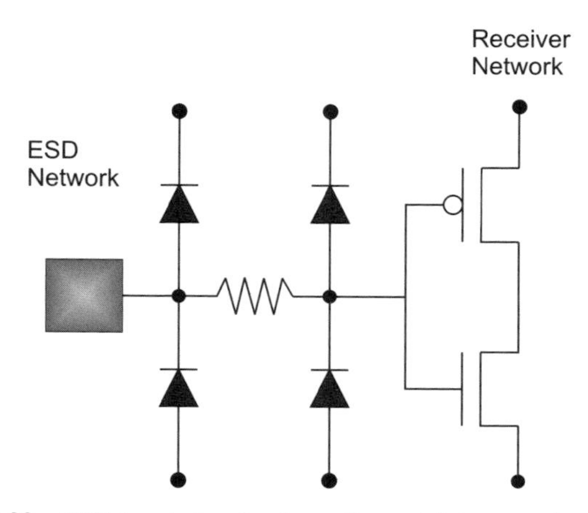

Figure 5.29 ESD input circuit schematic containing a resistor element

Figure 5.30 Polysilicon resistor element

polysilicon resistor element. The advantage of a non-diffused resistor element is that there is no parasitic diode element to the substrate or well regions. In the design synthesis, extra contacts are added to the two ends of the resistor as well as tapering in the region of the resistor region. Additional contacts on the ends lower the "end" resistance influence on the resistor region.

5.6.2 Diffusion Resistor Design Layout

Diffused resistor elements are also used in ESD input design. Diffused resistors have the advantage for ESD protection of assisting the discharge of ESD current. (A diffused resistor serves as a distributed element that contains a parasitic diode and the resistor characteristics [1–6,13–16].)

In the case of a two-stage dual diode – resistor – dual-diode network, the secondary stage is considerably smaller than the primary stage (see Figure 5.29). As a result, the resistor element can also serve as a diode for the secondary stage. A "dog bone" design resistor can be used as part of the CDM network. In this design synthesis, the octagonal ends of the dog bone resistor serve as a diode for the secondary stage. The physical placement of the resistor is local to the receiver element for CDM ESD protection.

5.6.3 P-diffusion Resistor Design Layout

P-diffusion resistors can be utilized in ESD networks, receiver networks, and OCD networks. P-diffusion resistors are used less often compared with n-diffusion resistor elements. The use of p-diffusion resistors in receiver networks is typically integrated as a single resistor element. Common use of p-diffusion resistors in networks can be as follows:

- Between the ESD network and the receiver network.
- Between the first and second stage of an ESD network.
- After a first ESD network stage, and serving as a resistor and part of the second stage of the ESD network.

The key design issues in the formation of the p-diffusion resistor structure are [1–6,13–16]:

- P-diffusion resistor end contact density.
- P-diffusion resistor contact-to-isolation space.
- P-diffusion resistor contact salicide-to-isolation space.

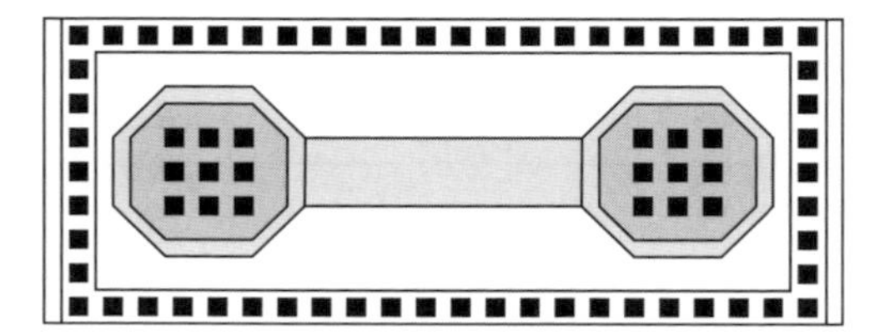

Figure 5.31 P+-diffusion dog-bone resistor layout

- P-diffusion resistor length.
- Separate n-well tub containing the p-diffusion resistor.

See Figure 5.31. In the design the resistor end contact density is needed to be adequate to prevent over-current conditions in the contact and via structure. Hence, the number of contacts must be of adequate number to prevent contact failure from ESD events. To provide good contact resistance it is necessary to provide silicide in the contact region. Hence, typically the resistor ends require silicide formation under the contacted area. The p-diffusion resistor length is defined by the desired resistor value conditions. The p-diffusion resistor length also plays a role in the onset of self-heating, avalanche, and second breakdown [1–6,13–16].

Integration of a resistor element into a multiple-stage ESD network can save loading capacitance, area, and improved ESD results. For example, in the construction of a two-stage ESD network for HBM and CDM issues, a network can be formed where there is a first-stage double-diode network, a p-type resistor, followed by the second-stage double-diode network (Figure 5.32). In this network, the first stage is significantly larger than the second stage. The first stage is to address HBM events, and the resistor and corresponding second stage are used to address CDM concerns. Using a "dog-bone" design style for the resistor element, the end of the resistor element itself closest to the receiver network can serve as a second-stage "CDM" p–n diode formed between the dog-bone end and the n-well tub (the n-well tub is electrically

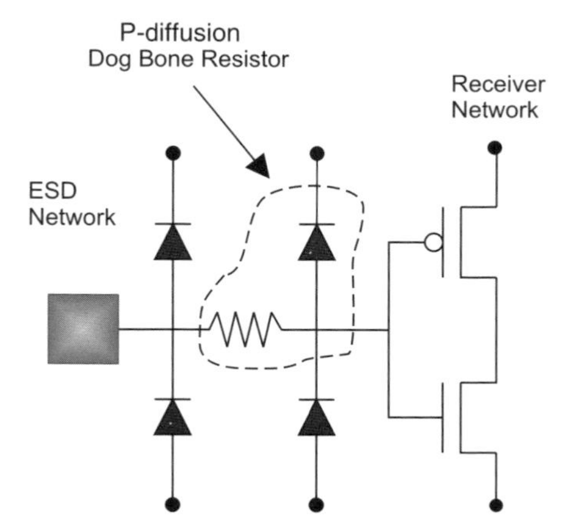

Figure 5.32 P-diffusion resistor element integrated into a two-stage HBM–CDM input node ESD network. Note the "end resistor" serves as the second-stage p–n diode to the V_{DD} power supply

connected to the highest supply voltage). In this implementation, the "resistor end" area must be adequate in size to address CDM events. As a result of integration of the resistor element and the second-stage CDM solution, additional loading capacitance and space is reduced. Note that the distinction in the ESD design practice is the area of the dog-bone end resistor, which must be adequately sized for the second stage. Also note that the resistor element itself serves as a buffering resistance for HBM events, and also participates in the CDM events associated with the semiconductor substrate. In the case of a HBM event, the resistor itself will be involved in the positive polarity ESD events. In this implementation, using a p-type resistor, a second n-diffusion will be needed to serve as part of the second stage to address the role of the second diode element.

5.6.4 N-diffusion Resistor Design

N-diffusion resistors can be utilized in ESD networks, receiver networks, and OCD networks [1–6,13–16]. The use of n-diffusion resistors in receiver networks is typically integrated as a single resistor element. Common uses of n-diffusion resistors in networks are the following:

- Between the ESD network and the receiver network.
- Between the first and second stage of an ESD network.
- After a first ESD network stage, and serving as a resistor and part of the second stage of the ESD network.

The key design issues in the formation of the n-diffusion resistor structure are:

- N-diffusion resistor end contact density.
- N-diffusion resistor contact-to-isolation space.
- N-diffusion resistor contact salicide-to-isolation space.
- N-diffusion resistor length.

Figure 5.33 shows an example of an n-diffusion resistor used for ESD protection. In the resistor "ends," the structure can be formed as a "dog-bone" design. The n-diffusion resistor length is

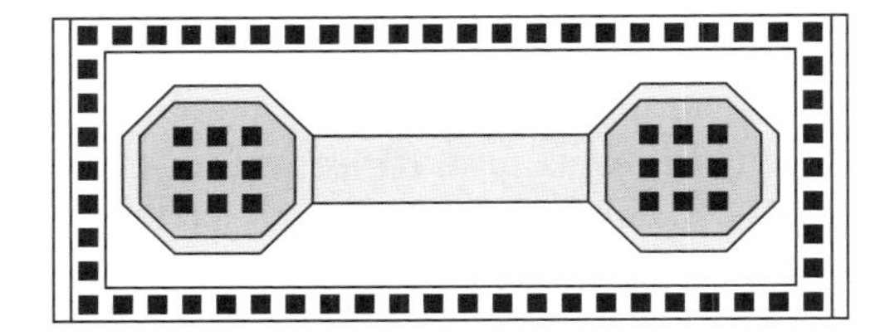

Figure 5.33 N-diffusion resistor element for ESD networks

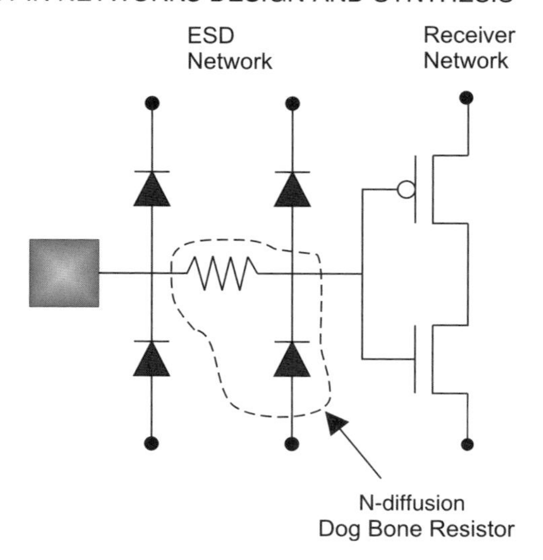

Figure 5.34 N-diffusion resistor element integrated into a two-stage HBM–CDM input node ESD network. Note the "end resistor" serves as the CDM diode to p-substrate

defined by the desired resistor value conditions. The n-diffusion resistor length also plays a role in the onset of self-heating, avalanche, and second breakdown.

As discussed in the previous section, as part of the ESD design synthesis, the end of the "dog-bone" design of the resistor can serve as a charged device model circuit "diode" element to the substrate. Figure 5.34 shows the integration of the dog-bone n-type resistor in the CDM circuitry.

5.6.5 Buried Resistors

In circuit design, a high-tolerance resistance element is desired for circuit design point accuracy and circuit matching [1,4]. Resistor passive elements are also needed for analog applications and ESD design. The buried resistor, also known as the gated diffusion resistor, has been used in CMOS design [1,4]. A buried resistor can be formed by utilizing the MOSFET gate structure as the silicide block mask. In CMOS technology, a buried resistor element is compatible with MOSFET devices. The polysilicon gate structure and isolation oxide layer, formed with the MOSFET structure, are compatible, scalable, and well-controlled features. The BR implant is typically a lower doped implant below the doping concentration of the MOSFET source and drain. Figure 5.35 shows an example of the BR element integrated with an n-channel MOSFET device.

Buried resistor elements can be used as an ESD element for negative polarity ESD events. For a BR placed in series with the signal pad, buried resistor elements can be utilized in a forward-bias diode mode of operation in bulk silicon. Buried resistors have an n+ diffusion region within the p-substrate. For negative polarity ESD events, the n+ buried resistor implant to the p-substrate will be forward biased, allowing hole and electron current flow across the metallurgical p–n diode junction.

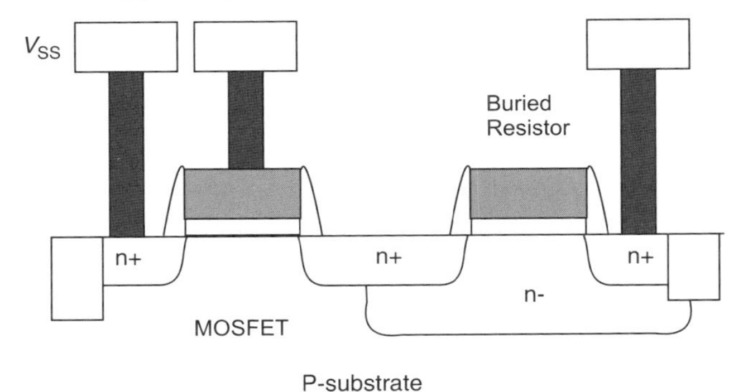

Figure 5.35 A BR structure with MOSFET integration

To utilize the BR element as an ESD diode element, there are some ESD design concepts to incorporate in order to achieve successful ESD results:

- **BR Width:** The width of the BR element on the input side must be of adequate width to provide a low current density per unit micron.
- **Metal Interconnect:** The metal bussing of the input side should be designed to introduce good voltage distribution and minimize the lateral voltage drops along the input of the resistor element.
- **BR Gate Connection:** The BR gate connection should not be electrically connected to V_{DD} or V_{SS} potential. The BR gate connection can be BR resistor input, output, or floated.
- **Substrate Contact:** For negative polarity ESD events relative to substrate (V_{SS} power rail), a p+ substrate contact local to the BR element is needed, and spaced an adequate distance to avoid current crowding.
- **Guard Ring:** For negative polarity ESD events relative to V_{DD} power rail, an n-diffusion region (e.g., n-well) or guard ring connection to V_{DD} is needed, and spaced a distance that provides the best ESD protection levels. For best results, the guard ring or n-diffusion shape should be placed at least on the input side.

5.6.6 N-well Resistors

In off-chip driver (OCD) networks, n-well resistors are used typically for ballasting elements in series with MOSFET structures [1–3,13–16]. Prior to silicide block masks, n-well resistors were implemented with the MOSFET networks. Only recently, has there been a renewed interest in the n-well resistor element. The implementation of n-well resistors can be integrated in different design styles.

- **Single Resistor:** Local and direct integration of a single n-well resistor with a single MOSFET drain in a multi-finger MOSFET device.
- **Parallel Resistors:** Local and direct integration of parallel n-well resistors with a single MOSFET drain in a multi-finger MOSFET device.

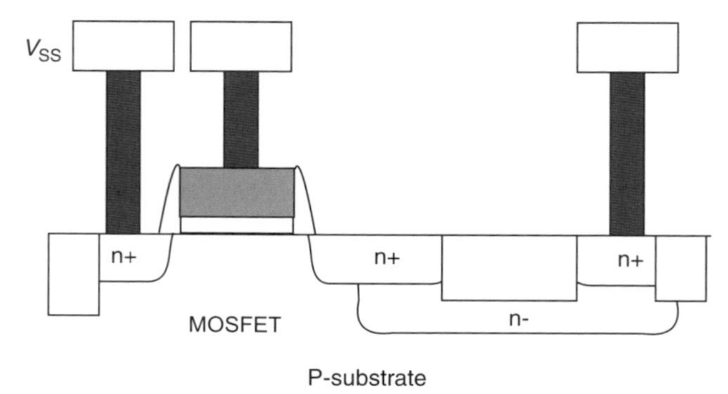

Figure 5.36 Cross-section of an n-well resistor with an n-channel MOSFET structure

- **Non-integrated Resistor Bank:** Non-local integration of a bank of n-well resistors with a plurality of MOSFET finger elements.

Figure 5.36 shows the cross section of a local integration of an n-well resistor with a n-channel MOSFET structure [13–16]. In the synthesis of the resistor element, the resistor output "end", n-well end contacts, and n-channel MOSFET drain contacts are eliminated. With the integration of the n-well resistor element and drain structure, the number of rows of contacts are eliminated allowing for compression of the design. The resistor input contacts serve as the MOSFET drain connection. In the integration, the n-well region is brought under the STI isolation, and under the n-channel MOSFET drain region. The shallow trench isolation over the n-well region serves as a means to block the salicide region between the resistor input and the MOSFET drain region. The key layout design issues in the formation of the structure is:

- *N-well resistor length.*
- *N-channel MOSFET drain length.*
- *N-well to n-channel MOSFET drain overlap.*
- *N-well to n-well space (in multi-finger MOSFET layout).*

N-well resistor length: The n-well resistor length is defined by the desired ballasting conditions. The n-well resistor length also plays a role in the onset of self-heating, avalanche, and second breakdown.

N-channel MOSFET drain length: The n-channel MOSFET drain length influences the silicided region of the MOSFET structure. The MOSFET drain length-to-width ratio influences how the current re-distributes in the silicided region.

N-well to n-channel MOSFET overlap: The n-well to n-channel MOSFET drain overlap is important in that when the n-well extends under the MOSFET region, the MOSFET current characteristics can be influenced.

To improve on the ESD robustness, it is possible to form a plurality of parallel n-well regions in parallel with the n-channel MOSFET drain. In the ESD design synthesis, a single MOSFET finger MOSFET drain is separated into parallel resistor-to-drain structures. In this fashion, given N segments, the ballast resistor in series with each MOSFET drain has increased by approximately by N. In this implementation, the spacing of the n-well-to-n-well influences the minimum lateral spacings between adjacent regions [1].

5.7 ESD DESIGN SYNTHESIS OF INDUCTORS

In the ESD design synthesis of radio frequency (RF) components, inductors are being introduced into the ESD signal pin design synthesis [4,5,26–38]. In RF ESD design, resistor elements are avoided and replaced with capacitor and inductors on the RF input and output ports. Inductors are used for matching filters, inductive isolation, and inductive degeneration in circuitry. In the ESD RF design synthesis, inductors are also used in the ESD input networks [28–36]. Inductor design layout are typically the following:

- *Rectangular layout.*
- *Octagonal layout.*
- *Circular layout.*

In RF design, the inductor structures are optimized for the highest quality factor (Q). For inductors to achieve a high Q, inductors are designed with low resistance, and far from the silicon substrate. Inductors are typically designed with thick metal layers. In RF design, ESD failure in inductors occur in the underpass wire connection [4,5]. Figure 5.37 shows an example of an inductor with the underpass connection. ESD failure occurred in the underpass connection, on the lower metal layer. Since in most technologies, the lower level layers are

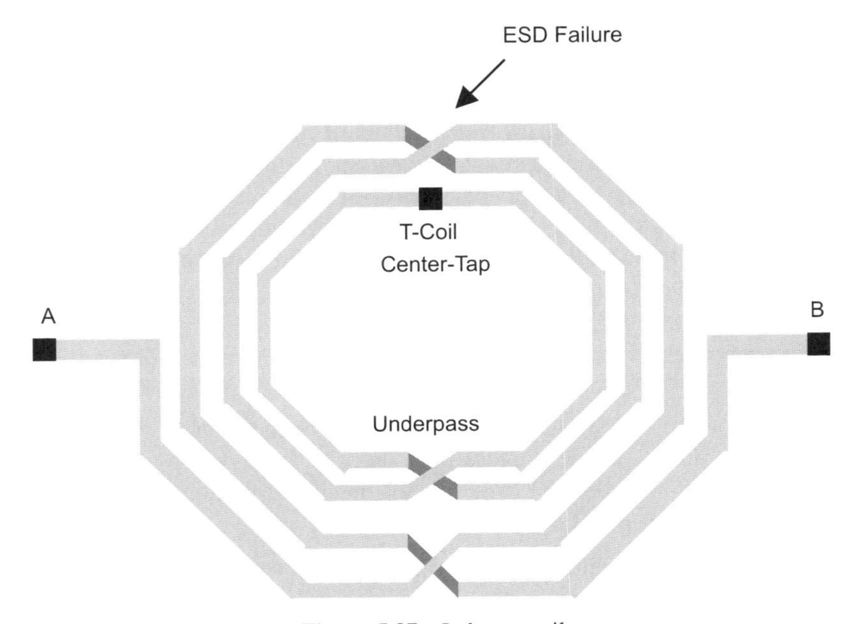

Figure 5.37 Inductor coil

thinner, the current density is higher if the underpass metal layer width is compensated to achieve the same current density.

For RF design, ESD robust inductors can be constructed for function circuits, matching and ESD elements [33–38]. For the functional circuits, the ESD failure can be eliminated with the following guidelines:

- **Thick Metal:** Design inductors with upper metal layers thick films.
- **Wide Wire Width:** Design inductors with adequate wire width.
- **Uniform Cross-sectional Area:** Maintain constant cross-sectional area through all metal layers.
- **Low Resistance:** Low-resistance inductors need to be effective as an ESD network element for shunting ESD current.

Inductors can be utilized in ESD networks if they maintain a low current density and have low resistance. ESD inductor elements used as inductive shunts to substrate must have a low series resistance, on the order of 1Ω.

One of the key issues with the integration of the inductors is the additional area. A solution for diode–inductor ESD networks is to synthesize the structure so the inductor element is placed over the silicon diode elements. Additionally, the integration of the inductor, bond pad, and silicon element can be co-integrated to reduce chip area [37,38].

5.8 SUMMARY AND CLOSING COMMENTS

In this chapter, the focus is on semiconductor active and passive elements that are used in CMOS digital, analog, and RF design. As part of the ESD design synthesis, the layout of these elements for ESD networks and in I/O circuitry is key to a successful design implementation.

In Chapter 6, the focus is on guard rings and guard ring placement. From a "top-down design" this subject could have been placed earlier in the design flow. But, in many technologies, the guard rings are either contained with the semiconductor active and passive elements or afterward, at the final stages of design integration. As part of the ESD design synthesis, the layout of these guard rings is key to a successful design implementation for both ESD and CMOS latchup.

PROBLEMS

5.1. Sketch a layout for an ESD dual-diode network. In the "up-diode," what design layout is to be made for the p+ anode to n+ cathode spacing? What decisions are to be made for the n+ contact to n-well edge? What are the metal distribution decisions? How much interconnects? What metal design layer?

5.2. In Problem 5.1, in the dual-diode network, create a "down-diode" using an n+ diffusion and n-well region. What are the spacing issues to be made between the n+ contact diffusion and the n-well edge? Is it dependent on the isolation type?

5.3. In Problem 5.1, add a guard ring that encloses both "up" and "down" diodes. Draw all the parasitic devices between the two diodes, and the guard ring structure. What spacing decisions are made between the guard ring and the diode structures?

5.4. A customer wants good voltage swing for both a positive and negative signal. Using a dual-diode network, the customer chooses to use a p + n-well for both the "up" and "down" diode in a p-substrate in a single-well process. Draw the physical design cross-section and layout. What guard rings are needed around this design? What are the advantages and disadvantages of this design from a capacitance loading, parasitics, and ESD perspective?

5.5. A design manager wants to map a design from single-well bulk CMOS to triple-well CMOS, and he wants a minimum number of ESD design changes. Given the ESD device in Problems 5.1 and 5.2, what changes are required? Note, a triple-well n-buried layer is placed in the technology. Given there is an RC-triggered MOSFET clamp network, what other changes are needed?

5.6. A design is mapped from a CMOS STI technology to a partially depleted SOI technology with a shallow silicon film. What modifications are needed in the ESD input network if it is a dual-diode network? What modifications are needed if the ESD input network is a grounded-gate n-channel MOSFET design?

5.7. A design team wants to add ballast resistors to an off-chip driver network. Show two implementations: (A) ballast resistors integrated into the MOSFET, (B) a separate array of ballast resistors in series with the MOSFET. Draw the layout of both implementations. Which one would be denser? What are the advantages and disadvantages of the different implementations?

5.8. Charged device model networks are added that include a resistor and diode element. Show the layout and schematic of a resistor element that consists of both the diode element and resistor section, for a p-doped and n-doped diffusion. Highlight both regions.

5.9. In early technology development, n-well resistors were used as ballast elements integrated with MOSFETs. Show the layout and cross-section of an n-well resistor integrated with a MOSFET. Highlight all design dimensions. What dimensions are limiting the usage of n-well resistors and why? What is the advantage of the n-well resistor compared to a silicide block mask of an n+ diffusion? What saturation current occurs in both elements?

5.10. Due to n-well to n-channel spacing rules, well scattering effects, minimum n-well width rules, and n-well sheet resistance control, n-well resistors were not used in advanced technologies as ballasting elements. But with the solving of many of these issues, they are being re-instituted. Why? What are the advantages and disadvantages of using n-wells instead of silicide block masks from a design layout, ESD, and cost perspective?

5.11. Draw a symmetric MOSFET that contains a silicide block mask on the drain structure of the MOSFET. The silicide block mask encloses the drain contact region. What choices are to be made on the dimensions of the silicide block mask? What are the spacings needed at the edges of the block mask? Why?

5.12. Inductors are used in ESD protection networks in RF circuits. If a "shunt inductor" is used between the signal pad and ground connection, what are the resistive and inductive characteristics that make it a good ESD network. What are the layout and design characteristics to make a robust inductor?

5.13. An inductor is used for a design to form an LC tank in an ESD network. The inductor forms an LC tank network which is tuned to the application frequency, so that the impedance is infinite at the resonant frequency. Show the equation that includes the LC tank equation. Include the bond pad capacitance, and all ESD capacitance elements. Assume all diode elements are capacitors.

REFERENCES

1. S. Voldman. *ESD: Circuits and Devices*. Chichester, UK: John Wiley and Sons, Ltd, 2005.
2. S. Voldman. *ESD: Physics and Devices*. Chichester, UK: John Wiley and Sons, Ltd, 2004.
3. A. Amerasekera and C. Duvvury. *ESD in Silicon Integrated Circuits*. Chichester, UK: John Wiley and Sons, Ltd, 1995.
4. S. Voldman. *ESD: Failure Mechanisms and Models*. Chichester, UK: John Wiley and Sons, Ltd, 2009.
5. S. Voldman. *ESD: RF Technology and Circuits*. Chichester, UK: John Wiley and Sons, Ltd, 2006.
6. V. Vashchenko and A. Shibkov. *ESD Design in Analog Circuits*. New York: Springer, 2010.
7. S. Voldman. *Latchup*. Chichester, UK: John Wiley and Sons, Ltd, 2007.
8. S. Voldman, V. Gross, M. Hargrove, J. Never, J. Slinkman, M. O'Boyle, T. Scott, and J. Delecki. Shallow trench isolation (STI) double-diode electrostatic discharge (ESD) circuit and interaction with DRAM circuitry. *Proceedings of the Electrical Overstress/Electrostatic Discharge (EOS/ESD) Symposium*, 1992; 277–288.
9. S. Voldman, V. Gross, M. Hargrove, J. Never, J. Slinkman, M. O'Boyle, T. Scott, and J. Delecki. Shallow trench isolation (STI) double-diode electrostatic discharge (ESD) circuit and interaction with DRAM circuitry. *Journal of Electrostatics*, **31** (2–3), 1993; 237–265.
10. S. Voldman and V. Gross. Scaling, optimization, and design considerations of electrostatic discharge protection circuits in CMOS technology. *Proceedings of the Electrical Overstress/Electrostatic Discharge (EOS/ESD) Symposium*, 1993; 251–260.
11. S. Voldman and V. Gross. Scaling, optimization, and design considerations of electrostatic discharge protection circuits in CMOS technology. *Journal of Electrostatics*, **33** (3), October 1994; 327–357.
12. S. Voldman. The impact of MOSFET technology evolution and scaling on electrostatic discharge protection. *Review Paper, Microelectronics Reliability*, **38**, 1998; 1649–1668.
13. S. Voldman, W. Anderson, R. Ashton, M. Chaine, C. Duvvury, T. Maloney, and E. Worley. Test structures for benchmarking the electrostatic discharge (ESD) robustness of CMOS technologies. *SEMATECH Technology Transfer Document, SEMATECH TT 98013452A-TR*, May 1998.
14. S. Voldman, W. Anderson, R. Ashton, M. Chaine, C. Duvvury, T. Maloney, and E. Worley. ESD technology benchmarking strategy for evaluation of the ESD robustness of CMOS semiconductor technologies. *Proceedings of the International Reliability Workshop (IRW)*, 1998; October 12–16.
15. S. Voldman, W. Anderson, R. Ashton, M. Chaine, C. Duvvury, T. Maloney, and E. Worley. A strategy for characterization and evaluation of the ESD robustness of CMOS semiconductor technologies. *Proceedings of the Electrical Overstress/Electrostatic Discharge (EOS/ESD) Symposium*, 1999; 212–224.
16. R. Ashton, S. Voldman, W. Anderson, M. Chaine, C. Duvvury, T. Maloney, and E. Worley. Characterization of ESD robustness of CMOS technology. *Tutorial Notes of the International Conference on Microelectronic Test Structure (ICMTS)*, Sweden, February 1999.

17. S. Voldman. ESD protection in a mixed voltage interface and multi-rail disconnected power grid environment in 0.5 and 0.25 μm channel length CMOS technologies, *Proceedings of the Electrical Overstress/Electrostatic Discharge (EOS/ESD) Symposium*, 1994; 125–134.
18. S. Voldman and G. Gerosa. Mixed voltage interface ESD protection circuits for advanced microprocessors in shallow trench and LOCOS isolation CMOS technology, *International Electron Device Meeting (IEDM) Technical Digest*, December 1994; 811–815.
19. S. Voldman, G. Gerosa, V. Gross, N. Dickson, S. Furkay, and J. Slinkman. Analysis of snubber clamped diode string mixed voltage interface ESD protection networks for advanced microprocessors. *Proceedings of the Electrical Overstress/Electrostatic Discharge (EOS/ESD) Symposium*, 1995; 43–62.
20. S. Voldman. The impact of technology evolution and scaling on electrostatic discharge (ESD) protection in high-pin-count high-performance microprocessors. *Proceedings of the International Solid State Circuits Conference (ISSCC)*, Session 21, WA 21.4, February 1999; 366–367.
21. S. Voldman. The state of the art of electrostatic discharge protection: physics, technology, circuits, design, simulation, and scaling. *IEEE Journal of Solid-State Circuits*, **34** (9), 1999; 1272–1282.
22. R. Countryman, G. Gerosa, and H. Mendez. Electrostatic discharge protection device, U.S. Patent No. 5,514,892, May 7, 1996.
23. S. Chittipeddi, W. Cochran, and Y. Smooha. Integrated circuit with active devices under bond pads, U.S. Patent No. 5,751,065, May 12, 1998.
24. W. Anderson. ESD protection under wire bond pads. Proceedings of the Electrical Overstress/Electrostatic Discharge (EOS/ESD) Symposium, 1999; 88–94.
25. S. Chittipeddi, W. Cochran, and Y. Smooha. Process for forming a dual damascene bond pad structure over active circuitry. U.S. Patent No. 6,417,087, July 9, 2002.
26. E. G. Gebreselasie, W. Sauter, S. St. Onge, and S. Voldman. ESD structures and circuits under bond pads for RF BiCMOS silicon germanium and RF CMOS technology. *Proceedings of the Taiwan Electrostatic Discharge Conference*, 2005; 73–78.
27. C. Richier, P. Salome, G. Mabboux, I. Zaza, A. Juge, and P. Mortini. Investigations on different ESD protection strategies devoted to 3.3 V RF applications (2 GHz) in a 0.18 μm CMOS process. *Proceedings of the Electrical Overstress/Electrostatic Discharge (EOS/ESD) Symposium*, 2000; 251–260.
28. S. Voldman, ESD Protection and RF Design, *Tutorial J, Tutorial Notes of the Electrical Overstress/Electrostatic Discharge (EOS/ESD) Symposium*, 2001.
29. C. Ito, K. Banerjee, and R.W. Dutton. Analysis and design of ESD protection circuits for high frequency/RF applications. *IEEE International Symposium on Quality and Electronic Design (ISQED)*, 2001; 117–122.
30. C. Ito, K. Banerjee, and R. Dutton. Analysis and optimization of distributed ESD protection circuits for high-speed mixed-signal and RF applications. *Proceedings of the Electrical Overstress/Electrostatic Discharge (EOS/ESD) Symposium*, 2001; 355–363.
31. P. Leroux and M. Steyart. High performance 5.25 GHz LNA with on-chip inductor to provide ESD protection. *IEEE Transactions of Electron Device Letters*, 37, (7), 2001; 467–469.
32. C.M. Lee and M.D. Ker. Investigation of RF performance of diodes for ESD protection in GHz RF circuits. *Proceedings of the Taiwan Electrostatic Discharge Conference (T-ESDC)*, 2002; 45–50.
33. M.D. Ker and C.M. Lee. ESD protection design for GHz RF CMOS LNA with novel impedance isolation technique. *Proceedings of the Electrical Overstress/Electrostatic Discharge (EOS/ESD) Symposium*, 2003; 204–213.
34. P. Leroux and M. Steyaert. RF-ESD co-design for high performance CMOS LNAs. *Proceedings of the Workshop on Advances in Analog Circuit Design*, Graz, Austria, 2003.
35. V. Vassilev, S. Thijs, P.L. Segura, P. Leroux, P. Wambacq, G. Groeseneken, M. I. Natarajan, M. Steyaert, amd H. E. Maes. Co-design methodology to provide high ESD protection levels in the advanced RF circuits. *Proceedings of the Electrical Overstress/Electrostatic Discharge (EOS/ESD) Symposium*, 2003; 195–203.

36. P. Leroux and M.Steyaert. *LNA-ESD Co-design for Fully Integrated CMOS Wireless Receivers.* Dordrecht, The Netherlands: Springer, 2005.
37. Z.X. He, R. M. Rassel, and S. Voldman. Integrated circuit structure incorporating an inductor, a conductive sheet and a protection circuit. U.S. Patent No. 7,750,408, July 6, 2010.
38. Z.X. He, R. M. Rassel, and S. Voldman. Integrated circuit structure incorporating an inductor, an associated design methodology and an associated design system. U.S. Patent Application 20100175035, July 8, 2010.
39. E. G. Gebreselasie, W.T. Motsiff, W. Sauter, and S. Voldman. Product and method for integration of deep trench mesh and structures under a bond pad. U.S. Patent No. 7,482,258, January 27, 2009.

6 Guard Ring Design and Synthesis

6.1 GUARD RING DESIGN AND INTEGRATION

Guard ring design and integration is a fundamental design synthesis practice in semiconductor design [1–64]. Guard rings are used in various places in semiconductor chip design to provide electrical isolation of circuit functions. Guard rings are a fundamental part of latchup physics, latchup characterization, and latchup analysis. Guard rings are used to prevent undesirable interaction between devices, circuits, sub-functions, and power domains. The guard ring prevents both current injection and potential perturbations that can lead to parasitic devices, noise, ESD failure, and latchup. In guard ring design, the key issues include the following:

- Guard ring placement.
- Guard ring effectiveness.

Guard ring placement addresses where the guard rings are to be placed around a given device, circuit, sub-function, or function. Guard rings are placed wherever concerns of parasitic interaction, noise, ESD, or latchup can occur.

In this chapter, the first section discusses the physics of a guard ring structure, and measurement of its effectiveness. This is followed by a "top-down" design approach instead of a "bottom-up" design method. This chapter will start with a full-chip domain isolation, I/O to I/O, within I/O, within ESD, and eventually within circuit elements. The chapter will then discuss special guard ring structures, such as high-voltage technology [35–42], deep trench [43–48], and through silicon via (TSV) structures [56–59]. The chapter will close on guard ring design methodologies and design rule checking [61–64].

ESD: Design and Synthesis, First Edition. Steven H. Voldman.

6.2 GUARD RING CHARACTERIZATION

Guard ring effectiveness is the degree to which the guard ring can isolate the specific element [1–11]. Guard ring efficiency is a metric to determine how effectively the guard ring provides the desired isolation.

Guard ring evaluation includes the metric known as the "injection ratio" [2]. Figure 6.1 is an example of a test structure to evaluate guard ring efficiency. An "injector" structure provides the source of the injection. A guard ring is placed between the "injector" and the "collector." A "collector" is a structure that collects the carriers outside the guard ring.

6.2.1 Guard Ring Efficiency

A metric for evaluation and characterization of the guard ring is the "Guard Ring Efficiency." The ratio of the captured electrons in the guard ring structure to the injected current is a measure of its guard ring efficiency. In this development, two metrics will be defined. Let us define first an injection metric F, where $F = I$ (injector)/I (collector) [2]:

$$F = \frac{I_{\text{inj}}}{I_{\text{coll}}}$$

Injection metric F is the ratio of the injected current into the structure inside the guard ring, divided by the current that is collected outside the guard ring [2,5].

If every minority carrier that is injected is also collected, the injection ratio would equal unity. As the number of collected minority carriers decreases, the factor F increases above

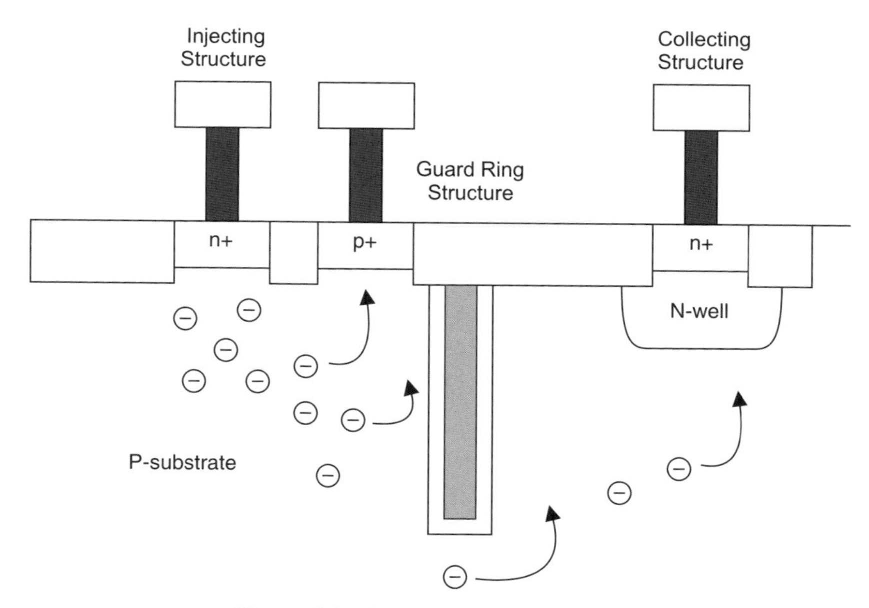

Figure 6.1 Guard ring and test structure

unity. This metric increase is a measure of the guard ring's ability to minimize the transport out of the guard ring to a region of interest where the carriers are collected. This can be quantified trivially on a test structure, or simulated to quantify the effective transport ratio. Since this term is not normalized to the injection current, the inverse relationship is better viewed from a probability perspective.

The inverse of F, $1/F$, will be defined as the "transport factor." The transport factor is the ratio of the collected current normalized to the injection current:

$$\frac{1}{F} = \frac{I_{\text{coll}}}{\text{I}_{\text{inj}}}$$

From the probability point of view, by normalizing to the injection carrier current, the collected current is the probability, given an injection current level, that the current reaches the structure of interest. The inverse term, $1/F$, is the escape probability (e.g., $P(E)$) in the approximation that the collected electrons outside the ring are significantly greater than the electrons that recombine outside the guard ring. This interpretation can change based on the structures between the injecting and collecting structure.

Hence, the "collected current" normalized to the "injected current" can be quantified using an "injecting structure" and the "collecting structure."

The escape probability, or $P(E)$, is then related to the following design parameters:

- **Injector Source Physical Dimensions:** Physical dimensions of the injection source (e.g., width, length, and depth).
- **Process:** Semiconductor process of the injection source (e.g., n+ diffusion, n-well, n-channel transistor).
- **Spacing:** Physical separation between the injection source and the collecting region.
- **Collecting Region Physical Dimensions:** Physical dimensions of the collecting region (e.g., width, length, and depth).
- **Bias Voltage:** Bias conditions on the collecting region (e.g., n-well to substrate bias).
- **Injection Current:** Current magnitude.
- **Structures:** All structures and guard ring structures between the injection source and the collection source.

In Figure 6.1, a guard ring test structure is shown as an example. In the structure there is an "injecting" structure and a "collecting" structure. In between the injecting and collecting structures are the structures that influence the transport of minority carriers.
J. Quinke [11] defined a guard ring efficiency:

$$\Psi_{\text{GRE}} = 1 - \frac{I_{\text{coll}}}{I_{\text{inj}}}$$

or

$$\Psi_{\text{GRE}} = 1 - \frac{1}{F}$$

In this case, if none of the current reaches the collecting structure, then the guard ring has captured all the current, and the guard ring efficiency is unity (e.g., 100%). If the guard ring is ineffective, and all of the current injected is collected by the outer structure, then the guard ring efficiency would be zero.

6.2.2 Guard Ring Theory – A Generalized Bipolar Transistor Perspective

The effectiveness of the guard ring structures to prevent CMOS latchup can be evaluated from different perspectives. Taking a first perspective, the problem can be viewed as a quantification of a generalized bipolar transistor [2]. First, there is an injection structure, a collecting structure, and a region of transport between; this forms a lateral bipolar transistor. From this perspective, the lateral bipolar characteristics can be quantified; the forward and reverse bipolar current gain can be evaluated as if this generalized region was actually a transistor structure [2]. The forward and reverse bipolar current gain is evaluated by switching the electrodes. Note that the n+ injector is enclosed inside the guard rings and n-well ring leading to asymmetrical results, just as in bipolar transistors having different forward and reverse characteristics. In this process, additional structures are placed in the "base" region between the emitter and collecting structure. From this perspective, the cumulative structures between the emitter and collector are evaluated as "spoilers" impacting the transport of electrons. In this fashion, a "bipolar" model can be used to quantify the relationship between the injection structure and the collecting structure [2].

6.2.3 Guard Ring Theory – A Probability of Escape Perspective

A second perspective is a more phenomenological formulation from a probability view. Minority carriers are injected into the substrate where either they are collected at a junction region, or recombine in the bulk or surface. The sum of the probability that an electron is collected plus the probability that an electron recombines equals unity:

$$P(\text{recombine}) + P(\text{collected}) = 1$$

Another perspective is that a minority carrier is either trapped, or escapes from the guard ring. The sum of the probability that an electron is trapped and the probability that an electron escapes equals unity [5]:

$$P(\text{trapping}) + P(\text{escape}) = 1$$

The probability that an electron is trapped is the probability of recombining, or collected within the guard ring structure (spatial region within the guard ring). As noted by Troutman, the probability of escape is the probability that an electron is collected, or recombines outside the guard ring structure [5]. The probability of a guard ring collecting an electron by a double guard ring structure is the current measured at the local p+ substrate ring and an n-type guard ring normalized to the injection current. The probability of an electron escaping from a guard ring, or series of guard rings, is the current measured at an additional ring outside the guard rings and the p+ substrate contact outside the guard ring normalized by the injection current.

6.2.4 Guard Ring – The Injection Ratio

The injection ratio can be defined as the ratio of the collected current on the "collector" over the injection current from the "injector." To evaluate the injection ratio, experimentally, the injector is swept as a function of voltage, or current. As the injector current is increased, the collected current can be measured on the collector structure. Figure 6.2 shows an example of the guard ring injection ratio.

In this development, two metrics will be defined. Let us define first an injection metric F, where

$$F = I(\text{injector})/I(\text{collector})$$

Injection metric F is the ratio of the injected current into the structure inside the guard ring, divided by the current that is collected outside the guard ring. If every minority carrier that is injected is collected, the injection ratio would equal unity. As the number of collected minority carriers decreases, the factor F increases above unity. This metric increase is a measure of the guard ring's ability to minimize the transport out of the guard ring to a region of interest where the carriers are collected. This can be quantified trivially on a test structure, or simulated to quantify the effective transport ratio. Since this term is not normalized to the injection current, the inverse relationship is better viewed from a probability perspective.

From the probability view, by normalizing to the injection carrier current, the collected current is the probability, given an injection current level, that the current reaches the structure of interest. The inverse term, $1/F$, is the escape probability in the approximation that the collected electrons outside the ring are significantly greater than the electrons that

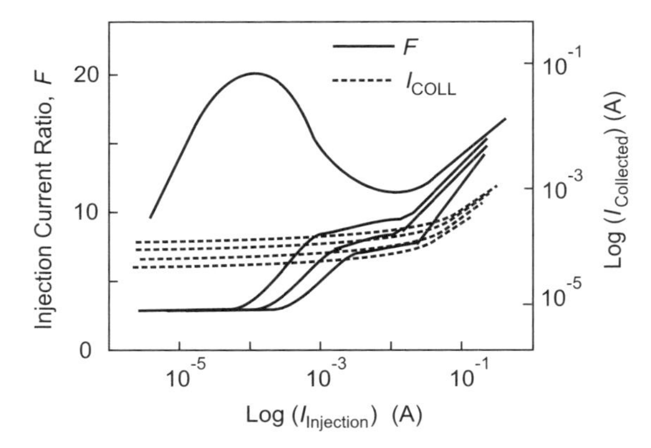

Figure 6.2 Guard ring injection ratio experimental results

recombine outside the guard ring. This interpretation can change based on the structures between the injecting and collecting structure. Let us refer to $1/F$ as the guard ring efficiency (GRE), where GRE $= 1/F$ (e.g., associated with the collected charge to the outer structure). From electrical measurements, the collected minority carriers in a metallurgical junction (e.g., an n-well region containing a p-channel transistor) form the structure of interest. Hence, the "collected current" normalized to the "injected current" can be quantified using an "injecting structure" and the "collecting structure."

The escape probability, or GRE, is then related to the following design parameters:

- Physical dimensions of the injection source (e.g., width, length, and depth).
- Semiconductor process of the injection source (e.g., n+ diffusion, n-well, n-channel transistor).
- Physical separation between the injection source and the collecting region.
- Physical dimensions of the collecting region (e.g., width, length, and depth).
- Bias conditions on the collecting region (e.g., n-well to substrate bias).
- Current magnitude.
- All structures and guard ring structures between the injection source and the collection source.

As a result, mathematically, we can establish the partial derivatives of the guard ring metric F. For example, the partial derivatives of interest are the change in the efficiency as a function of the injected current. A second partial derivative of interest is the change in the F factor as a function of the bias on the collecting structure. Both these terms are evident experimentally in the testing of these guard ring test structures.

6.3 SEMICONDUCTOR CHIP GUARD RING SEAL

In semiconductor chip design, it is a requirement to have a semiconductor chip guard ring seal on the perimeter of the semiconductor chip (Figure 6.3). In the semiconductor chip guard ring seal, metal and via structures are stacked to prevent migration of mobile ions or moisture from the outside of the chip penetrating into the semiconductor chip. The guard ring seal extends to a doped region of the same doping concentration as the substrate wafer. There are two states of the substrate region:

- Seal ring and substrate floating.
- Seal ring and the substrate non-floating and electrically connected to a power rail.

In the majority of applications, the seal ring and substrate are biased to an electric potential (e.g., zero volts). In the case that the seal ring is electrically biased, this can influence ESD protection.

In some applications, the seal ring and the substrate are left "floating" in silicon on insulator technology, or power applications.

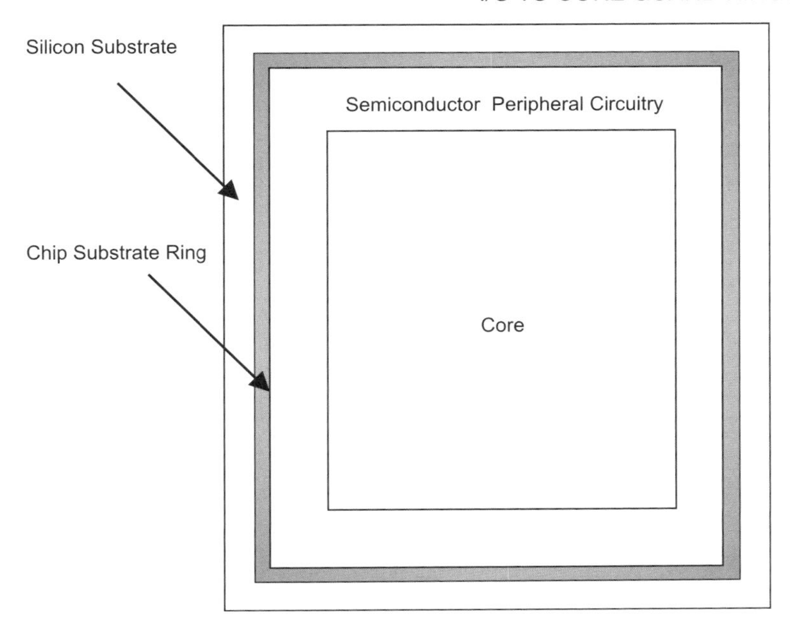

Figure 6.3 Placement of semiconductor chip guard ring

6.4 I/O TO CORE GUARD RINGS

In peripheral I/O design, the I/Os are physically separated from the core circuitry [3]. Core circuitry can contain sensitive circuitry which must be isolated from the peripheral circuits through separate electrical power domains, but also spatial separation. The core region can contain decoupling capacitors, memory, analog and digital circuitry. To spatially isolate the I/O circuitry from the core regions, additional guard rings are added to isolate the physical regions.

Guard rings are placed between peripheral I/O circuitry and core circuitry. To avoid interactions between the peripheral circuits and the core, a guard ring is placed between all the circuitry on the periphery and internal circuits.

Figure 6.4 shows an example of a chip floorplan, with guard rings between the peripheral circuitry and the interior of the semiconductor chip. In the design decision process, the following parameters are chosen for the technology:

- Guard ring type.
- Guard ring width.
- Guard ring resistance.
- Guard ring spacing: peripheral I/O to peripheral I/O guard ring structure.
- Guard ring spacing: core circuitry and the peripheral I/O guard ring structure.
- Spacing between the peripheral circuit elements and the core circuitry.

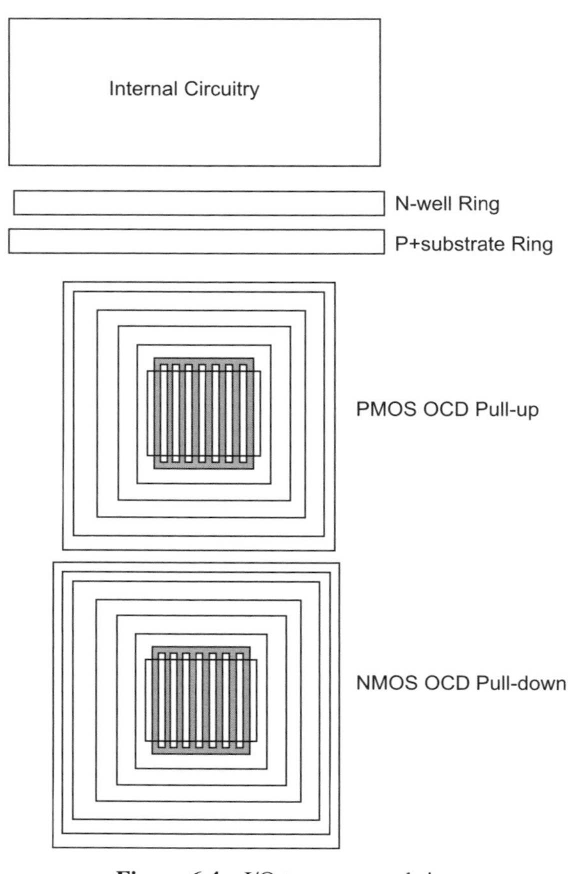

Figure 6.4 I/O to core guard ring

6.5 I/O TO I/O GUARD RINGS

I/O to I/O guard rings are important to avoid CMOS latchup and parasistic interaction in both peripheral I/O and array architectures. I/O to I/O guard rings are of three forms:

- Placement of p+ substrate contact guard ring between adjacent I/O standard cell circuits.
- Placement of n + -based guard rings between the two adjacent I/O standard cell circuits.
- Placement of trench isolation structures between adjacent I/O cells.

In the placement of a p+ substrate contact guard ring, the physical space, contact, and wire density are critical. With the lack of an n+ region, minority carrier injection is not collected by the guard ring between the two I/O cells.

In the placement of an n+ -based guard ring, the guard ring resistance is critical. The guard ring resistance is critical to the effectiveness of the guard ring. Figure 6.5 contains two adjacent I/O standard cells with I/O-to-I/O guard rings.

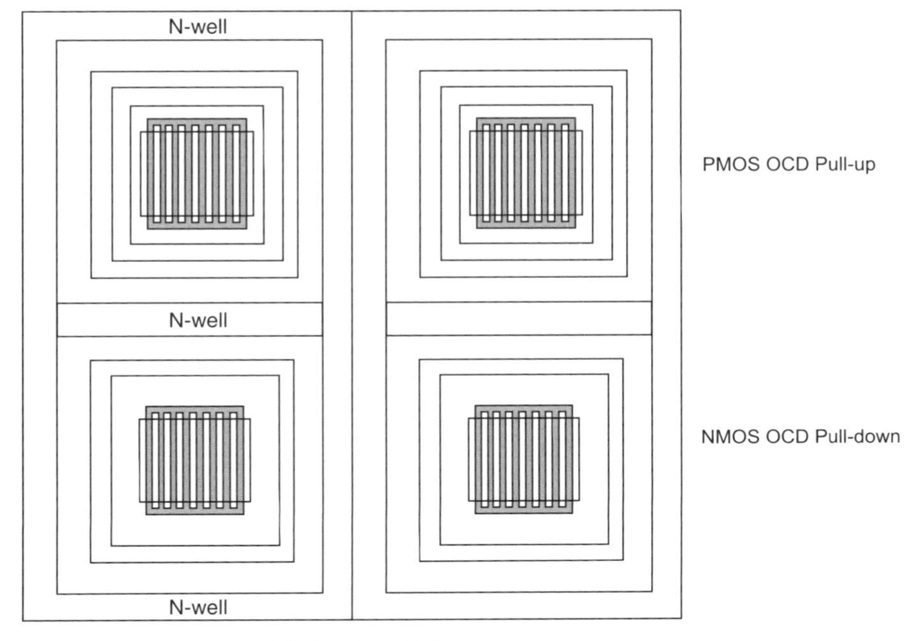

Figure 6.5 I/O to I/O guard rings

6.6 WITHIN I/O GUARD RINGS

In standard cell libraries, the off-chip driver (OCD) network contains a driver output stage, pre-drive circuitry and signal pin ESD networks. An OCD network output stage contains a large p-channel MOSFET, and a large n-channel MOSFET. In the standard cell design, these output MOSFETs are adjacent to the signal pin ESD network.

6.6.1 Within I/O Cell Guard Ring

To avoid CMOS latchup, guard rings are needed to separate the p-channel MOSFET and the n-channel MOSFET. Figure 6.6 is an example of a guard ring structure to isolate the PFET, the NFET, and the pre-driver circuitry. In this guard ring structure, the p-channel MOSFET is separated from the n-channel MOSFET to avoid CMOS latchup. A "figure 8" is formed around the OCD p-channel MOSFET pull-up and n-channel MOSFET pull-down.

6.6.2 ESD-to-I/O OCD Guard Ring

In standard cell methodologies, the ESD networks are inherently integrated with the input, output, and I/O circuitry. In standard cell libraries, the OCD network contains a driver output

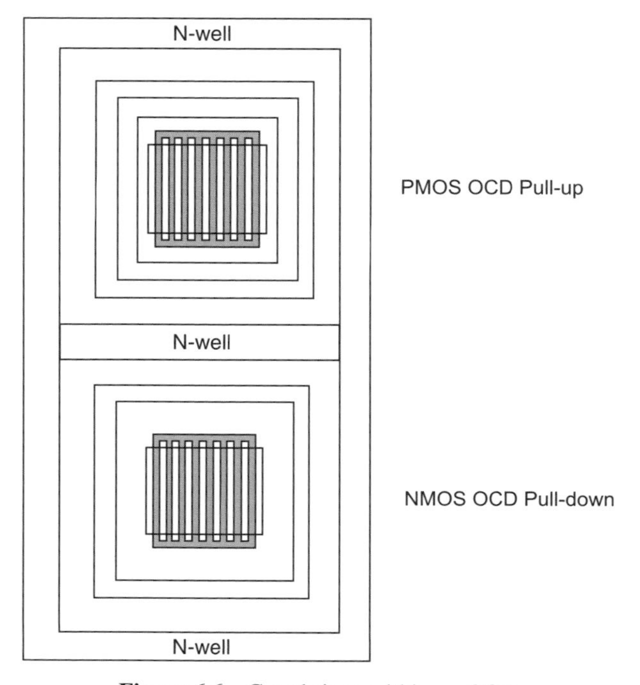

Figure 6.6 Guard rings within an OCD

stage, pre-drive circuitry, and signal pin ESD networks. In the standard cell design, these output MOSFETs are adjacent to the ESD network. To avoid ESD concerns, and CMOS latchup, guard rings are needed to separate the p-channel MOSFET, the n-channel MOSFET, and the ESD network elements. Figure 6.7 is an example of a guard ring structure to isolate the PFET, the NFET, and the ESD network elements. In each case, a guard ring encloses all the elements in the standard cell, self-enclosed for all elements. To save space, the rings are merged. In this guard ring structure, the p-channel MOSFET, the n-channel MOSFET, and the ESD network are separated to avoid CMOS latchup. In some implementations, the placements of the p-channel MOSFET and the n-channel MOSFET are separated, with the ESD network placed between them.

6.7 ESD SIGNAL PIN GUARD RINGS

In ESD design synthesis of semiconductor chips, ESD devices on signal pins are needed to avoid injection of minority carriers into adjacent circuitry. Signal pins experience voltage and current transients from both signals, noise, and ESD events. Voltage and current excursions, whether in a powered or unpowered state, can lead to minority carrier injection. Signal pin ESD networks are adjacent to the following:

- I/O circuits within the same signal pad and network.
- I/O circuits in adjacent signal pads.

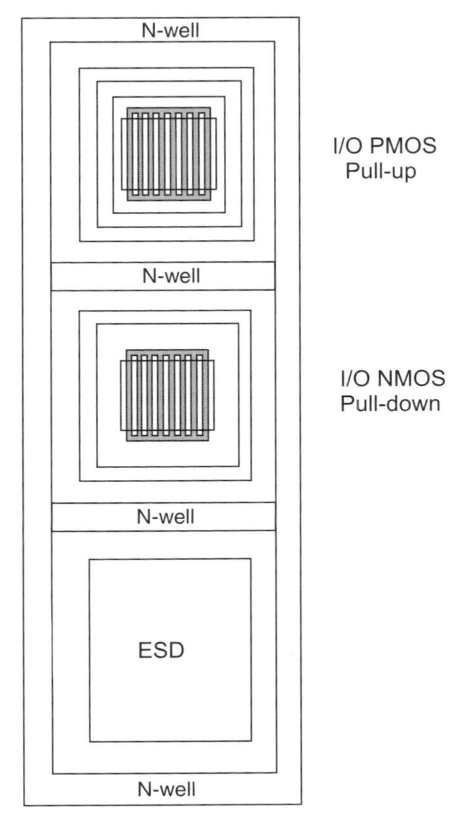

Figure 6.7 ESD to OCD guard rings

- ESD networks in adjacent signal pads.
- ESD networks in adjacent power pads.
- Decoupling capacitors.
- Core circuitry.

Figure 6.8 is an example of a guard ring structure around an ESD network. In low-voltage CMOS, guard rings can be constructed with CMOS implants. P+ guard ring structures and substrate contacts are formed with p+ diffusions, and p-well implants. The n+ guard rings are constructed with n+ diffusions and n-well regions. P-wells and n-wells can be either diffused or retrograde well profiles. For p+ diffusions, guard rings are placed around the n-well which contains the p-diffusion-based device. A separate n-well ring can surround the p+ device within the n-well to avoid interaction with adjacent structures. An n-well guard ring can be placed around an n-diffusion structure to prevent the injection of the n-diffusion minority carrier electrons to other devices. The n-diffusion is to provide a low resistance back to the power supply rail.

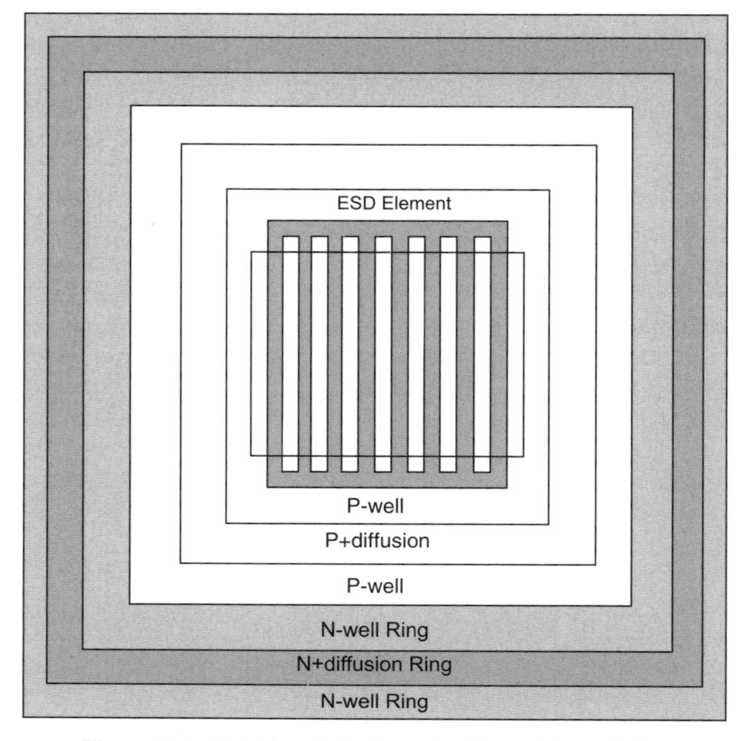

Figure 6.8 ESD signal pin element with n-well guard ring

6.7.1 ESD Signal Pin Guard Rings and Dual-Diode ESD Network

As an example, an ESD dual-diode signal pin network is shown that integrates a guard ring structure into a low-voltage CMOS technology (Figure 6.9). A continuous guard ring is formed around the entire ESD dual-diode network. An n-well diode is used for negative mode discharge to V_{SS} (substrate), and a p+/n-well diode is used for a positive mode discharge to V_{DD} (power). An ESD solution utilizes the n-well diode and the guard ring structure to provide a current path from the input node to the V_{DD} power supply for negative pulse events. The n-well diode and n-well guard ring provide a lateral npn transistor to the V_{DD} power supply. On the other side, the n-well diode and n-well tub of the p+/n-well diode form a second lateral npn transistor to the V_{DD} power supply. In the design synthesis, the well-to-well spacing is chosen to provide symmetrical discharge of both lateral npn transistors. For ESD networks, during normal operation, the guard ring serves as a means to collect minority carriers; during ESD events, it is utilized to provide ESD protection relative to the V_{DD} power supply for negative pulse ESD events. What is unique about this structure is that the guard ring is intentionally integrated and optimized as an ESD solution for negative pulse events.

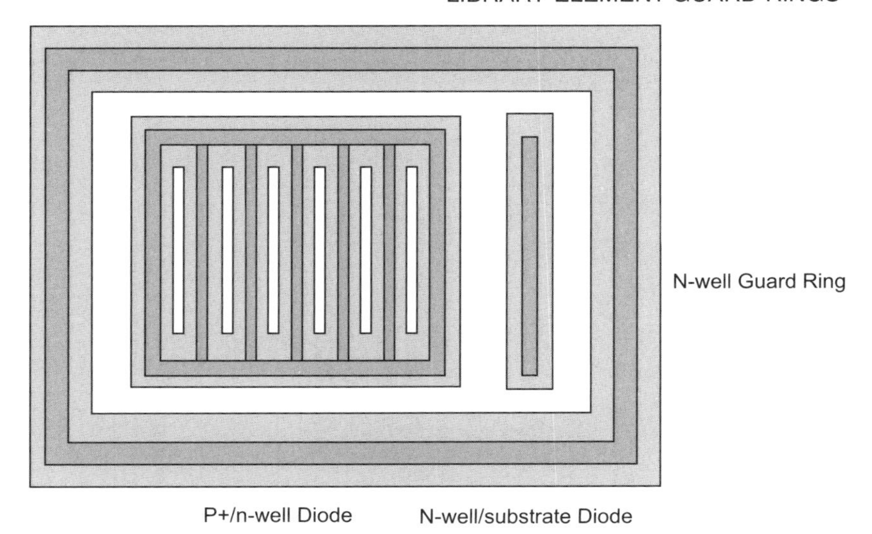

Figure 6.9 ESD dual-diode network with integrated guard ring

6.8 LIBRARY ELEMENT GUARD RINGS

Guard rings are needed for any element in the technology library that will be connected to signal pins. In the ESD design synthesis, DRC can include a rule that any physical element electrically connected to a signal pad must have a guard ring structure. As a result, guard rings may be needed for any physical element in the technology library. These can include both passive and active elements, such as MOSFETs, diodes, resistors, and diffused capacitors.

6.8.1 N-channel MOSFET Guard Rings

Figure 6.10 shows an example of an n-channel MOSFET with a guard ring structure.

Figure 6.11 shows an example of an n-channel transistor with an n-well guard ring. The n-well guard ring junction depth is deeper than the n-channel MOSFET device. Electron injection from the n-channel MOSFET can be collected by the n-well region. For negative polarity events, a lateral npn is formed between the n-diffusion and the n-well region. The n-diffusion to n-well spacing must be separated from the n-well guard ring to avoid ESD failures from negative polarity ESD events.

Figure 6.12 shows an example of an n-channel transistor with a p+ substrate guard ring. Providing a low-resistance path to the substrate, minority carrier collection can be carried out at the substrate potential. Collection of carriers prevents the lowering of the substrate potential near the MOSFET structure. With a low resistance to the p- substrate, forward biasing of the MOSFET source-to-substrate metallurgical junction can be prevented, preventing MOSFET snapback.

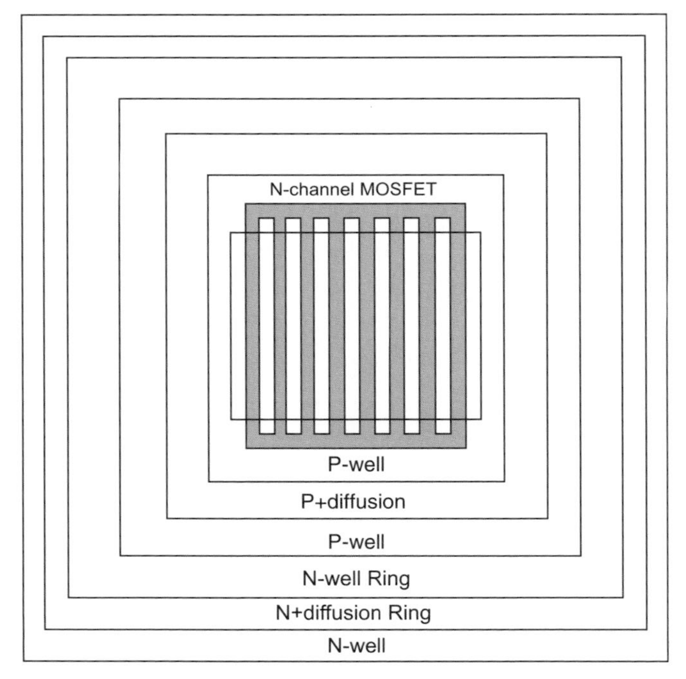

Figure 6.10 N-channel transistor with guard ring

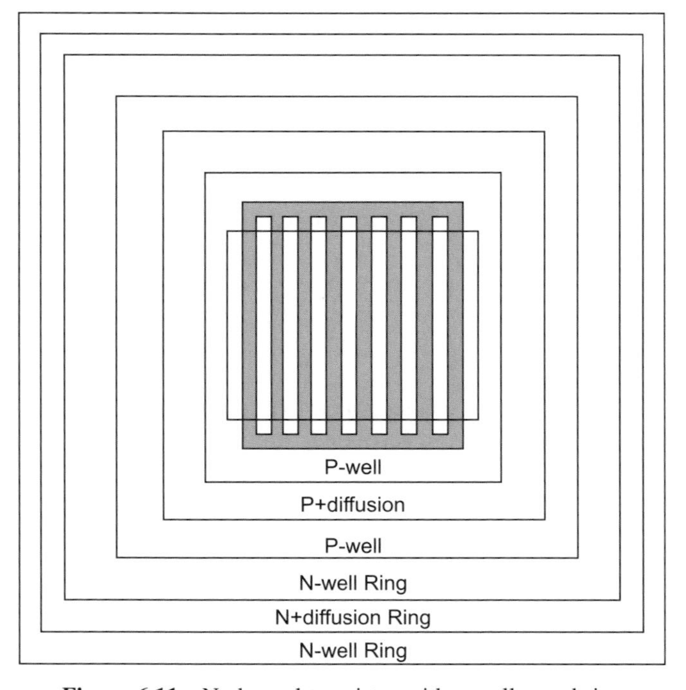

Figure 6.11 N-channel transistor with n-well guard ring

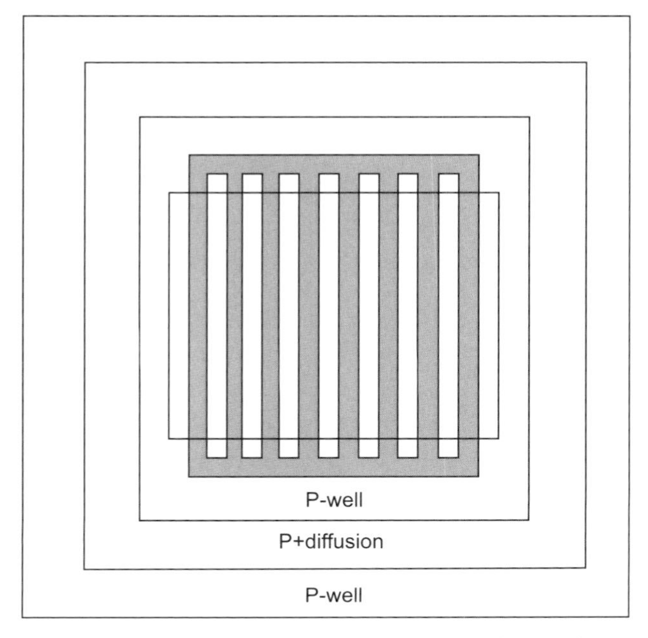

Figure 6.12 N-channel transistor with p+ substrate ring

Figure 6.13 shows an example of an n-channel transistor with a single p+ substrate stripe. For the case of a single p+ stripe, the resistance between the p+ stripe and the different MOSFET fingers can lead to non-uniform turn-on of the MOSFET device. When MOSFET snapback occurs, the current distribution is non-uniform.

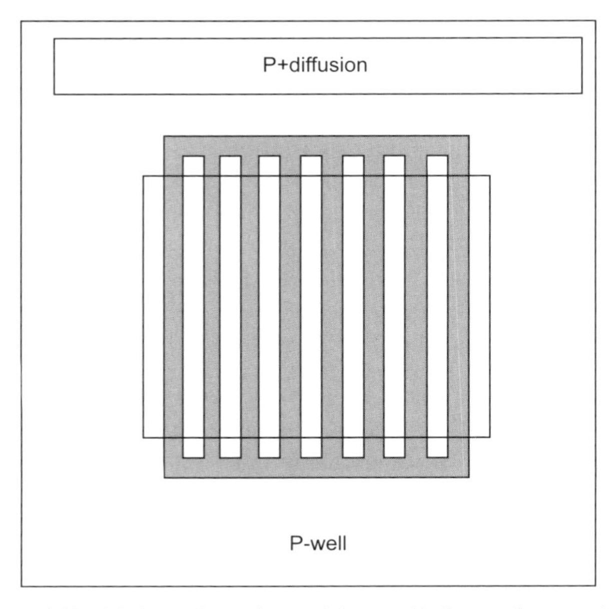

Figure 6.13 N-channel transistor with one-sided p+ substrate pickup

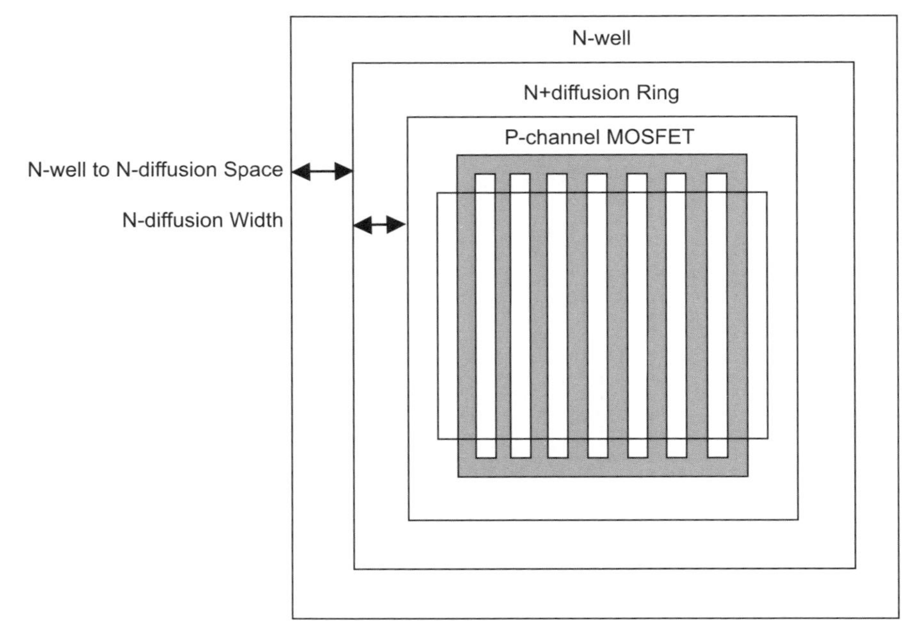

Figure 6.14 P-channel transistor with n+ ring within n-well

6.8.2 P-channel MOSFET Guard Rings

Figure 6.14 shows an example of a p-channel transistor inside an n-well. The n-well contains an n+ diffusion within the n-well region. The n+ diffusion provides a uniform low-resistance region within the n-well. This is important to avoid forward biasing of the p+ to n-well junction. In a LOCOS technology, the n+ ring increases the surface recombination along the edge of the p+ diffusion, lowering the lateral pnp current gain. In shallow trench isolation, the n+ implant serves as a low-resistance short in the n-well region. The n+ diffusion serves as a means to prevent forward biasing of the p+ diffusion.

Figure 6.15 shows an example of a p-channel transistor with a separate n-well. The independent n-well ring prevents CMOS latchup by separating the p-channel MOSFET from the n-channel MOSFET. In the well of the p-channel transistor, the n-well contains an n+ diffusion within the n-well region. The n+ diffusion provides a uniform low-resistance region within the n-well. For the independent n-well guard ring, it provides a low resistance to the V_{DD} power supply rail; this minimizes concerns over CMOS latchup and parasitic pnp turn-on from external sources outside the n-well guard ring.

Figure 6.16 shows an example of a p-channel transistor with a one-sided n+ well pickup. In this case, the distance between the n+ pickup and the p-channel MOSFET varies across the structure. The n+ diffusion provides a non-uniform resistance region within the n-well; this does not minimize concerns over CMOS latchup and parasitic pnp turn-on.

Figure 6.17 shows an example of a p-channel transistor with a separate p+ substrate ring. The independent p+ substrate ring provides low resistance to the substrate ground plane. In the well of the p-channel transistor, the n-well contains an n+ diffusion within the n-well region.

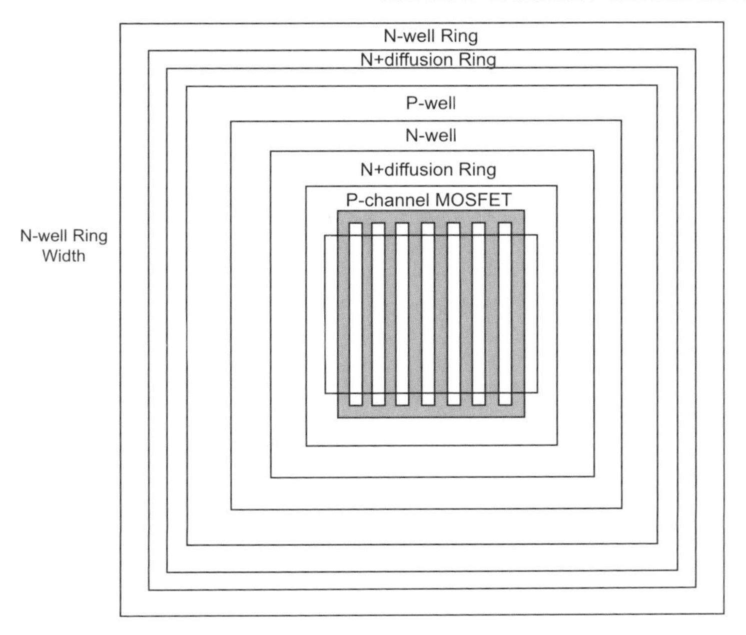

Figure 6.15 P-channel transistor with n-well guard ring

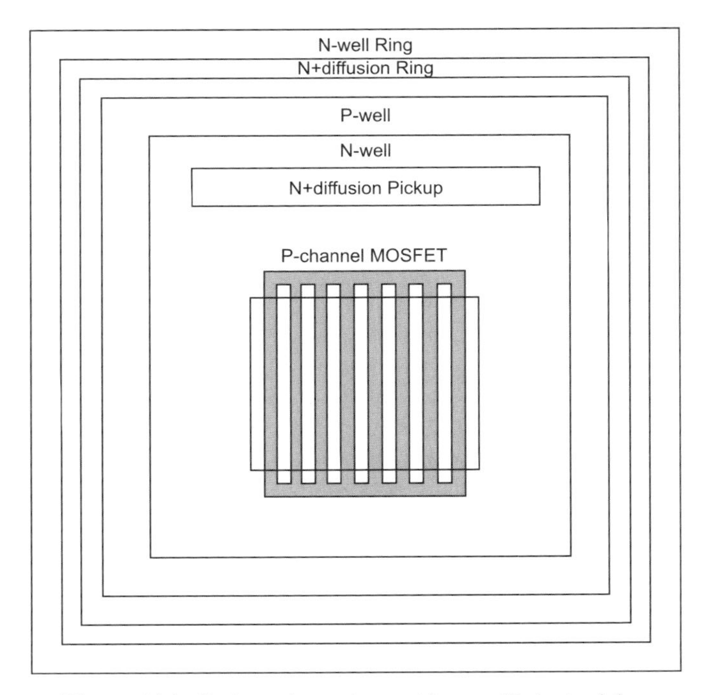

Figure 6.16 P-channel transistor with one-sided n+ pickup

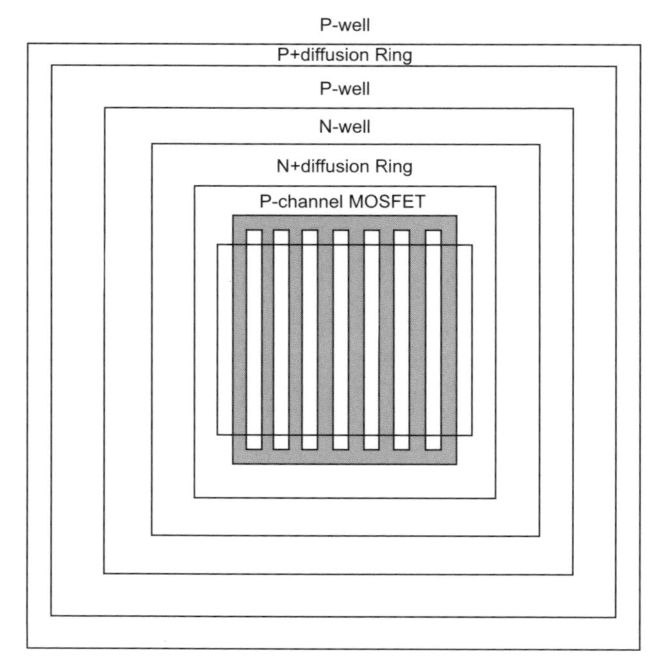

Figure 6.17 P-channel transistor with p+ substrate pickup ring

The n+ diffusion provides a uniform low-resistance region within the n-well. For the independent p+ guard ring, it provides a low resistance to the V_{SS} power supply rail; this does not minimize concerns over CMOS latchup and parasitic pnp turn-on from external sources outside the guard ring.

6.8.3 RF Guard Rings

For RF applications, the guard ring structures are integrated with the physical devices. RF guard rings are designed to minimize noise injection in devices, and to provide accurate RF models. Typically, radio frequency technologies have very accurate RF models for all the devices in the released library. All library elements are growable elements and provided as parameterized cell (p-cell) elements. The guard rings are integrated into the p-cell elements.

The dilemma that occurs for ESD development is that the guard rings being integrated with the devices prevents usage of parasitic devices between RF elements for ESD protection, or to provide a dense physical layout of ESD networks [27,28].

6.9 MIXED-SIGNAL GUARD RINGS – DIGITAL TO ANALOG

Digital noise can affect analog circuitry in a mixed-signal chip with digital and analog circuits on a common substrate [36]. The solution to this is to have separate power domains between

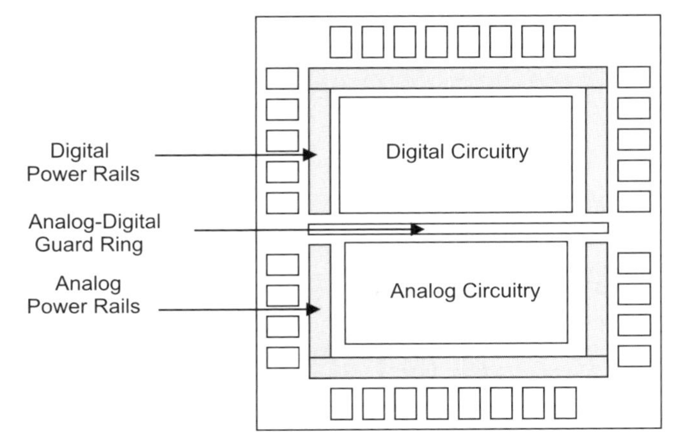

Figure 6.18 Digital-to-analog guard rings

digital and analog circuits, triple-well isolation (isolated p-well regions), and spatial separation of the digital and analog circuitry.

Guard rings can also be placed in a mixed-signal chip that separates the digital and analog regions. These structures are referred to as "guard rings" or "moats." The different structures placed between can be as follows:

- CMOS design layers: wide n + /n-well region.
- LDMOS design layers: deep n-well/HV n-well/n-well/n + .
- BiCMOS design layers: deep trench.
- Through silicon vias (TSV).

The guard ring regions can be "passive guard rings" or "active guard rings" [2,36–42]. Passive guard rings typically have a fixed bias connected to the guard ring structure. Active guard rings were first developed by mixed-signal circuit designers to minimize the effect of substrate injection on the analog circuits. Active guard rings can be electrically connected to circuitry that compensates for the injection [36–42].

Figure 6.18 shows the integration of guard rings between the digital and analog domains within a semiconductor chip. In designs that can physically separate the two domains, a wide guard ring can be placed to separate the digital noise from impacting the analog circuitry.

6.10 MIXED-VOLTAGE GUARD RINGS – HIGH VOLTAGE TO LOW VOLTAGE

High-voltage LDMOS-based circuits can inject electrons into the substrate, which can affect both digital and analog circuitry in a mixed-signal chip with digital and analog circuits on a common substrate [35–38]. The solution to this is to have separate high-voltage power domains between the digital and analog circuits, triple-well process [31],

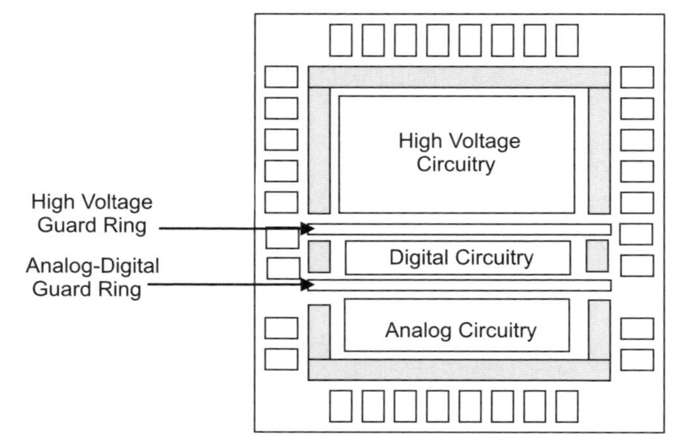

Figure 6.19 High-voltage LDMOS to low-voltage circuitry guard rings

isolated epitaxy, and spatial separation of the LDMOS transistors from the digital and analog circuitry.

Guard rings can also be placed in a mixed-signal chip that separates the LDMOS transistors from the digital and analog domains. These structures are referred to as "high-voltage guard rings" or "moats." The different structures placed between can be as follows:

- LDMOS design layers: deep n-well/HV n-well/n-well/n + .
- Deep trench (DT).

For the high-voltage LDMOS devices, the two solutions for co-synthesis are to have guard rings around the high-voltage elements, and independent guard rings between the high-voltage and low-voltage circuitry. In the design synthesis, design rules are needed to address this:

- **Space Rule:** Distance between LDMOS circuit and low-voltage CMOS.
- **Guard Ring Rule:** Specified guard ring design, and width.

Figure 6.19 shows an example of a floorplan that separates the high-voltage domain from the digital and analog domain; in this case, spatial separation of the domains is possible, and enclosed rings may not be required. A linear stripe or "moat" can be created and spatial separation can be used. In some cases, the separation may only require a good electrical ground using a p+ diffusion substrate contact; in other implementations, isolation structures, wells, or trenches can improve the guard ring efficiency.

6.10.1 Guard Rings – High Voltage

In high-voltage CMOS and LDMOS, guard rings can be constructed with both CMOS and LDMOS implants [35]. P+ guard ring structures and substrate contacts are formed with p+ diffusions, and p-well implants. From the LDMOS transistor, p-body implants and high-

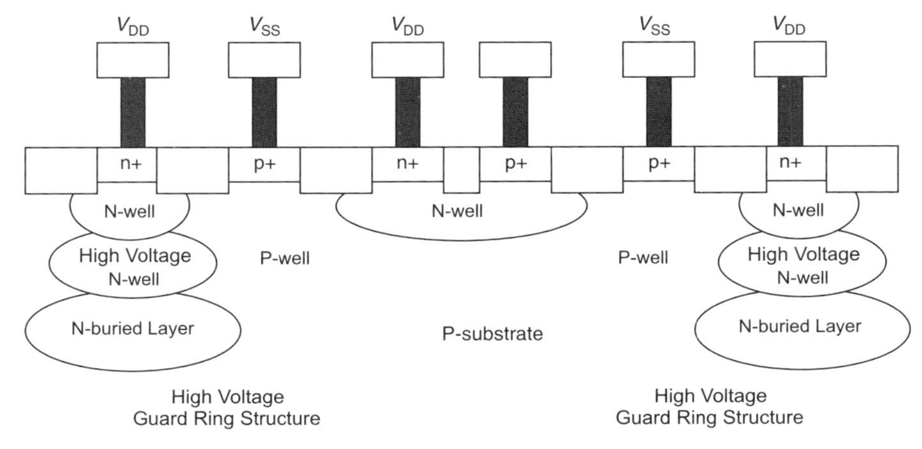

Figure 6.20 Cross-section of high-voltage guard rings

voltage p-well implants can be utilized. The n+ guard rings are constructed with CMOS n+ diffusions and n-well regions. In addition, from the high-voltage LDMOS transistors, high-voltage n-well and n+ buried layers can be used for guard ring structures. P-wells and n-wells can be either diffused or retrograde well profiles. Figure 6.20 shows an example of guard rings for high-voltage CMOS/LDMOS technology.

6.11 PASSIVE AND ACTIVE GUARD RINGS

In the definition of guard rings, guard rings can be classified as "passive" versus "active" guard rings. Passive guard rings are typically guard ring structures that collect minority carrier power supplies of opposite polarity. N-type guard rings that are connected to the V_{DD} power supply, that collect minority carrier electrons in the substrate, are "passive guard ring" structures.

6.11.1 Passive Guard Rings

In the definition of passive guard rings, guard rings can be classified as guard ring structures that block the flow of minority carriers to a collecting structure. Passive guard rings can include the following [2]:

- Trench isolation.
- Deep trench isolation.
- Through silicon vias.
- Buried layers.
- Buried layers with connecting implants.
- N-type guard rings that are connected to the positive power supplies.
- P-type guard rings that are connected to negative power supplies or ground potential.

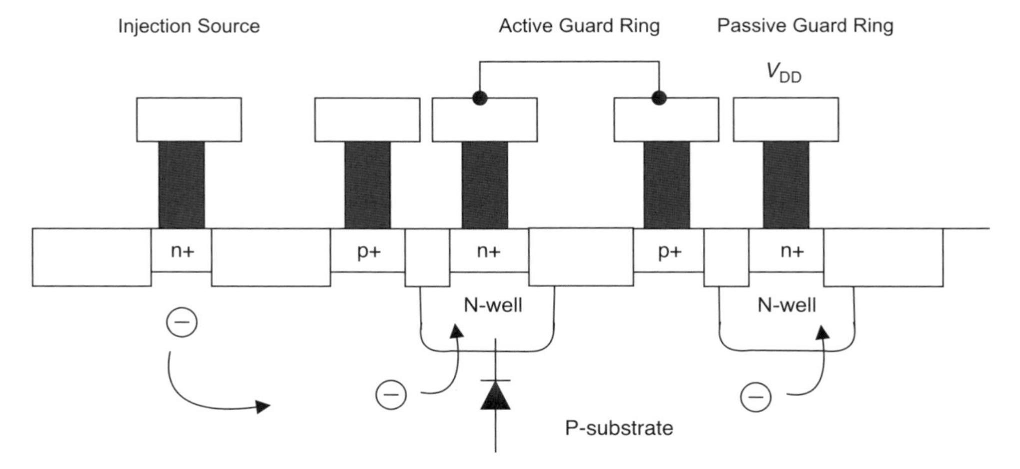

Figure 6.21 Active guard rings

6.11.2 Active Guard Rings

In contrast to passive guard rings, "active guard rings" absorb and process the current or voltage response from injection. In some applications, it is not desirable to inject the minority carriers into the V_{DD} power rail or V_{SS} ground substrate power rails. Active guard rings are guard rings which collect the minority carriers, process the "signal," and return it to the physical region [2,36–42]. Figure 6.21 is an example of an "active guard ring" structure. Minority carrier electrons are injected into the p- substrate. An "active guard ring" is formed by electrically connecting the n-well to a p+ diffusion (instead of connection to a V_{DD} power supply). The electrons are collected by the n-well region. With electrical attachment to the p+ region, the substrate pulse is sent into the p-substrate. The lowering of the substrate potential establishes a lateral electric field in opposition to the flow of minority carriers.

Active guard rings can also have active circuitry and process the signal prior to re-injection into the substrate. Figure 6.22 is an example of an active guard ring with circuitry connected.

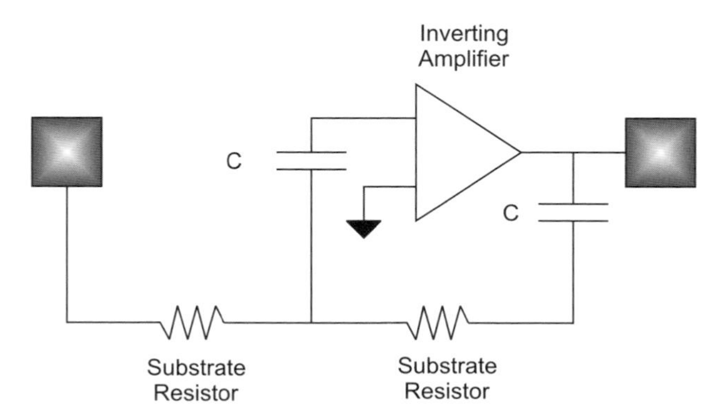

Figure 6.22 Active guard rings and circuitry

Minority carrier injection response is sent into the circuitry, where the signal is processed. The output is inverted and re-sent into the substrate wafer.

6.12 TRENCH GUARD RINGS

Trench structures that are available in BiCMOS technology or SOI technology can be used to create guard ring structures. In a BiCMOS technology, there exist deep trench isolation structures used for the bipolar transistor. The deep trench of a bipolar transistor can be utilized as a guard ring structure to improve CMOS latchup (Figure 6.23). The DT structure can be placed between the n-well and p-well regions in the CMOS section of a BiCMOS semiconductor technology. Deep trench guard ring effectiveness is superior to n-well (NW) guard ring structures [2,31–35,43–48].

Figure 6.24 shows the usage of a DT structure as an independent guard ring. The trench structure can be placed around any physical elements to prevent the flow of carriers either into or out of a structure. Deep trench structures used from a BiCMOS technology (typically used for trench-defined bipolar transistors) are very effective in preventing the flow of carriers laterally in the substrate [2,31–35,43–48].

In Figure 6.25, the deep trench guard ring injection ratio, F, is shown [43,44]. A. Watson and S. Voldman measured the guard ring efficiency of a deep trench structure as a function of deep trench depth. In Figure 6.26, the inverse injection ratio, $1/F$, is shown. As the trench is deeper, the guard ring efficiency of the structure improves.

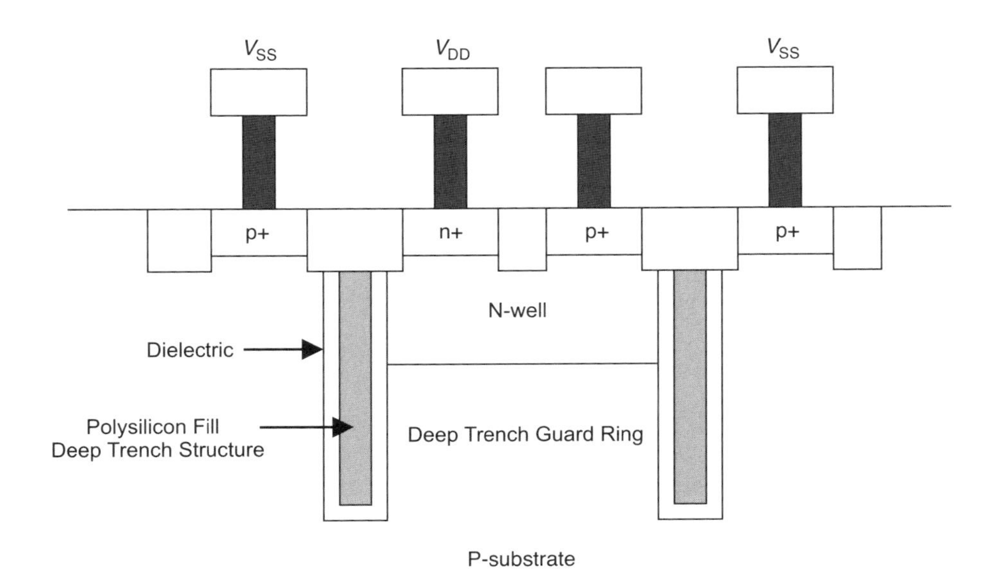

Figure 6.23 Deep trench guard ring structure

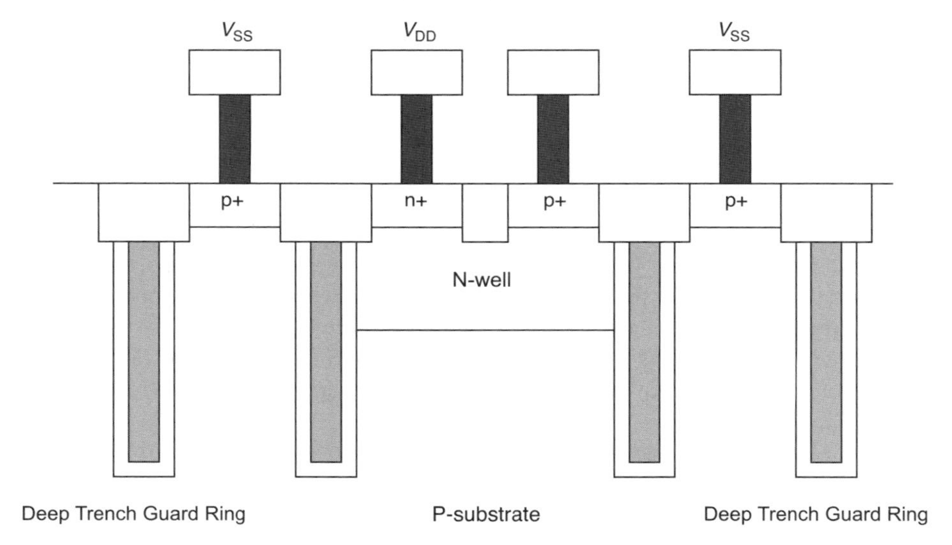

Figure 6.24 Independent deep trench guard ring structure

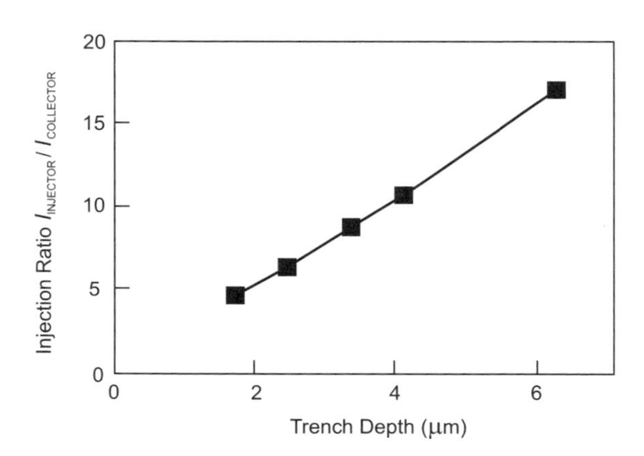

Figure 6.25 Deep trench guard ring injection ratio as a function of trench depth

6.13 TSV GUARD RINGS

In modern integrated circuits, I/O circuits are placed within a region of the integrated circuit chip containing logic circuits. By placing I/O circuits in such close proximity to logic circuits, CMOS FETs in the circuits have been found to be susceptible to latchup. Latchup causes FETs to consume large amounts of current, overheating and destroying the integrated circuit in which latchup occurs. Existing methods for reducing this

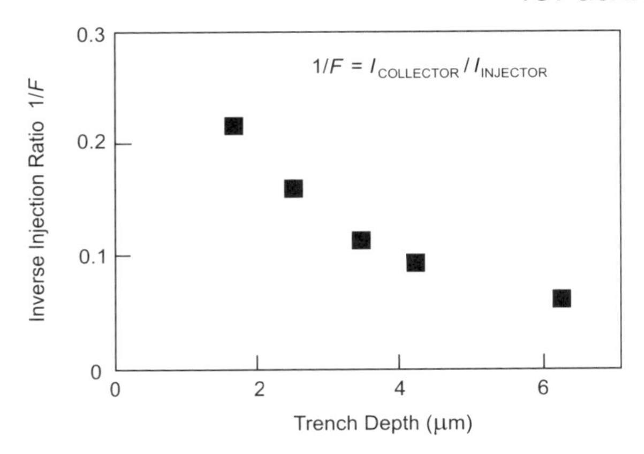

Figure 6.26 Deep trench guard ring inverse injection ratio as a function of trench depth

propensity to latchup have become increasingly less effective as doping levels of the substrates of integrated circuits have decreased. Therefore, there is a need in the industry for more robust latchup preventive structures and methods for preventing latchup for integrated circuits having I/Os embedded in the logic circuit regions of integrated circuit chips.

In multi-chip integration, and high-bandwidth applications, it is desirable to have TSV structures. TSV structures are etched through a silicon wafer, and electrically connected to interconnects, backside regions, or the substrate wafer. Some TSV structures have an insulator sidewall to allow the signal to be electrically separated from the substrate. TSV structures can also be used to serve as low-resistance substrate contacts.

TSV structures can be placed in many locations within a semiconductor chip to eliminate CMOS latchup. TSV structures can be placed in the following locations:

- Between p-channel MOSFETs and n-channel MOSFETs [57,58].
- Within ESD networks [59].
- Between I/O and ESD networks [59].
- Between I/O and core circuitry [57,58].
- Between I/O and memory arrays.
- Between power domains (e.g., digital, analog, RF, and power).

TSVs can be utilized as guard ring structures. Figure 6.27 shows an example of using a TSV as a guard ring structure. TSV structures can prevent the transport of minority carriers, and provide isolation between devices, circuits, sub-functions, or functions within a semiconductor chip [56–58]. TSV structures can be placed between p-channel and n-channel MOSFET devices to avoid forming a parasitic pnpn (preventing CMOS latchup). TSV structures can be placed in the I/O circuit, ESD networks, or core circuitry. These TSVs can be placed between digital, analog, and power domains to prevent noise

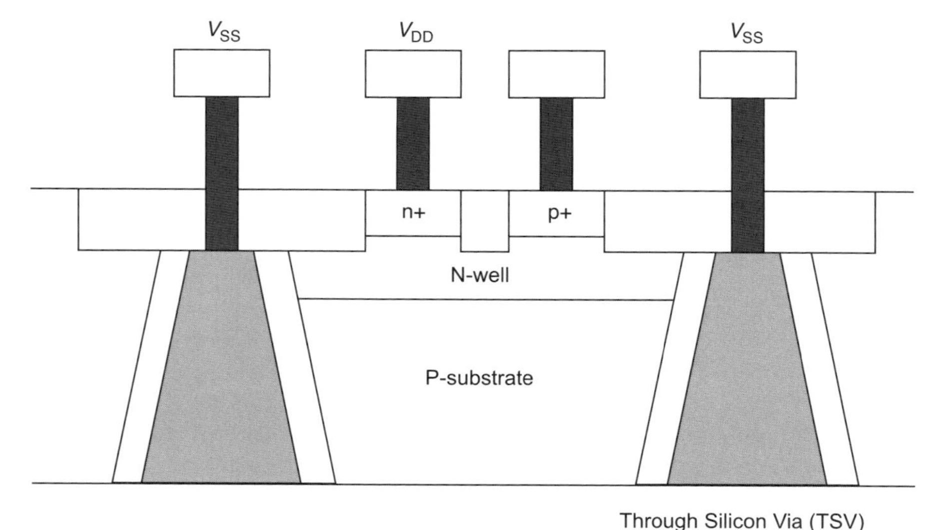

Figure 6.27 Through silicon via guard ring

or injection between functional circuit blocks. This concept was first proposed by S. Voldman, D. Collins, and P. Chapman [56–59].

6.14 GUARD RING DRC

Guard ring design rules exist to minimize ESD and latchup concerns in semiconductor chip design [2]. DRC processes are initiated to prevent non-compliance with the design objectives. Guard ring DRC rules can consist of the following:

- **Guard Ring Existence:** Existence of guard rings around a desired structure or device.
- **Guard Ring to Device Spacing:** Spacing of the guard ring relative to the edge of a device or structure.
- **Placement:** Placement of guard rings between structures or circuit types.
- Guard ring contact density.
- Guard ring resistance.
- Guard ring resistance between an injection source and a power supply connection.

In the ESD design synthesis, using DRC rules, checking and verification of the guard ring design can be achieved.

DRC and verification can consist of both physical spaces and electrical characteristics. DRC CAD methods typically evaluate only geometric spacings. There also exist checking methods that evaluate both the electrical conditions and the geometric spaces. For latchup, there are fundamental rules established in most semiconductor corporations. These latchup rules are universal and associated with the fundamental equations of latchup physics.

6.14.1 Internal Latchup and Guard Ring Design Rules

Guard ring rules and latchup rules are intertwined in the ESD and latchup design synthesis [2]. In latchup, the physical dimensions associated with the parasitic pnpn network are checked and verified. Fundamental design rule checks include the following:

Minimum p+ to n-well Space: The physical space between the p+ diffusion and the n-well edge is set to some minimum value based on the desired p + /n+ minimum rule.

Minimum n+ to n-well Space: The physical space between the n+ diffusion and the n-well edge is set to some minimum value based on the desired p + /n+ minimum rule.

Maximum n-well Resistance Requirement: The maximum n-well resistance is established based on the maximum allowed well shunt resistance. This is typically represented as a physical distance between the n-well contact and the p-channel MOSFET or any p-doped element in an n-well region.

Maximum p-substrate Resistance Requirement: The maximum substrate resistance is established based on the maximum allowed substrate shunt resistance. This is typically represented as a physical distance between the p-well contact and the n-channel MOSFET or any n-doped element in a p-well region.

Guard Ring Type Rule: Design rules require guard rings for all elements electrically connected to an external node. The type of guard ring is a function of whether an element is p-doped or n-doped, and the technology requirements.

Minimum Guard Ring Space Rule: Typically, the guard rings are spaced relative to the physical diffusion to allow electrical biasing without interaction. In addition, the spacing is optimized so as not to be too close to elements to establish interaction, but at the same time not too far to collect minority carrier injection.

Minimum Guard Ring Width Rule: Guard ring width influences the guard ring efficiency of a guard ring structure. Hence, many technology guidelines will define the guard ring width, or minimum width.

Maximum Guard Ring Resistance Rule: Guard ring design is either defined, or a maximum guard ring resistance rule is established.

Butted Contact Rules: In many technologies, butted contacts are desired to minimize the resistance between a contact and the device, recommending that butted contacts should be utilized to minimize latchup concerns.

6.14.2 External Latchup Guard Ring Design Rules

External latchup rules can be established when the location of the external source can be identified. In the case where the DRC system can locate and define an injection source, a design rule can be established based on the relative distance and the injection source magnitude [2,52–55]. The injection source can be local or global. In CMOS semiconductor chips, the injection source can include the following:

- I/O circuit.
- ESD circuit.
- N-wells connected to V_{DD}.
- Triple-well regions connected to V_{DD}.

Injection conditions are established for evaluation of the impact on adjacent circuitry. There is a physical relationship between the relative distance between the injection source and victim circuit, and the current required to initiate latchup. Hence, rules can be established as follows:

Well and Substrate Contact Spacing vs. the Injector-to-Circuit Distance: In the case that a sensitive circuit is close to an injection source, the spacing of the well and the contacts can be adjusted to avoid satisfying the differential latchup criterion in the presence of an external source [2]. As a result, the well and substrate contact spacing can be established based on distance from the external injection source.

6.15 GUARD RINGS AND COMPUTER AIDED DESIGN METHODS

In the following sections, a few examples of guard ring computer aided design (CAD) methods will be discussed.

6.15.1 Built-in Guard Rings

An ESD design synthesis methodology to ensure guard rings are contained in the designs is to release customized elements (e.g., all designs are released from a set of pre-designed elements with built-in guard rings) [2]. This can be achieved on the following levels of design:

- Guard rings within primitive semiconductor devices in a released library.
- Guard rings within released circuit books.
- Guard rings within the design methodology itself through GUIs.

The problem with this methodology is that not all physical elements require guard rings. Secondly, the use of the guard ring is a function of the placement within the chip; the choice of the guard ring structure may not be suitable for both internal and external devices. Hence, the integration of a p-cell with a defined and fixed guard ring may not be advantageous.

Guard rings can be integrated into larger circuits or design books. For example, in an ASIC methodology, guard rings can be inherently integrated into the released peripheral I/O books. In this fashion, the peripheral book must be qualified in satisfying the latchup specification.

6.15.2 Guard Ring Parameterized Cells

In a third method, the guard ring itself can be a p-cell, which is identifiable by the design environment [27,28,49–51,64]. With the establishment of a hierarchical Cadence™-based ESD design methodology, the opportunity to integrate guard rings into the design methodology provided both built-in compliance, checking, and verification. As a design methodology, D. Jordan, S. Strang, and S. Voldman constructed a graphical layout, and schematic and symbolic cell view representations of ESD networks that were hierarchical, which allowed both variable design size and circuit topology [49,50]. The elements are constructed of primitive order "1" O[1] devices, which were standard kit library items as well as ESD optimized elements.

Guard ring p-cells can be constructed which are compiled into the network to provide a higher-order circuit, where the guard ring is detectable by the design environment in the layout, schematic, and symbol "cell views." In this fashion, the checking of the guard ring is evaluated by the identification of the guard ring p-cell in the net listing. To address design integration of guard rings in a Cadence-based p-cell system, C.N. Perez and S. Voldman developed an independent guard ring parameterized cell [51,64]. The guard ring p-cell consists of a plurality of guard rings which can be integrated with the primitive O[1] device elements, or the higher-order O[*n*] hierarchical parameterized cells [27,28,51,64]. The guard ring p-cells are designed such that the guard ring structures can be turned "off" using switches in the GUI. In the guard ring p-cell, a large combination of rings can be switched "on" or "off," allowing significant design flexibility and co-synthesis of RF design and latchup optimization. An independent guard ring p-cell was defined which contains a plurality of consecutive ring structure types and number. This allows for the following:

- Independent design of a guard ring structure.
- Choice of a guard ring based on guard ring efficiency requirements.
- A growable guard ring which expands based on identification of the element type.
- A growable guard ring which expands based on the circuit or function block design input parameters.

This concept also allows for the following:

- Design of a guard ring structure with generated virtual design levels (e.g., guard ring virtual level, I/O virtual level, ESD virtual level).
- Graphical and schematic representation view of a guard ring structure.
- Design of a guard ring structure symbology and symbol.
- Design of a guard ring structure symbology and symbol hierarchy which integrates the existing circuit design symbol view with the guard ring symbol view.

- Checking of a guard ring structure by p-cell identification.
- Verification of a guard ring structure by p-cell identification.

Figure 6.28 is an example of a parameterized cell guard ring. Figure 6.29 is an example of the guard ring p-cell symbol cell view.

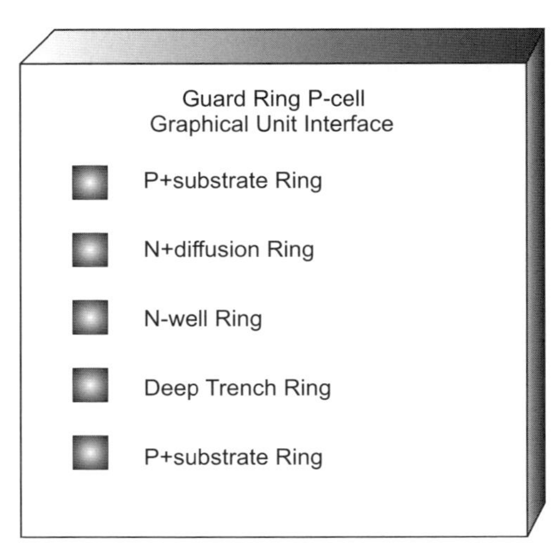

Figure 6.28 Guard ring p-cell

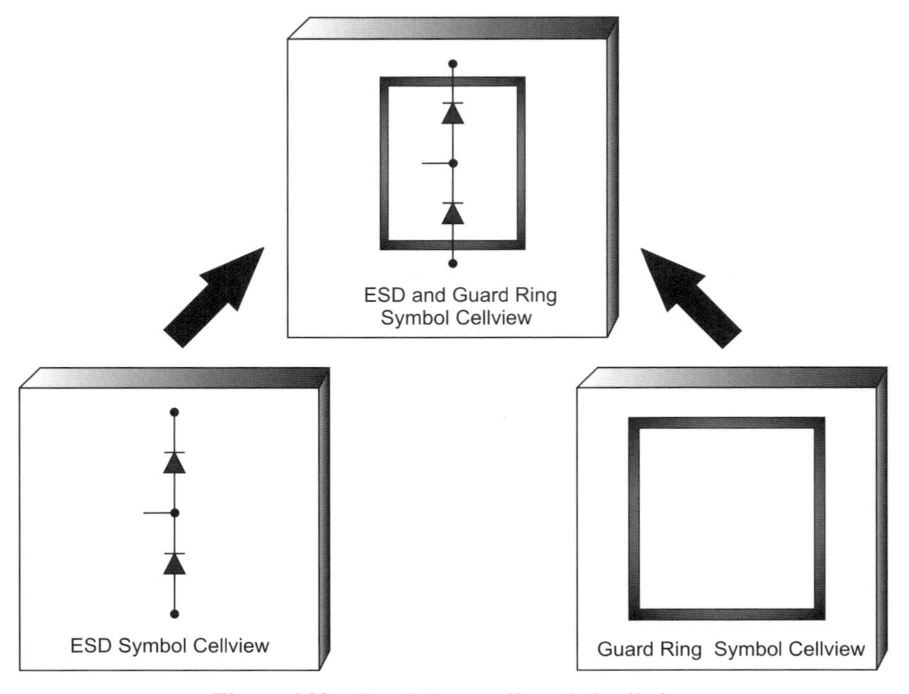

Figure 6.29 Guard ring p-cell symbol cell view

6.15.3 Guard Ring p-Cell SKILL Code

Guard ring generation can be a conversion of the guard ring structure as a parameterized cell [64]. Guard ring p-cells can be established using graphical or code-based implementation. A guard ring p-cell code can be used to generate a structure with a plurality of concentric rings where the dimensional information can be stored. In this section, the following SKILL code was generated:

```
;;Skill procedure to create ESD guardring pcell.
```

In this SKILL code, the procedure must first be established to generate the library element and cell:

```
procedure( create_guardring(LIBRARY CELL)
 prog( (libId cellId)
  unless( libId = ddGetObj( LIBRARY )
   return()
  )
  unless(cellId = dbOpenCellViewByType(libId CELL "layout" "maskLay-
out" "a")
   return()
  )
   dbSave(cellId)
   dbClose(cellId)
```

In the SKILL code for the guard ring p-cell structure, the cell defines which guard ring structures are desirable for the given design implementation. In this implementation, the series of rings consists of p+ substrate rings, n-well rings, and deep trench ring structures:

```
pcDefinePCell( list( ddGetObj( LIBRARY ) CELL "layout" )
   (
    (l          string "10u") ;; length (Y) of opening for innermost ring
    (w          string "10u") ;; width (X) of opening for innermost ring
    (dtRing1?      boolean t)    ;; toggle first DT ring (Deep Trench)
    (subRing?      boolean t)    ;; toggle substrate ring (P+ diffusion)
    (subRingW      string "1.0u") ;; substrate ring path width
    (dtRing2?      boolean t)    ;; toggle second DT ring (Deep Trench)
    (nwRing?      boolean t)    ;; toggle NW ring (N-well)
    (nwRingW      string "1.0u") ;; NW ring path width
    (dtRing3?      boolean t)    ;; toggle third DT ring (Deep Trench)
    (wire?      boolean t)    ;; toggle wiring
   )
```

```
let( ( cv mfgGrid techId CA RX BP DT M1 NW grCaWidth grCaSpace
  grCaWithinRx grCaWithinM1 grRxpWithinBp grDtOutsideCornerToRx
  length offset OL grRxNwcToBp grRxToDt grDtWidth grDtSpace width
  subRingWidth nwRingWidth rodFig1 pts n subRect subShapes
  grNwToDt)

cv = pcCellView
techId = techGetTechFile(cv~>lib)
mfgGrid = techGetMfgGridResolution(techId)

;;Layer variables
```

To construct the guard ring p-cell structure, the level names of the guard rings must be defined in the listing. Additionally, spatial ground rules must also be defined:

```
CA = '("CA" "drawing")
RX = '("RX" "drawing")
BP = '("BP" "drawing")
NW = '("NW" "drawing")
DT = '("DT" "drawing")
M1 = '("M1" "drawing")
OL = '("OUTLINE" "drawing")

;;Groundrules

grCaWidth          = 0.20
grCaSpace          = 0.25
grCaWithinRx         = 0.15
grCaWithinM1         = 0.05
grRxpWithinBp         = 0.15
grRxNwcToBp         = 0.15
grRxToDt           = 0.25
grNwToDt           = 1.0
grDtWidth           = 1.0
grDtSpace           = 1.5
grDtOutsideCornerToRx = 2.00

;;Evaluate strings and set dimensions to microns

width = cdfParseFloatString(w)*1e6
length = cdfParseFloatString(l)*1e6
subRingWidth = cdfParseFloatString(subRingW)*1e6
nwRingWidth = cdfParseFloatString(nwRingW)*1e6

;;Create outline shape. First ring is drawn with inner
;;edge coincident with outline shape.
```

```
rodFig1 = rodCreateRect(
        ?name  "outline"
        ?layer OL
        ?width width
        ?length length
        ?origin 0:0)

;;Create first DT ring.

when( dtRing1?

 offset = 0.0

 pts = list(rodAddToY(rodFig1~>lL -offset)
        rodFig1~>uL
        rodFig1~>uR
        rodFig1~>lR
        rodAddToX(rodFig1~>lL -grDtWidth-offset))

 rodFig1 = rodCreatePath(
        ?name      "dtRing1"
        ?layer     DT
        ?width     grDtWidth
        ?pts      pts
        ?justification "right"
        ?offset    offset)

) ; when dtRing1?

;;Create substrate ring.

when( subRing?

 if(dtRing1? then offset = grDtOutsideCornerToRx else offset =
grRxpWithinBp)

 pts = list(rodAddToY(rodFig1~>lL -offset)
        rodFig1~>uL
        rodFig1~>uR
        rodFig1~>lR
        rodAddToX(rodFig1~>lL -subRingWidth-offset))

 rodFig1 = rodCreatePath(
        ?name     "subRing"
        ?layer    RX
        ?width    subRingWidth
        ?pts     pts
        ?offset   offset
        ?justification "right"
```

```
        ?encSubPath  list(list(
                     ?layer    BP
                     ?enclosure  -grRxpWithinBp
                     ?beginOffset -grRxpWithinBp
                     ?endOffset  grRxpWithinBp )))
) ; when subRing?

;;Create second DT ring.

when( dtRing2?

 offset = cond(
        (subRing? grRxToDt)
        (dtRing1? grDtSpace)
        (t 0.0))

 pts = list(rodAddToY(rodFig1~>lL -offset)
       rodFig1~>uL
       rodFig1~>uR
       rodFig1~>lR
       rodAddToX(rodFig1~>lL -offset-grDtWidth))

 rodFig1 = rodCreatePath(
        ?name       "dtRing2"
        ?layer      DT
        ?width      grDtWidth
        ?pts       pts
        ?justification "right"
        ?offset     offset)

) ; when dtRing2?
```

The n-well ring is also constructed using the rod commands:

```
Create nwell ring.

when( nwRing?

 offset = cond(
       (dtRing2? grDtOutsideCornerToRx)
       (subRing? grRxNwcToBp+grRxpWithinBp)
       (dtRing1? grDtOutsideCornerToRx)
       (t 0.0))

 pts = list(rodAddToY(rodFig1~>lL -offset)
       rodFig1~>uL
       rodFig1~>uR
       rodFig1~>lR
```

```
        rodAddToX(rodFig1~>lL -offset-nwRingWidth))

    rodFig1 = rodCreatePath(
          ?name       "nwRing"
          ?layer      RX
          ?width      nwRingWidth
          ?pts       pts
          ?offset     offset
          ?justification "right"
          ?encSubPath  list(list(?layer NW ?enclosure 0.0)))

   ) ; when nwRing?

   ;;Create third DT ring.

   when( dtRing3?

    offset = cond(
          (nwRing? grNwToDt)
          (dtRing2? grDtSpace)
          (subRing? grRxToDt)
          (dtRing1? grDtSpace)
          (t 0.0))

    pts = list(rodAddToY(rodFig1~>lL -offset)
         rodFig1~>uL
         rodFig1~>uR
         rodFig1~>lR
         rodAddToX(rodFig1~>lL -offset-grDtWidth))

    rodFig1 = rodCreatePath(
          ?name       "dtRing3"
          ?layer      DT
          ?width      grDtWidth
          ?pts       pts
          ?justification "right"
          ?offset     offset)

   ) ; when dtRing3?

   ;;Add first level metal wiring to substrate ring.

   when( wire?
    when( subRing?
   n = max( fix( (subRingWidth - 2*grCaWithinRx + grCaSpace) / (grCaWidth +
grCaSpace)) 1 )
     subRect = nil
     for( i 1 n
      subRect = cons(list(
```

```
                ?layer        CA
                ?width        grCaWidth
                ?length       grCaWidth
                ?space        grCaSpace
                ?sep        minus( (i-1) * grCaSpace + i*grCaWidth
                            + grCaWithinRx )
                ?justification  "right"
                ?endOffset     -grCaSpace
                ?beginOffset    -grCaSpace) subRect)
     ) ; for i

     rodCreatePath(
      ?name    "subRingM1"
      ?layer   M1
      ?width   subRingWidth
      ?pts    rodGetObj( "subRing" )~>dbId~>points
      ?subRect subRect)



    ) ; when subRing?

    ;;Add first level metal wiring to nwell ring.

    when( nwRing?
   n = max( fix( (nwRingWidth - 2*grCaWithinRx + grCaSpace) / (grCaWidth +
grCaSpace)) 1 )
     subRect = nil
     for( i 1 n
      subRect = cons(list(
                ?layer        CA
                ?width        grCaWidth
                ?length       grCaWidth
                ?space        grCaSpace
                ?sep        minus( (i-1) * grCaSpace + i*grCaWidth
                            + grCaWithinRx )
                ?justification  "right"
                ?endOffset     -grCaSpace
                ?beginOffset    -grCaSpace) subRect)
     ) ; for i

     rodCreatePath(
      ?name    "nwRingM1"
      ?layer   M1
      ?width   nwRingWidth
      ?pts    rodGetObj( "nwRing" )~>dbId~>points
      ?subRect subRect)
    ) ; when nwRing?
   ) ; when wire?
```

```
      ;;convert all paths to polygons
    foreach(path setof(shape rodGetNamedShapes() shape~>dbId~>objType
== "path")
      if( subShapes = path~>subShapes then
        dbConvertPathToPolygon( path~>dbId )
      foreach( subShape setof( shape subShapes shape~>objType == "path")
          dbConvertPathToPolygon( subShape)
        )
      else
        dbConvertPathToPolygon( path~>dbId )
      ) ; if subShapes
    ) ; foreach path

    ) ; let
  ); pcDefinePCell
 ) ; prog
) ; procedure create_guardring
```

6.15.4 Guard Ring Resistance CAD Design Checking

In the integration of guard rings into a semiconductor chip design, CAD methods are being used for latchup. Guard rings are placed around injection sources that can trigger latchup. Injection sources can be n+ diffusions, n-type resistors, and n-wells (e.g., in a p-type substrate wafer). One of the primary design issues with guard rings is the effectiveness of collecting the minority carrier injection. The guard ring effectiveness is dependent on the guard ring type, guard ring depth, guard ring physical width, and relative spacing from the injection source. Guard ring resistance is also a key criterion [60]. The reason this is a growing issue is that the CMOS and BiCMOS dimensional scaling increases in the peripheral I/O density (e.g., I/O book width scaling). With the technology dimensional scaling, the physical dimensions of the p+, n+, and n-well have been reduced. With the scaling of the minimum n-well widths, the width of the n-well guard ring has been scaled. As a result, the resistance along the length has increased. With the increase in circuit density, the I/O circuit density has increased. ASIC environments have focused on reducing the width of the I/O peripheral book to allow more I/O circuits on the periphery of a semiconductor chip. In this process, the peripheral I/O length has been increased to compensate the reduction of the peripheral I/O book width.

With the placement of an n-type guard ring in the p-type substrate, a metallurgical junction is formed which can collect the minority carrier electrons injected into the substrate. As an example, an n-type guard ring is biased to the power supply voltage, collecting the injection current.

At low injection currents, the electrons are collected by the reverse-biased metallurgical junction formed between the substrate and the n-type guard ring. But, at very high injection currents, the series resistance between the power supply voltage and the guard ring is a key latchup design factor.

The injection source serves as an "emitter," the substrate serves as a "base" region, and the guard ring serves as a "collector." When the emitter–base junction is forward active, the electrons are injected into the substrate region. When the collector is biased positive at the power supply voltage, the collector-to-emitter voltage is positive. In this state, the parasitic transistor formed between the injection source and the guard ring is forward active. When the resistance of the guard ring increases, a voltage drop occurs in the guard ring. The voltage drop is equal to the product of the guard ring resistance and the injection current:

$$\Delta V = I_{\text{inj}} R_{\text{GR}}$$

where I_{inj} is the injection current and R_{GR} is the guard ring resistance between the point of injection and the power supply voltage. At the location of the injection, the voltage at the guard ring is equal to

$$V_{\text{GR}} = V_{\text{DD}} - \Delta V_{\text{GR}} = V_{\text{DD}} - I_{\text{inj}} R_{\text{GR}}$$

As the voltage drop increases due to the injection current, the guard ring voltage at the point of injection will decrease. When the effectiveness of the guard ring at collecting the current is minimized as a result of de-biasing, the minority carrier electron current will flow to alternative structures (e.g., outside the guard ring). Design parameters that influence the resistance are the following:

- Guard ring sheet resistance (e.g., n-well sheet resistance, or plurality of implants in the guard ring).
- Guard ring width.
- Guard ring contact density.
- Guard ring contact resistance.
- Guard ring silicide resistance.
- Metal bus resistance.
- Distance between the injection location and the power supply voltage source.

Historically, guard ring resistance was not a critical issue due to the technology, the ground rule dimensions, and the I/O density. From the 1980s to 2000, the guard rings used were typically n-well regions. In this time frame, the ground rules for both diffused and retrograde wells prevented narrow-width n-well regions. As a result, the ground rules prevented scaling of the guard ring widths below some minimum dimension (e.g., typically wells could not be scaled below 3–7 μm). With the utilization of n-diffusion and silicides (e.g., titanium silicide and cobalt silicide), and large contact dimensions, the resistance was very low. Additionally, due to wide "wiring tracks" and peripheral I/O design, the power bus width was wide (e.g., 10–30 μm). In addition, the I/O density was low.

In this millennium, the vertical semiconductor process profile was scaled, leading to higher well sheet resistance. In addition, vertical scaling allowed for a decreased minimum well width requirement, allowing a narrower guard ring structure in I/O design. In each technology generation, the number of I/Os increases, leading to high aspect ratio I/O books that are long and narrow. In this case, the guard ring width is reduced, as well as increased length between injection sources and the power supply voltage. In addition, with the metal scaling, the wire widths are reduced to allow a higher density of wire tracks. With all the scaling issues for both the semiconductor process and the semiconductor chip layout design, the resistance issue is more critical.

A CMOS latchup CAD system can be developed that addresses the guard ring resistance [60]. The CMOS latchup CAD evaluation must address a maximum resistance requirement for the guard ring resistance. The guard ring resistance can be evaluated as follows:

- Identify injection source.
- Identify the location near the guard ring structure.
- Calculate the total resistance to the power supply V_{DD}.
- Evaluate the maximum resistance allowed for the guard ring for the given conditions.

6.15.5 Post-Processing Methodology of Guard Ring Modification

CAD methods to integrate guard ring structures (and contact density) can also be implemented in an integrated semiconductor chip design as a post-processing methodology [61]. Given a semiconductor foundry environment, the peripheral "I/O books" can be pre-defined; this may include the OCD, the receiver, the ESD circuit, and the guard ring structures. A given pre-defined I/O book may be mapped into different chip architectures, and different technologies. Latchup sensitivity is influenced by the semiconductor technology and the integration into the power bus architecture; each design methodology may have different integration placement and practices. As a result, the placement of the guard rings relative to the power bus, and the adequacy of the guard rings to minimize latchup requirements, may need to be modified as a post-processing of the peripheral I/O design or upon full-chip integration.

M.D. Ker, H.C. Jiang, J.J. Peng, and T.L. Shieh developed a CAD methodology that addresses this capability [61]. A design solution is a procedure that can be established, which allows for the placement of additional guard rings within a given design. To address satisfying the latchup guard rings for given latchup foundry requirements, additional guard rings can be added to the pre-existing design. To address integration with the power rails, additional shapes can be added under the power rails to electrically connect to the power rails (e.g., V_{DD} and/or V_{SS}).

In this latchup CAD design practice, the design method allows for the auto-generation of diffusion shapes serving as a guard ring under the power bus for V_{DD} and V_{SS}. In this method, the designer has the ability to create n+ or p+ arrays under the power rail. For the case of a V_{DD} power bus, n+ diffusion shapes and electrical contacts, and vias, are generated to serve as an n+ guard ring attached. For the case of a V_{SS} power bus, p+ shapes, contacts, and vias are generated under the V_{SS} power bus. To integrate the overlay of the guard ring, the shapes, and the power bus, design layers are used to coordinate the integration. The auto-generate guard ring procedure is as follows [61]:

- Load the program.
- Select to run V_{DD} or V_{SS} guard ring.
- Select the guard ring type.

- Input spacings and define variables.
- Run simulation.

6.16 SUMMARY AND CLOSING COMMENTS

In this chapter, a "top-down" design synthesis approach for guard rings was shown for a semiconductor chip, starting with the seal ring, to domains, standard cell-to-standard cell, within standard cell, and down to the individual devices. A "bottom-up" approach starts with the individual devices and works its way up to the full-chip implementation. Special structures and cases were shown as examples of how to further isolate both domains and devices. A small taster was also given of what is possible with the guard ring design synthesis, and integration with both devices to full-chip implementations.

In Chapter 7, the focus will be on examples of design synthesis in full-chip implementations. Examples of DRAM, SRAM, microprocessors, mixed-voltage, mixed-signal, and RF applications will be shown. As part of the ESD design synthesis, the layout is key to a successful design implementation for both ESD and CMOS latchup. These examples will provide some understanding of the challenges in the ESD full-chip integration issues.

PROBLEMS

6.1. Draw an n+ well contact in an n-well. Place a single stripe p-channel transistor in the n-well region. Estimate the resistance from n+ contact to the p-channel transistor to its nearest edge from the n+ stripe, as a function of the width of the transistor. Estimate the resistance to the farthest edge from the n+ stripe. Draw the resistor model for the p-channel MOSFET and its well resistance. Derive an equation for the resistance as a function of the n+ stripe width, the n-well sheet resistance, and physical distances.

6.2. Draw two n+ well contact stripes in an n-well in parallel. Place a single stripe p-channel transistor in the n-well region between the two stripes. Estimate the resistance from n+ contact to the p-channel transistor to its nearest edge from the n+ stripe, as a function of the width of the transistor. Estimate the resistance to the farthest edge from the n+ stripe. Draw the resistor model for the p-channel MOSFET and its well resistance. Derive an equation for the resistance as a function of the n+ stripe width, the n-well sheet resistance, and physical distances.

6.3. Create the complete n+ ring around the p-channel transistor in an n-well. Draw the resistor model for the p-channel MOSFET and its well resistance. Derive an equation for the resistance as a function of the n+ stripe width, the n-well sheet resistance, and physical distances. Break the ring into four separate stripes for estimation.

6.4. Draw a p-channel MOSFET contained within an n-well. Draw a separate guard ring consisting of an n+ contact and n-well region, placed around the n-well containing the p-channel MOSFET. Create a high-current Ebers–Moll model for the parasitic npn transistor model based on the design dimensional parameters, where the npn transistor is created by the two n-well regions. The model should be based on the design parameters in all dimensions. Create a small-signal model of the lateral npn whose parameters are based on the design parameters.

6.5. An OCD guard ring is formed by a "figure 8" guard ring around a p-channel MOSFET and an n-channel MOSFET. The p-channel MOSFET is contained within its own n-well and surrounded by the separate OCD guard ring. The n-channel MOSFET is contained in a p-substrate region, and also surrounded by the n-well ring (the second part of the n-well ring). Derive the current model for the first npn formed between the p-channel pull-up n-well and the guard ring (e.g., Problem 6.4), and the second npn formed between the n-channel pull-down and the n-well ring.

6.6. As in Problem 6.5, the n-well ring is extended to form a third region for the ESD network. The n-well ring is now formed of three regions. Add an additional model for an npn transistor between an ESD n-well diode and the I/O cell guard ring.

6.7. An OCD I/O cell contains a p-channel pull-up and an n-channel pull-down aligned with the bond pad. The "width" of the I/O cell is equal to the bond pad width. A space is defined between the two bond pads for bond wire placement rules. No guard ring is placed between the two I/O cells. Between the two I/O cells, the two n-channel pull-downs and two p-channel transistors are adjacent to each other. Derive a lateral npn model between the two pull-up n-wells, as a function of the design parameters and bond pad parameters.

6.8. As Problem 6.7, with two n-channel pull-down diffusions. Assume the orientation of the n-well stripes is parallel to each other (e.g., polysilicon gate structures are continuous from the PFET to the NFET). Derive a lateral npn model between the two pull-down n-channel MOSFETs, as a function of the design parameters and bond pad parameters.

6.9. As Problems 6.7 and 6.8, with two I/Os side by side. In this problem, an n-well guard ring stripe was placed in the symmetric center of the I/O cells. Derive an npn model for the transistor formed between the n-well guard ring stripe and the p-channel OCD n-well, as well as a second npn model transistor formed between the n-well guard ring stripe and the n-channel pull-down. How do these compare to the models of Problems 6.7 and 6.8?

6.10. In the I/O n-well guard ring, evaluate the resistance assuming electrical connection to the V_{DD} power supply at the top of the n-well ring. First, evaluate the resistance without the "figure 8" but just as a rectangular ring. Second, evaluate with the "figure 8" connection between the p-channel FET and n-channel FET. Add the third ring, and evaluate the resistance to the ESD region. Parameters should include the distance of the V_{DD} connection to any distance along the n-well ring, the n-well guard ring sheet n-well resistance, n-well guard ring width.

6.11. Given a guard ring, derive a model for a lateral npn of an injecting n-well region, and a guard ring consisting of a plurality of n-well regions of different depths. Assume the following: case (A) n-well stripe, depth t_{NW}; case (B) high-voltage n-well, depth $t_{HV\ NW}$; case (C) n-buried layer, t_{NBL}.

6.12. An active guard ring of an n-well is placed between an injecting source and a collecting region. The active guard ring is an n-well electrically connected to a local p+ substrate contact. Show the lateral potential, and estimate the lateral electric field introduced by the active guard ring. Evaluate the impact of the lateral electric field on the carrier transport. (*Hint*: evaluate the drift-diffusion current equation.) At what electric field will the active guard ring counteract the diffusion transport?

6.13. Given a guard ring stripe, evaluate the series resistance assuming the following: case (A) n-well stripe; case (B) n-well and high-voltage n-well; case (C) n-well, high-voltage n-well and n-buried layer.

6.14. A trench region is used as a guard ring structure. As the trench becomes deeper, the guard ring efficiency improves. Establish a relationship for trench depth vs. carrier transport assuming that the trench sidewall is "absorbing" the carriers, preventing them from passing beyond the trench region.

6.15. A through silicon via is placed between two circuits. Show how this can serve as a guard ring structure. Assume the TSV can not be formed in a complete ring but must leave a space $W_{TSV\text{-}TSV}$ between the adjacent vias.

REFERENCES

1. R. R. Troutman. *Latchup in CMOS Technology: The Problem and the Cure*, Boston: Kluwer Academic Publishers, 1986.
2. S. Voldman. *Latchup*. Chichester, UK: John Wiley and Sons, Ltd, 2007.
3. M. D. Ker and S.F. Hsu. *Transient-Induced Latchup in CMOS Integrated Circuits.* Singapore: John Wiley and Sons, Pte Ltd, 2009.
4. D.B. Estreich. *The Physics and Modeling of Latch-up and CMOS Integrated Circuits.* Technical Report G-201-9, Integrated Circuits Laboratory, Stanford University, Stanford, CA, November 1980.
5. R.R. Troutman. Epitaxial layer enhancement of n-well guard rings for CMOS circuits. *IEEE Electron Device Letters*, **4** (12), 1983; 438–440.
6. R. R. Troutman and H.P. Zappe. Layout and bias considerations for preventing transiently triggered latchup in CMOS. *IEEE Transactions on Electron Devices*, **ED-31**, March 1984; 279–286.
7. R.R. Troutman and H.P. Zappe. A transient analysis of latchup in bulk CMOS. *IEEE Transactions on Electron Devices*, **ED-30**, February 1983; 170–179.
8. H.P. Zappe. *A Transient Analysis and Characterization of Latchup in Bulk CMOS.* B.S. and M.S. Thesis, Massachusetts Institute of Technology, Cambridge, MA, February 1983.
9. IEEE Latchup Standards Committee. Latchup Test Method for Process Characterization, 1988.
10. JEDEC Standard No. 17. *Latchup in CMOS Integrated Circuits*, Joint Electron Device Engineering Council, August 1988.
11. J. Quinke. Novel test structures for the investigation of the efficiency of guard rings used for I/O latchup prevention, *Proceedings of the International Conference on Microelectronic Test Structures (ICMTS)*, 1990; 35–40.
12. T. Cavioni, M. Cecchetti, M. Muschitiello, G. Spiazzi, I. Vottre, and E. Zanoni. Latch-up characterization in standard and twin-tub test structures by electrical measurements, 2-D simulations and IR microscopy, *Proceedings of the International Conference on Microelectronic Test Structures (ICMTS)*, 1990; 41–46.
13. C. Mazure, W. Reczek, D. Takacs, and J. Winnerl. Improvement of latching hardness by geometry and technology tuning, *IEEE Transactions on Electron Devices*, **ED-35** (10), 1988; 1609–1615.

14. Y. Song, J.S. Cable, K.N. Vu, and A.A. Witteles. The dependence of latchup sensitivity on layout features in CMOS integrated circuits. *IEEE Transactions of Nuclear Science*, **NS-33** (6), 1986; 1493–1498.
15. R. Lohia, and A. Ali. Parametric formulation of CMOS latchup as a function of chip layout parameters. *IEEE Journal of Solid State Circuits*, **23** (1), February 1988; 245–250.
16. R. Menozzi, L. Selmi, E. Sangiorgi, G. Crisenza, T. Cavioni, and B. Ricco. Layout dependence of CMOS latchup. *IEEE Transactions on Electron Devices*, **ED-35** (11), 1988; 1892–1901.
17. W. Reczek, F. Bonner, and B. Murphy. Reliability of latchup characterization procedures. *Proceedings of the International Conference on Microelectronic Test Structures (ICMTS)*, 1990; 51–54.
18. C. Cane. *Optimitzacio de pous n per una tecnologia CMOS. Aplicacio a la prevencio del latch-up.* Ph.D. Thesis, Polytechnics University of Catalonia, Barcelona, 1989.
19. C. Cane, M. Lozano, E. Cabruja, E. Lora-Tamayo, and F. Serra-Mestres. A new test structure to characterize the latchup effect. *Proceedings of the International Conference on Microelectronic Test Structures (ICMTS)*, 1990; 47–51.
20. T. Aoki. A new latchup test structure for practical design methodology for internal circuits in standard-cell-based CMOS/BiCMOS LSIs. *Proceedings of the International Conference on Microelectronic Test Structures (ICMTS)*, 1992; 18–23.
21. S. Voldman. Test structures for analysis and parameter extraction of secondary photon-induced leakage currents in CMOS DRAM technology. *Proceedings of the International Conference on Microelectronic Test Structures (ICMTS)*, 1992; 39–43.
22. H. Momose, T. Maeda, K. Inoue, Y. Urakawa, and K. Maeguchi. Novel test structures for the characterization of latchup tolerance in a bipolar and MOSFET merged device. *Proceedings of the International Conference on Microelectronic Test Structures (ICMTS)*, 1991; 225–230.
23. T. Kessler and F.W. Wulfert. Diagnosing latch-up with backside emission microscopy, *Semiconductor International*, **7** (23), July 2000; 313–316.
24. W.K. Chim, *Semiconductor Device and Failure Analysis: Using Photon Emission Microscopy*. Chichester, UK: John Wiley and Sons, Ltd, 2000.
25. S. Liao C. Niou, W.T.K. Chien, A. Guo, W. Dong, and C. Huang. New observance and analysis of various guard ring structures on latchup hardness by backside photoemission image. *Proceedings of the International Reliability Physics Symposium (IRPS)*, April 2003; 92–97.
26. D. Tremouilles, M. Bafluer, G. Bertrand, and G. Nolhier. Latch-up ring design guidelines to improve electrostatic discharge (ESD) protection scheme efficiency. *IEEE Journal of Solid-State Circuits*, **39** (10), 2005; 1778–1782.
27. S. Voldman, C.N. Perez, and A. Watson. Guard rings: theory, experimental quantification, and design, *Proceedings of the Electrical Overstress/Electrostatic Discharge (EOS/ESD) Symposium*, October 2005; 131–140.
28. S. Voldman, C.N. Perez, and A. Watson. Guard rings: structures, design methodology, integration, experimental results, and analysis for RF CMOS and RF mixed signal silicon germanium technology. *Journal of Electrostatics*, **64**, 2006; 730–743.
29. D. Tremouilles, M. Scholz, G. Groeseneken, M.I. Natarajan, N. Azilah, M. Bafluer, M. Sawada, and T. Hasebe. A novel method for guard ring efficiency assessment and its applications for ESD protection design and optimization. *Proceedings of the International Reliability Physics Symposium (IRPS)*, 2007; 606–607.
30. S. Voldman, E. G. Gebreselasie, D. Hershberger, D. S. Collins, N. B. Feilchenfeld, S. A. St. Onge, A. Joseph, and J. Dunn. Latchup in merged triple well technology. *Proceedings of the International Reliability Physics Symposium (IRPS)*, 2005; 129–136.
31. S. Voldman and E.G. Gebreselasie. The influence of merged triple well, deep trench and subcollector on CMOS latchup. *Proceedings of the Taiwan Electrostatic Discharge Conference (T-ESDC)*, 2006; 49–52.

32. S. Voldman, Latchup Physics and Design, *Tutorial Notes, ESD Tutorials of the Electrical Overstress/Electrostatic Discharge (EOS/ESD) Symposium*, September 20, 2004.
33. S.Voldman, ESD and Latchup in Advanced Technologies, *Tutorial Notes of the International Reliability Physics Symposium (IRPS)*, April 25, 2004.
34. S. Voldman, CMOS Latchup, *Tutorial Notes of the International Reliability Physics Symposium (IRPS)*, April 17, 2005.
35. S. Voldman. Guard ring structures for high voltage CMOS/low voltage CMOS technology using LDMOS (lateral double-diffused metal oxide semiconductor) device fabrication. U.S. Patent No. 7,541,247, June 2, 2009.
36. W. Winkler, and F. Herzl, Active substrate noise suppression in mixed-signal circuits using on-chip driven guard rings. *Proceedings of the IEEE 2000 Custom Integrated Circuits Conference (CICC)*, May 2000; 356–360.
37. O. Gonnard and G. Charitat. Substrate current protection in smart power IC's. *Proceedings of the International Symposium on Power Semiconductor Devices (ISPSD)*, 2000; 169–172.
38. O. Gonnard, G. Charitat, P. Lance, M. Susquet, M. Bafluer, and J.P. Laine. Multi-ring active analogic protection (MAAP) for minority carrier injection suppression in smart power IC's. *Proceedings of the International Symposium on Power Semiconductor Devices (ISPSD)*, 2001; 351–354.
39. M. Schenkel, P. Pfaffli, W. Wilkening, D. Aemmer, and W. Fichtner. Transient minority carrier collection from substrate in smart power design. *Proceedings of the European Solid State Device Research Conference (ESSDERC)*, 2001; 411–414.
40. R. Singh and S. Voldman. Method and apparatus for providing ESD protection and/or noise reduction in an integrated circuit. U.S. Patent No. 6,826,025, November 30, 2004.
41. R. Singh and S. Voldman. Method and apparatus for providing ESD protection and/or noise reduction in an integrated circuit. U.S. Patent No. 7,020,857, March 28, 2006.
42. W. Horn. *On the Reverse-Current Problem in Integrated Smart Power Circuits*. Ph.D. Thesis, Technical University of Graz, Austria, April 2003.
43. A. Watson, S. Voldman, and T. Larsen. Deep trench guard ring structures and evaluation of the probability of minority carrier escape for ESD and latchup in advanced BiCMOS SiGe technology, *Proceedings of the Taiwan Electrostatic Discharge Conference (T-ESDC)*, 2003; 97–103.
44. S. Voldman and A. Watson. The influence of deep trench and substrate resistance on the latchup robustness in a BiCMOS silicon germanium technology. *Proceedings of the International Reliability Physics Symposium (IRPS)*, 2004; 135–142.
45. S. Voldman, R. A. Johnson, L.D. Lanzerotti, and S.A. St Onge. Deep trench-buried layer array and integrated device structures for noise isolation and latch up immunity. U.S. Patent No. 6,600,199, July 19, 2003.
46. S. Gupta, S.L. Kosier, and J. C. Beckman. Guard ring structure for reducing crosstalk and latch-up in integrated circuits. U.S. Patent No. 6,747,294, June 8, 2004.
47. S. Voldman and A. Watson. The influence of polysilicon-filled deep trench and sub-collector implants on latchup robustness in RF CMOS and BiCMOS SiGe technology. *Proceedings of the Taiwan Electrostatic Discharge Conference (T-ESDC)*, 2004; 15–19.
48. S. Voldman. The influence of a novel contacted polysilicon-filled deep trench (DT) biased structure and its voltage bias state on CMOS latchup. *Proceedings of the International Reliability Physics Symposium (IRPS)*, 2006; 151–158.
49. S. Voldman, S. Strang, and D. Jordan. A design system for auto-generation of ESD circuits. *Proceedings of the International Cadence Users Group (ICUG)*, September 2002.
50. S. Voldman, S. Strang, and D. Jordan. An automated electrostatic discharge computer-aided design (CAD) system with the incorporation of hierarchical parameterized cells in BiCMOS analog and RF technology for mixed signal applications. *Proceedings of the Electrical Overstress/Electrostatic Discharge (EOS/ESD) Symposium*, October 2002; 296–305.

51. C.N. Perez, S. Voldman, Method of forming a guard ring parameterized cell structure in a hierarchical parameterized cell design, checking and verification system. U.S. Patent Application 20040268284, December 30, 2004.
52. Y. Huh, K. Min, P. Bendix, V. Axelrad, R. Narayan, J.W. Chen, L.D. Johnson, and S. Voldman. Chip level layout and bias considerations for preventing neighboring I/O cell interaction-induced latchup and inter-power supply latchup in advanced CMOS technologies. *Proceedings of the Electrical Overstress/Electrostatic Discharge (EOS/ESD) Symposium*, 2005; 100–107.
53. A. Weger, S. Voldman, F. Stellari, P. Song, P. Sanda, and M. McManus. A transmission line pulse (TLP) pico-second imaging circuit analysis (PICA) methodology for evaluation of ESD and latchup. *Proceedings of the International Reliability Physics Symposium (IRPS)*, 2003; 99–104.
54. S. Voldman. Latchup and the domino effect. *Proceedings of the International Reliability Physics Symposium (IRPS)*, 2005; 145–156.
55. N.P. Sanda, S. H. Voldman, and A.J. Weger. Method and application of PICA (picosecond imaging circuit analysis) for high current pulsed phenomena. U.S. Patent No. 6,943,578, September 13, 2005.
56. P. Chapman, D. Collins, and S. Voldman. Latchup robust gate array using through wafer via. U.S. Patent No. 7,498,622 March 3, 2010.
57. P. Chapman, D. Collins, and S. Voldman. Structure for a latchup robust gate array using through wafer via. U.S. Patent No. 7,696,541, April 13, 2010.
58. P. Chapman, D. Collins, and S. Voldman. Latchup robust array I/O using through wafer via. U.S. Patent No. 7,741,681, June 22, 2010.
59. S. Voldman. ESD network circuit with a through wafer via structure and a method of manufacture. U.S. Patent Application No. 20100244187, September 30, 2010.
60. P. Chapman, D. Collins, and S. Voldman. Design methodology of guard ring design resistance optimization for latchup prevention. U.S. Patent No. 7,549,135, June 16, 2009.
61. M.D. Ker, H.C. Jiang, J.J. Peng, and T.L. Shieh. Automatic methodology for placing the guard rings into chip layout to prevent latchup in CMOS IC's. *International Electron Device Meeting (IEDM) Technical Digest*, 2001; 113–116.
62. S. Voldman. Methodology for placement based on circuit function and latchup sensitivity. U.S. Patent No. 7,401,311, July 15, 2008.
63. A.E. Watson, and S. Voldman. Methodology of quantification of transmission probability for minority carrier collection in a semiconductor chip. U.S. Patent No. 7,200,825, April 3, 2007.
64. C.N. Perez, and S. Voldman. Method of displaying a guard ring within an integrated circuit. U.S. Patent No. 7,350,160, March 25, 2008.

7 ESD Full-Chip Design Integration and Architecture

7.1 DESIGN SYNTHESIS AND INTEGRATION

ESD protection is fundamental to semiconductor chip design [1–5]. In this chapter, examples of ESD design synthesis and integration will be discussed for different chip designs. The chapter will begin by discussing architectures for memory [6–21], microprocessors [22–42], ASIC standard cell [43–49], analog and radio frequency [50–59]. The sections will provide examples of digital, analog, RF, mixed-signal, and mixed-voltage interface chip integration and architectures.

7.2 DIGITAL DESIGN

ESD design synthesis in digital design typically does not require co-synthesis of the ESD solutions and the digital circuitry for digital design applications below 1 GHz [1–49]. But, the ESD design synthesis must be integrated into the floorplan of a semiconductor chip.

Integration of the ESD networks is built into the architecture of the chip to ensure achieving ESD protection for all pin combinations specified in the ESD standards. ESD protection must be provided between signal pins and power rails, power rail-to-power rails, and pin-to-pin. In semiconductor designs, circuitry interfacing with other semiconductor chips or systems can be at the same voltage or different voltage levels. Peripheral circuits and power rails are separated from internal core circuitry and other sensitive functions. Guard rings are placed to minimize interaction and injection between the different chip domains. To address these considerations, the semiconductor design synthesis must have the following:

- Full-chip guard ring.
- Separate power rails for different power domains.

ESD: Design and Synthesis, First Edition. Steven H. Voldman.

- ESD networks within each domain.
- ESD networks between each chip domain.

7.3 CUSTOM DESIGN vs. STANDARD CELL DESIGN

In semiconductor chip design, ESD design synthesis is a function of whether the design is custom [1–42], or in a standard cell architecture [43–49]. In custom design, the floorplan engineer has the freedom to place different domains in different locations in the semiconductor chip. Power bus placement and ESD can be placed as desired within the ESD design rules, and design manual rules for the technology. ESD designs can be customized to the area or region for optimized placement. ESD designs can be pre-established, parameterized cells, or customized design layers.

In a standard cell design integration, specific rules are established for placement of the bond pads, peripheral circuits, power bus, ground bus, guard rings, and ESD elements. In a standard cell design methodology, ESD designs are integrated with the standard cell circuitry. ESD input and output circuitry are placed in the standard cell footprint. ESD power clamps are also pre-defined for the customer. With this architecture, only limited changes in the design can be integrated.

7.4 MEMORY ESD DESIGN

In this section, ESD design synthesis for different memory architectures will be shown as examples.

7.4.1 DRAM Design

ESD protection and design synthesis is challenging in dynamic read access memory (DRAM) [6–15]. Area and capacitance loading of ESD networks are severe constraints on the high-speed memory input receiver networks. Additionally, advanced DRAMs are typically the first product in a technology generation, leading to new ESD concerns and learning in a new technology. To add to this, the native voltage of the DRAM is lower than the products it is interfacing with; this requires mixed-voltage interface (MVI) circuitry that must drive higher voltages.

In a DRAM design floorplan, the memory arrays are segmented into bits of DRAM cells, bit lines, word lines, and sense amplifier logic. All other functions are contained within a given region of the semiconductor chip. There are three architecture floorplans:

- **Peripheral Exterior:** Bond pads and I/O on the periphery of the semiconductor chip.
- **Vertical Spine:** Bond pads and I/O in the symmetrical center in the long dimension of the semiconductor chip (e.g., height).
- **Horizontal Spine:** Bond pads and I/O in the symmetrical center in the short dimension of the semiconductor chip (e.g., width).

For floorplans with peripheral I/O, the bond pad, power rails, and ESD placement are similar to many other standard products. In this case, the edges of the I/O bond pads have a short wire bond length to the package pins. The peripheral I/Os are separated from the DRAM arrays with a guard ring, so I/O circuitry and ESD circuitry does not interfere with the DRAM cells.

For the other cases, with a vertical or horizontal "spine," the configuration is significantly different. The "spine" is the section of the DRAM that contains the bond pads, I/O circuitry, and ESD networks. In this region, the I/O circuit power bus and I/O ground are defined. The area constraints used for the spine are limited to avoid wasting too much area for non-memory function. In the spine region, there exist the following:

- Bond pads.
- V_{DD} (I/O) power rail.
- V_{SS} (I/O) power rail.
- V_{DD} core power rail.
- V_{SS} substrate power rail.
- Voltage regulators V_{DD} (I/O) to V_{DD} (core).
- I/O circuitry.
- ESD signal pin ESD networks.
- ESD V_{DD} (I/O) to V_{SS} (I/O) networks.
- ESD V_{SS} (I/O) to V_{SS} networks.
- Decoupling capacitors.

Figure 7.1 shows an example of a DRAM chip floorplan with a vertical spine. A vertical spine was required in a 4 Mb DRAM.

Figure 7.2 shows an example of a DRAM chip floorplan with a horizontal spine. A horizontal spine was utilized in a 16 Mb DRAM [6–8].

In the case of an internal spine, the package must interface with the internal bond pads. One method to interconnect the package to the internal spine is using a "transfer wire," which is a thick layer that is part of the packaging and molding process. The transfer wire is connected to a short wire bond between the transfer wire and the internal bond pads.

Some of the challenges in ESD are as follows:

- Mixed-voltage interface I/O circuits.
- Multiple power rails (e.g., V_{DD} (I/O), V_{DD}, V_{SS} (I/O), and V_{SS}).
- Low-capacitance external power rail (e.g., V_{DD} (I/O)).
- Regulated core V_{DD}.
- Limited area in the spine region.

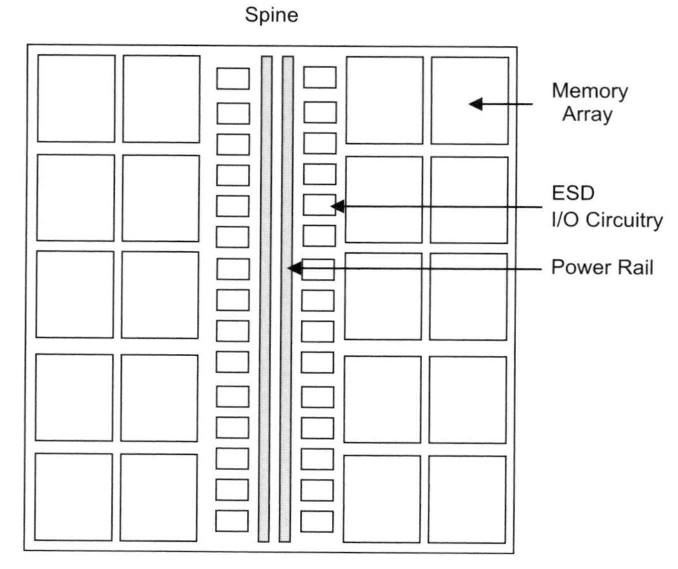

Figure 7.1 DRAM design with a vertical spine

Mixed-voltage interface circuits: To address the mixed-voltage interface concern, OCDs introduce resistors in series with the NFET transistors. Typically, no integrated ballast resistors are used to save space. Using an independent resistor element is more efficient from an area perspective. Using the PFET drain-to-n-well diode, this assists in providing good ESD protection with a reduced area for the ESD element.

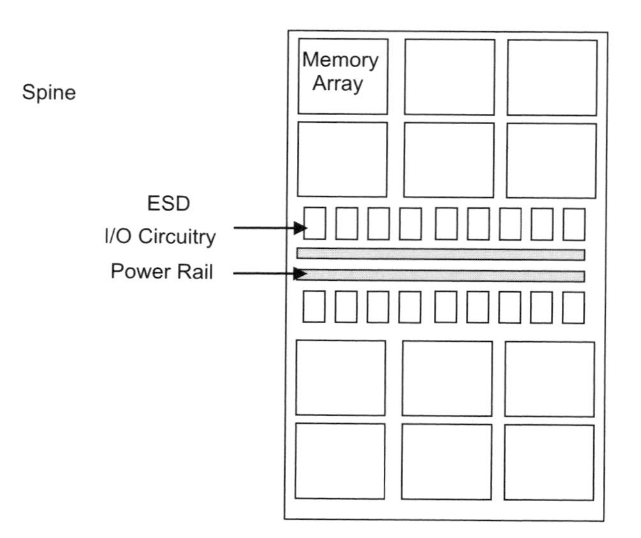

Figure 7.2 DRAM design with a horizontal spine

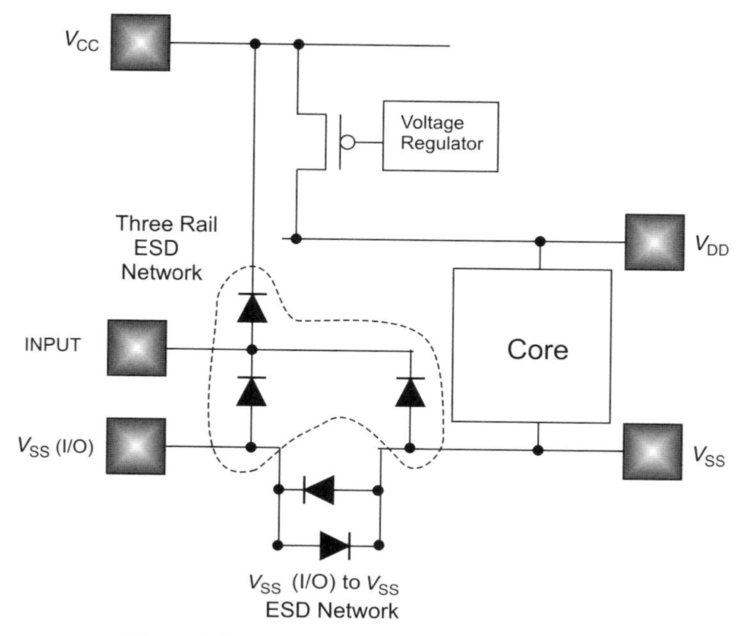

Figure 7.3 DRAM ESD network with three rails

Multiple-power ESD signal pin networks: Unique ESD networks are used at the bond pad that have direct current paths to three or four rails [7,11]. These "three-rail" ESD networks can provide a forward-biased diode discharge path to V_{DD} (I/O), V_{SS} (I/O), and V_{SS}. These have been utilized from 16 Mb to 1 Gb DRAM chips. Figure 7.3 shows an example of a three-rail ESD network [11,12].

ESD regulator bypass network: In DRAMs, the voltage regulator is a p-channel transistor whose well and source are connected to the V_{DD} (I/O) rail (Figure 7.4). As a result, no current flows from the I/O power rail to the regulated V_{DD} core power rail. In early DRAM development, an ESD regulator bypass network was used in parallel with the regulator to allow the ESD current flow to the high-capacitance rail of the semiconductor chip [13]. An n-channel MOSFET was distributed along the length of the spine, where the n-channel gate and source was connected to V_{DD}, and whose drain was connected to V_{DD} (I/O).

ESD V_{DD}(I/O) power clamp: In advanced technology, concerns over the discharge into the core memory led to not adding an ESD regulator bypass. As a result, a local ESD power clamp was added between the V_{DD} (I/O) rail and V_{SS} (I/O) (or directly to V_{SS}). In advanced 1 Gb DRAMs, a DTSCR was used for the ESD power clamp [14,15]. The disadvantage of this method is the limited area for external ESD power clamps, as well as not taking advantage of the natural capacitance and size of the DRAM chip. Figure 7.5 shows an example of a DTSCR.

With technology scaling, one of the challenges of the ESD design is in the spine region. With the scaling of the bond pad size, and power buses, it is more difficult to achieve HBM, MM, and CDM results in the advanced DRAM products. But still today, 2000 V HBM, 200 V MM, and 1000 V CDM are achievable with 1–2 Gb DRAM products in 45 nm technology [5].

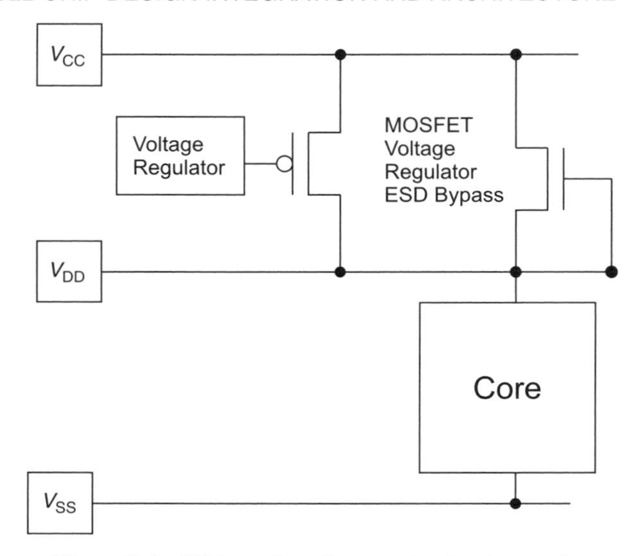

Figure 7.4 ESD regulator bypass circuitry integration

7.4.2 SRAM Design

In a static read-access memory (SRAM) design, the architecture is established to provide the highest utilization for the memory array, and the highest speed [17–20]. ESD design synthesis of a SRAM design typically requires peripheral pad design. The floorplan contains all the bond pads on the periphery of the semiconductor chip, away from the core SRAM cell arrays (Figure 7.6). For speed, it is advantageous to have the OCD circuitry on the periphery with the

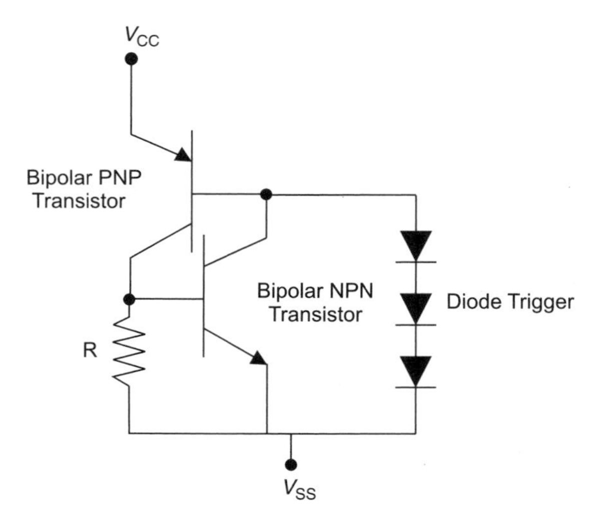

Figure 7.5 1 Gb DRAM diode-triggered SCR for ESD V_{DD}-to-V_{SS} power clamp

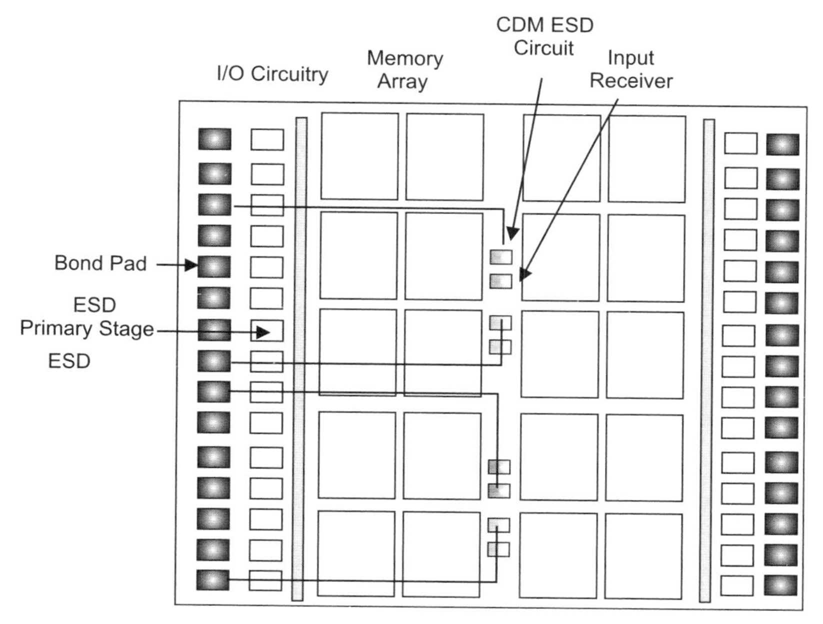

Figure 7.6 SRAM design floorplan

bond pad, power rails, and ESD network. To provide fast memory, it is an advantage to have the receiver networks in the interior of the core region.

As an example of a SRAM ESD architecture, the receiver networks were such that the bond pad and primary ESD network were on the periphery of the chip, but the receiver itself was on the interior of the design.

On the periphery, ESD p + /n-well diodes and n-well to substrate diodes were used between the power rail and ground. The core capacitance was large, so no additional ESD power clamps were used. ESD HBM results exceeded 5 kV HBM [18].

Figure 7.7 shows the ESD circuitry for the SRAM design. A wide interconnect extended from the periphery to the interior, forming an 8Ω resistor in series with the receiver. Since the placement of the receiver network was a significant distance from the periphery, a secondary stage CDM network was placed local to the receiver. The interconnect wire was utilized as a CDM resistor; no additional resistor element placed in series. A 40 μm wide grounded-gate n-channel MOSFET was placed near the receiver inverter network. CDM test simulation results showed that without the CDM MOSFET near the receiver, the CDM results were 200 V CDM. With the CDM MOSFET, the CDM results exceeded 1200 V.

This implementation demonstrates the following ESD design synthesis concepts:

- Usage of peripheral ESD networks on the exterior of the semiconductor chip.
- Utilization of a secondary ESD CDM stage placed local to the inverter receiver.
- Utilization of the natural resistance of the wire interconnect from the exterior of the chip to the internal receiver for CDM protection.

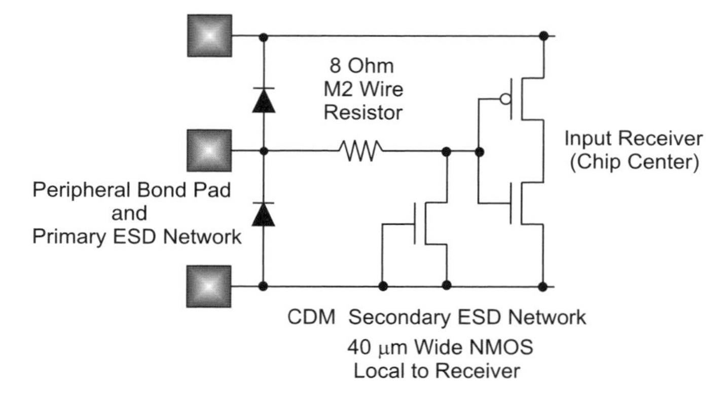

Figure 7.7 SRAM design ESD circuitry

7.4.3 Non-Volatile RAM ESD Design

In non-volatile RAM (NVRAM) applications, semiconductor chips may have a 3.3 V native power rail, a 5 V power rail, and a 12 V programming pin for the NVRAM circuitry. Some of the unique issues with the NVRAM ESD protection are as follows:

- **Program Pin Latchup:** 12 V power pin can lead to CMOS latchup and electrical overstress.
- **Programming Flash-Induced Injection:** Programming flash can lead to photon-induced leakage events in functional circuits and ESD networks.

Programming pins can lead to injection of minority carriers. Programming pins must be sufficiently isolated and have guard rings that satisfy high-voltage circuit solutions. Forward biasing of ESD elements, electrical overstress, and mis-sequencing is an issue with the programming pin.

In a NVRAM technology, the MOSFET gate oxide can use a dual-gate thin oxide [21]. Figure 7.8 shows a sequence-independent ESD network using a triple-oxide p-channel

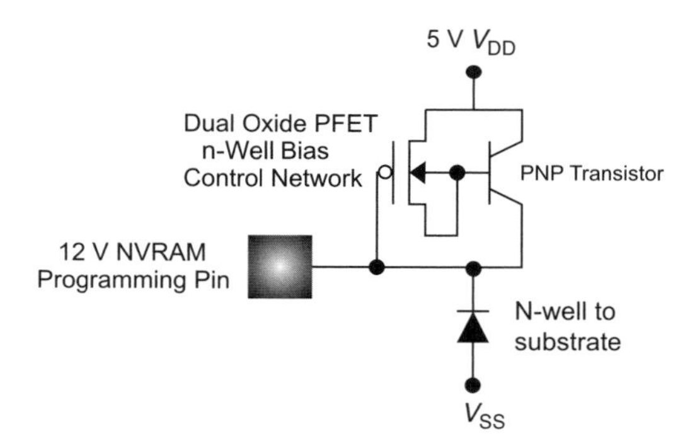

Figure 7.8 NVRAM chip design

MOSFET. The ESD network has a p-channel MOSFET as a "well bias control"; a lateral pnp is contained within the well. In one aspect, there is a dual-oxide ESD protective network for non-volatile memory in which ESD protection is provided using a thick oxide PFET in a thick epitaxial layer with sequence-independent circuitry. The dual-oxide gate of the p-channel MOSFET is connected to the 12 V input programming node, the source is connected to the 5 V power rail, and the drain is connected to its own n-well region. When the programming pin is low, the p-channel MOSFET is "on" and the n-well is biased at 5 V. When the programming pin is high (e.g., 12 V), the n-well floats. This provides ESD protection and avoids electrical overstress.

The dual-oxide ESD network includes a high-voltage PFET ESD network for 12 V to 5 V applications, as well as a low-voltage PFET network for 5 V to 3 V applications taking advantage of dual oxides supported by the disclosed technology. The circuit saves space, is migratable, improves reliability, and is voltage differential independent.

7.5 MICROPROCESSOR ESD DESIGN

In a microprocessor, the architecture and chip size are important for both microprocessor performance, yield, and net profits. Microprocessors utilize advanced CMOS technology [22–24]. Additionally, SOI technology is used for mainstream microprocessor development [38–42]. In 25-wafer lots of microprocessors, each lot can be worth millions of US dollars. Significant focus in a microprocessor is on the placement of the cache, and other critical circuitry that optimizes the chip performance metrics.

7.5.1 3.3 V Microprocessor with 5.0 V to 3.3 V Interface

As an example of one microprocessor architecture, in a 0.35 μm channel length technology, a PowerPC microprocessor was constructed with 140 MHz performance [30–35]. The area needed for ESD protection represented 8% of the entire chip area. It was decided – for chip size, performance, and net cost – to place the ESD circuitry under the bond wire pads (Figure 7.9). The bond pad size, at this technology node, was 130 μm × 130 μm. The microprocessor had a 3.3 V power supply, and needed to interface with 5 V. The floorplan contained all the bond pads on the periphery of the semiconductor chip. The package was a quad flat pack (QFP) package.

The off-chip driver was a mixed-voltage interface circuit containing an n-channel MOSFET series cascode pull-down, and a "floating well" p-channel pull-up. For the mixed-voltage interface, the ESD design needed to be compatible with the I/O bi-directional circuit.

The ESD design chosen was a series of p+/n-well diodes between the signal pads and the 3.3 V to receive a 5 V signal. On the periphery, five ESD p+/n-well diodes and n-well to substrate diodes were used between the power rail and ground. The core capacitance was so significant that no additional ESD power clamps were used. Since no other circuits were allowed under the bond pads except diode networks, the ESD network filled the area under the bond wire pads. The ESD network was placed adjacent to the bi-directional off-chip driver signal pins. With the spatial placement of the ESD network local to the bi-directional OCD, no additional CDM secondary network was needed for CDM protection.

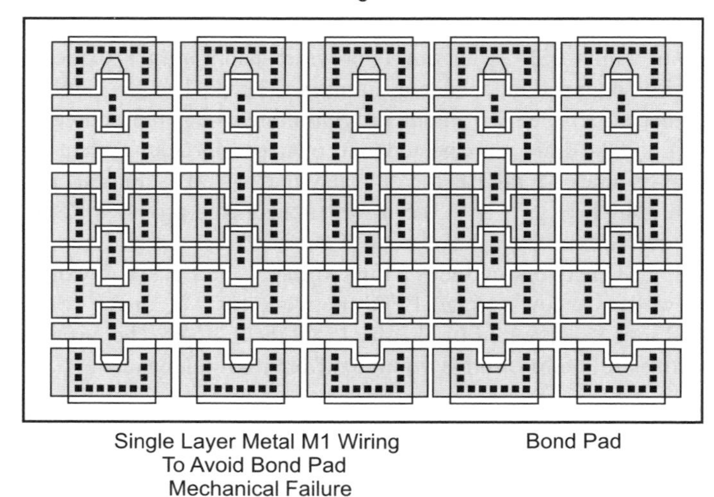

Figure 7.9 ESD under bond pad chip design – five diode string

A first concern was the design of the metal layer pattern under the wire bond pad. In the ESD design layout synthesis, different metal patterns were used to verify the integrity of the bond pad after the bond pad stress test.

A second concern was with the implementation of five p+/n-well diodes in series; a parasitic bipolar Darlington effect was discovered due to the common collector parasitic pnp transistors. This common collector amplification of the leakage was eliminated by a "snubber diode" [30–32]. Figure 7.10 shows the ESD network with the "snubber diode" to eliminate the Darlington amplification.

ESD results of the CPU design (after optimization of the perimeter of the diodes) of 8000 V HBM were achieved after two successive microprocessor design passes. Perimeter-to-area

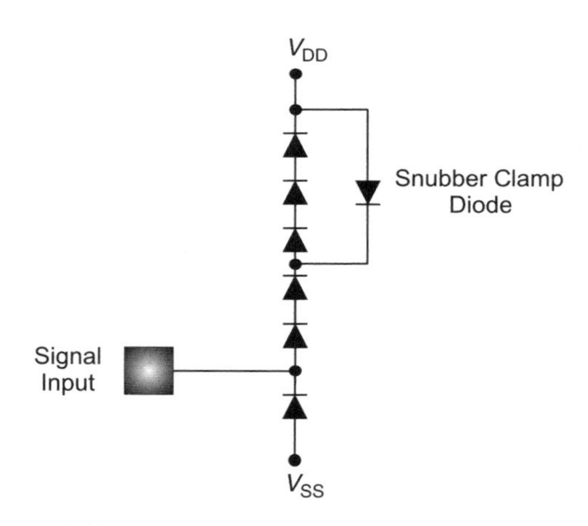

Figure 7.10 ESD schematic with snubber diode element

optimization of the STI p–n diodes increased the HBM ESD results from 800 V to 8000 V with no change in the size of the ESD network [30–32].

This implementation demonstrates the following ESD design synthesis concepts:

- **ESD Under Bond Pads:** Usage of peripheral ESD networks under bond wire pads.
- **Metal Pattern Optimization:** Optimization of metal wire pattern under bond pad was required to avoid insulator cracking and bond pad integrity.
- **Mixed-Voltage Interface ESD:** Utilization of a mixed-voltage interface ESD network.
- **Darlington Effect:** Elimination of the parasitic Darlington effect in the ESD network using a snubber diode between successive diode elements.
- **ESD Diode Perimeter-to-Area Optimization:** Perimeter-to-area optimization can lead to significant increase in the ESD robustness of a semiconductor product with no process or area increase of the ESD design.

7.5.2 2.5 V Microprocessor with 5.0 V to 2.5 V Interface

The first 2.5 V power supply microprocessor was required to interface with 5.0 V power supply semiconductor chips [3,5]. The first 2.5 V power supply PowerPC chip, at 100 MHz, replaced a 5 V/3.3 V 33 MHz microprocessor. This implementation was the first 0.25 μm CMOS technology in the semiconductor industry. This microprocessor integrated two power supply rails in the design; an external power rail at 5 V, and an internal core native power supply voltage at 2.5 V power supply (Figure 7.11). For the external I/O circuitry, the circuitry must receive and drive 5 V signals. The input circuitry contained an n-channel half-pass transistor to lower the voltage on the MOSFET gate of the receiver. The n-channel MOSFET half-pass transistor was followed by a BR element and grounded-gate MOSFET prior to the input gates.

An ESD network connected a p–n diode between the signal pad and the 5 V power supply pin. A "diode string" of seven p–n diodes was placed between the 5 V power rail and the 2.5 V power rail [3,5]. In this time period, no ESD power clamps were placed between the V_{DD} and the V_{SS} power rail. The ESD event was distributed to the 2.5 V power rail, and the native capacitance of the chip was used to discharge the current to the ground plane.

7.5.3 1.8 V Microprocessor with 3.3 V to 1.8 V Interface

In the next technology node, another microprocessor chose to place the OCDs in the interior of the microprocessor. The OCDs were formed in OCD banks, where each grouping contained a service module. The service module contained power and the ESD power clamps for the set of local off-chip drivers. Eight I/Os were placed in each OCD bank, and these OCD banks were strategically placed for both performance and packaging considerations. To allow for sequence independence between the 3.3 V power rail and a 1.8 V power rail, a sequence-independent ESD network was utilized; this allowed for each power rail to be brought up independently

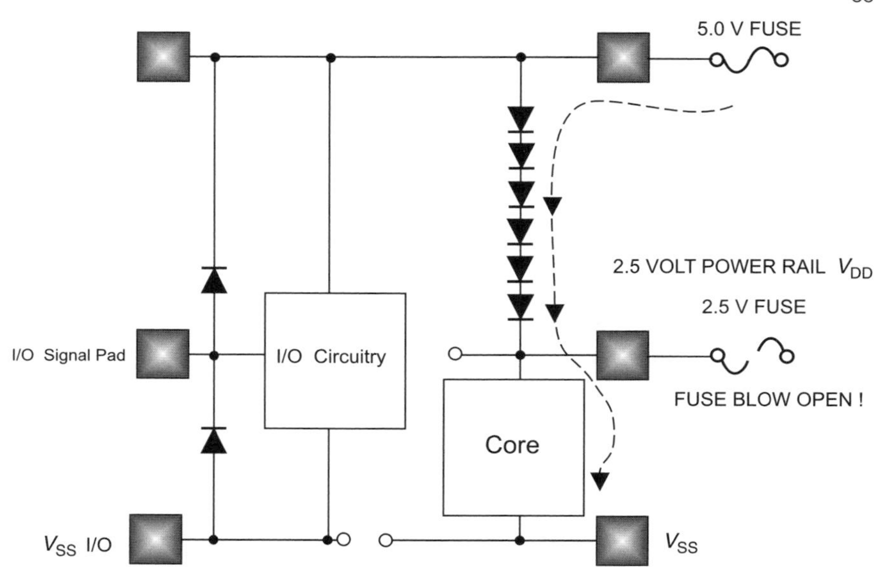

Figure 7.11 5 V to 2.5 V CMOS microprocessor architecture

without forward biasing of an ESD network. The sequence-independent network used a lateral pnp element in a "floating well"; the floating well was biased by a floating well control network using a p-channel MOSFET [11,12].

With technology scaling, as the power supply differential voltages were reduced, the number of elements between the power supplies was reduced. At the same time, there was an increase in the need for independence of the power supply rails. In this period, innovation introduced different circuits to allow this to occur. The design practice also eventually introduced ESD power clamps between the V_{DD} and V_{SS} power rails, eliminating the V_{DD}-to-V_{DD} ESD networks in most applications; this avoided sequencing issues between power rails, improved the domain-to-domain isolation, and at the same time eliminated redundancy of current paths.

7.6 APPLICATION-SPECIFIC INTEGRATED CIRCUITS

In this section, ESD implementation and design synthesis in ASIC applications will be discussed.

7.6.1 ASIC ESD Design

Application specific IC (ASIC) design synthesis must incorporate the ESD design synthesis in the formation of the methodology from the first definition of the "ASIC methodology" [44]. In an ASIC methodology, the floorplan and the ASIC "rule set" are integrated for customer usage.

The ASIC methodology must be a co-synthesis with the ESD design process. The ASIC method that influences ESD must incorporate the following:

- Power rail placement.
- Power rail bus widths.
- Bond pad width and spacing (pitch).
- Standard cell form factor (width and height).
- Power bond pad placement and "power book."
- Signal pin bond pad placement and standard cell.
- Input placement of I/O standard cells.
- Power rail placement over the I/O standard cells.
- Signal pin standard cell percentage of area for ESD networks.
- Signal pin standard cell guard rings.
- Power pad standard cell ESD networks.

In the ASIC methodology, all these design variables must be established to define the ESD implementation [48].

In a full ASIC methodology, there may be as many as 400–500 different I/O standard cells. These I/O standard cells contain different receiver and off-chip circuitry.

7.6.2 ASIC Design Gate Array Standard Cell I/O

In the ASIC design methodology, the functional and ESD requirements must be co-synthesized. With a large number of standard cells, methods are needed to reduce the number of circuits using creative methods, rules, and CAD techniques. In a standard cell design integration, specific rules are established for placement of the bond pads, peripheral circuits, power bus, ground bus, guard rings, and ESD elements. In a standard cell design methodology, ESD designs are integrated with the standard cell circuitry. ESD input and output circuitry is placed in the standard cell footprint. ESD power clamps are also pre-defined for the customer. With this architecture, only limited changes in the design can be integrated.

In one methodology, to provide different standard cell impedance, resistors are formed in a "gate array" method. In this methodology, the same standard cell can be used for different OCD strengths. These resistor elements serve as both series resistors for the correct series impedance and ballasting resistor.

Figure 7.12 shows an example of a standard cell with an array of different resistors. One set is used for each different desired impedance. The unused resistors are "grounded" and the row of incorporated resistors wired in series with the MOSFET OCD. This methodology was incorporated into a 0.35 μm CMOS technology generation. In the integration of the different circuits, each case of the circuit is tested and verified for functionality and ESD protection levels.

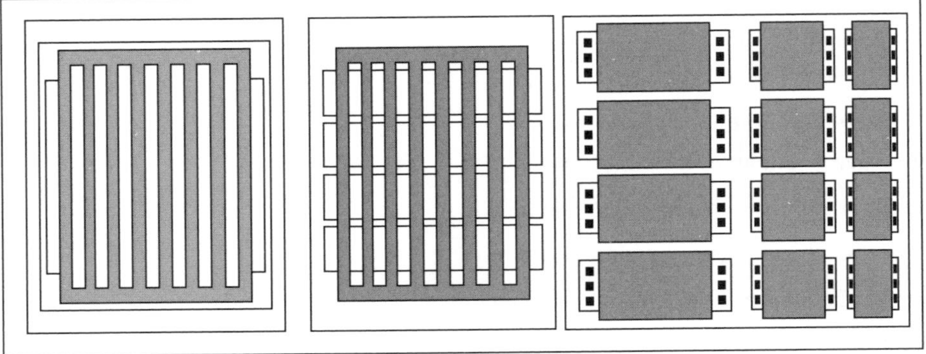

Figure 7.12 ASIC chip design with resistor "bank" for different I/O OCD strength

7.6.3 ASIC Design System with Multiple Power Rails

In an ASIC system, multiple power supplies are needed due to mixed-voltage environments. ESD protection must exist in ASIC standard cells, and as part of the ASIC architecture. Sequence independence between power rails is key to avoid limiting customers to sequencing requirements.

A novel ESD protection circuit for multiple power supplies, having both inventive inter-rail ESD circuitry and inventive single-rail ESD circuitry, was implemented by D.W. Stout, S. Voldman, J. Sloan, J. Pequignot, and T. Rahman [45]. The inter-rail ESD circuitry is scalable and comprises one or more diode strings for interconnecting a pair of power rails. The ESD trigger voltage for a diode string is set by the number of diodes within the diode string and preferably a sufficient number of diodes are provided within each diode string for power-up and power-down sequence independence. The single-rail ESD circuitry is connected to a level-shifter and may comprise an RC discriminator comprising two NFET transistors connected in series. The RC discriminator may be connected to a clamping transistor via a buffering circuit, such as an inverter stage, that isolates the gate capacitance of the clamping transistor from the RC discriminator so that the RC characteristics of the RC discriminator are unaffected by the choice of the clamping transistor. The ESD protection circuit may be constructed from a selection of user-selectable discrete circuit elements formed on the chip [45–47].

7.6.4 ASIC Design System with Voltage Islands

In advanced technologies, power consumption is a large concern in big semiconductor chips and systems. Power management techniques were incorporated into microprocessors to reduce the power consumption.

In an ASIC environment, a concept known as "voltage islands" is used to reduce power consumption. Figure 7.13 shows an example of a voltage island. In a voltage island, an independent power domain is established internally by forming a "dummy bus" within the voltage island. The voltage island incorporates a "header," a "header control," a dummy power

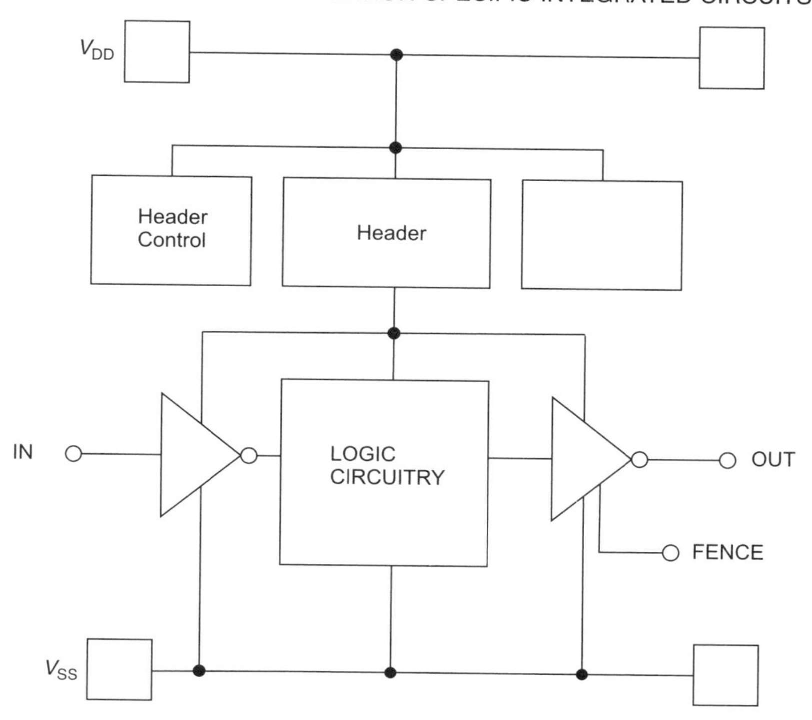

Figure 7.13 ASIC chip design with voltage islands

bus, and a "fence." In this implementation, what distinguishes this concept from microprocessors is that in this method, some signals are still "powered" whereas all other circuits within the "voltage island" are shut down. In the past, microprocessors may have powered down completed chip sectors; in this method, some signals remain powered.

The ESD and latchup issues influenced are as follows:

- Segmentation of the power grid into separated domains from the V_{DD} through header and header controls.
- Placement of active circuit-powered networks adjacent to shut-down networks.

Many customers or corporations have different preferences in the ESD power clamp design. One "gate array" ESD power clamp concept is to have a standard cell which contains a plethora of elements for the design team to construct their own RC-triggered ESD power clamp. Figure 7.14 shows an example of a "ESD gate array kit" for design teams to develop their own design [46,47]. The kit would contain a multi-finger MOSFET, inverters, resistors, and capacitor arrays. This concept was proposed by S. Voldman and J. Pequignot for ASIC development. In this fashion, the customer can provide his own design synthesis within the parameters and models of a given standard cell environment. This is a method for automatically generating a custom ESD network for an integrated circuit. When a user provides chip size and chip capacitance for the integrated circuit, components for the customized ESD network are

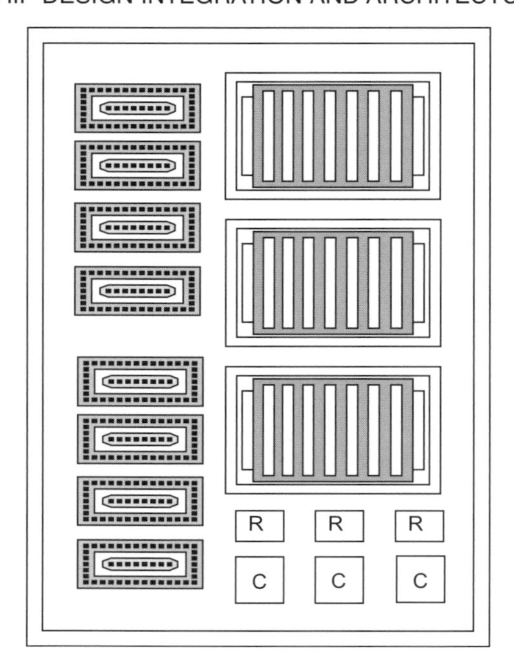

Figure 7.14 ASIC gate array ESD design

automatically selected based on the user-provided chip size and chip capacitance, and the adequacy of the ESD behavior of an ESD network employing the selected components is evaluated.

An ASIC book comprising a gate-array format of ESD components is provided. A customized, optimized, and tuned ESD network can be constructed from the ASIC book. Novel ESD circuitry having inter-rail ESD circuitry and single-rail ESD circuitry can be constructed. The inter-rail ESD circuitry is scalable and comprises one or more diode strings for interconnecting a pair of power rails. The ESD trigger voltage for a diode string is set by the number of diodes within the customized diode string, and preferably a sufficient number of diodes are provided within each diode string for power-up and power-down sequence independence. The single-rail ESD circuitry is connected to a level-shifter and may comprise an RC discriminator comprising a customizable plurality of NFET transistors connected in series. The RC discriminator may be connected to a clamping transistor via a buffering circuit, such as an inverter stage, that isolates the gate capacitance of the clamping transistor from the RC discriminator. In a second aspect of the invention, an ASIC book comprising a gate-array format of ESD components is provided. A customized, optimized, and tuned ESD network can be constructed from the ASIC book.

7.7 CMOS IMAGE PROCESSING CHIP DESIGN

Image processing chips have seen considerable growth in the last few years, with the usage of digital cameras and cell phones. Image processing chips are used in digital cameras, where both high-quality images and low power consumption are important. Noise and power are important to utilize these chips in digital cameras, where battery life limitations are a concern. CMOS imaging chips are advantageous over CCD chips due to low power consumption, making it ideal for battery power applications. Additionally, it does not introduce "blooming." CMOS image processing chips can incorporate analog-to-digital converters (A/DC) for direct digital output.

Small imaging device are also used in cellular phones. Specifically, solid state imaging devices are needed that can achieve high levels of micro-miniaturization, low cost, and high performance.

A metric for an image processing semiconductor array is the number of pixels and pixel densities. It is desirable to have the size of the imaging array as large as possible. For example a CMOS sensor may have a 1.3 Mega-pixel imaging array with 1280×1024 resolution. As a result, semiconductor imaging chips have large "core" area, and are "core-limited" design architectures. The imaging area must also be free of circuits to allow capturing of the image through the entire physical area. To achieve this, the digital logic circuits, I/O and ESD networks must exist on the periphery of the semiconductor image processing chip.

In a "core-limited" ESD design synthesis for CMOS imaging chips, the architecture contains the following:

- Peripheral power rails.
- Peripheral ground rails.
- Peripheral bond pads.
- Many power domains for isolation (e.g., 8 power domains).
- Breaker cells between the power domains.
- ESD power clamps between each power and ground domain.
- ESD V_{SS}-to-V_{SS} cells between the ground power rails.

A unique feature of the CMOS image processing chips is that the area on the periphery is large, and the pin count is low for the size of the chip. A CMOS image chip may be as large as a large ASIC design, but may have less than 50 pins (e.g., 48-pin package). In a standard ASIC cell system, the design is "pad-limited" with long/narrow standard cells on the periphery. In a CMOS image processing chip core-limited design, the spacing between signal pads is large, allowing for a large number of ESD resources. In these designs, foundry long/narrow standard cells would waste significant area. The solution for this is a foundry cell known as a short/wide standard cell, or rotation of the long/narrow standard cells perpendicular to the normal orientation.

7.7.1 CMOS Image Processing Chip Design with Long/Narrow Standard Cell

Figure 7.15 shows a CMOS image processing chip using a long/narrow I/O standard cell. The standard cell is rotated relative to the normal orientation of the design. This then requires

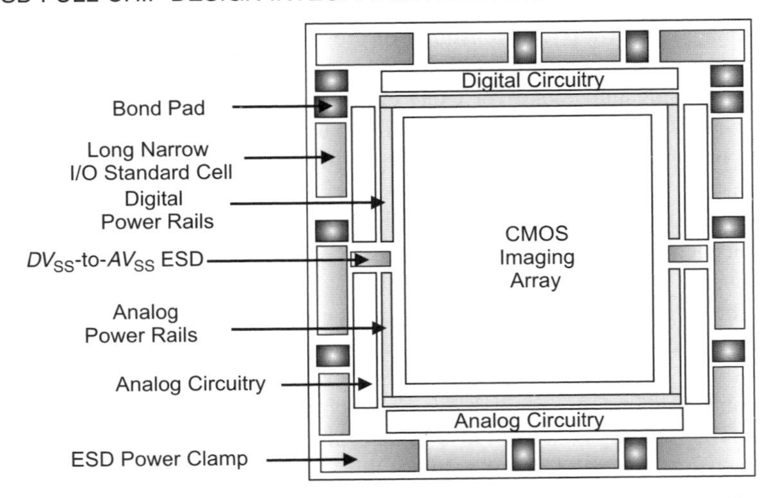

Figure 7.15 CMOS image processing chip design with long/narrow I/O

additional connections to the power bus, which is parallel to the long dimension of the long/narrow I/O cell. ESD power clamps must also be rotated since, in a standard cell library, these also conform to the same form factor as the standard cell itself.

7.7.2 CMOS Image Processing Chip Design with Short/Wide Standard Cell

Figure 7.16 shows a CMOS image processing chip with the short/wide I/O standard cell, and a short/wide ESD power clamp. In this case, the peripheral area is better utilized. In these image

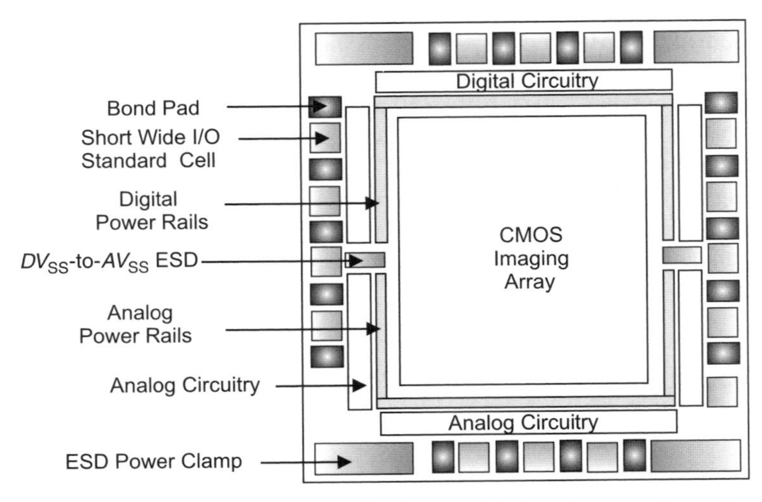

Figure 7.16 CMOS image processing chip design with short/narrow I/O standard cells

processing chips, significant area is available to add ESD power clamps and breaker cells between the power domains. In a CMOS image processing chip, there are a significant number of separated power domains due to the noise considerations. A second key consideration for the number of ESD power clamps is the total chip I_{DD} leakage current. In the chip design synthesis, the total chip I_{DD} leakage specification is significantly lower than an ASIC-type application. As a result, consideration of the leakage on the V_{DD} is a major ESD design limitation for the number of ESD power clamps utilized in the design (not the silicon area).

7.8 MIXED-SIGNAL ARCHITECTURE

In an MS architecture, digital, analog, and RF circuitry can be contained on the same semiconductor chip. In the following sections, chip design synthesis and ESD issues will be discussed.

7.8.1 Mixed-Signal Architecture – Digital and Analog

In an MS architecture, the digital and analog circuitry is separated into different power domains. Analog design requires unique layout and design characteristics [50–53]. Power electronics also require unique layout, design, and guard ring requirements [54,55]. Figure 7.17 shows an example of a semiconductor chip with digital and analog domains. To avoid ESD failures in an MS semiconductor chip, ESD protection networks are placed between the analog ground (AV_{SS}) and the digital ground (V_{SS}). Typical architectures contain a separate ESD power clamp in each domain. An ESD power clamp exists in the digital domain, between V_{DD} and V_{SS}, and a second ESD power clamp exists in the analog domain, between analog V_{DD} (AV_{DD}) and analog ground (AV_{SS}).

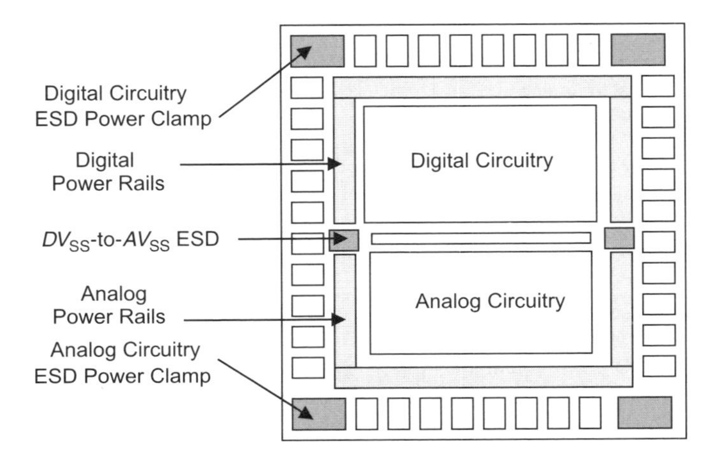

Figure 7.17 Mixed-signal architecture – digital and analog architecture

7.8.2 Mixed-Signal Architecture – Digital, Analog, and RF

In an MS architecture, the digital, analog, and RF circuitry is separated into different power domains. Radio frequency ESD design synthesis differs significantly from both analog and digital design [56–59]. To avoid ESD failures in an MS semiconductor chip, ESD protection networks are placed between the analog ground (AV_{SS}), the digital ground (V_{SS}), and the RF ground. Typical architectures contain a separate ESD power clamp in each domain. An ESD power clamp exists in the digital domain, between V_{DD} and V_{SS}; a second ESD power clamp exists in the analog domain, between analog V_{DD} (AV_{DD}) and analog ground (AV_{SS}); and a third ESD power clamp exists between RF V_{DD} (RFV_{DD}, or V_{CC}) and RF ground (RFV_{SS}, or V_{EE}). In these mixed-signal chips, the RF application voltage is typically higher than the analog and digital application voltage.

Figure 7.18 shows an example of a mixed-signal chip with RF, analog, and digital circuitry. To separate the analog circuitry from the digital noise, separate power rail domains exist. Additionally, a guard ring "moat" separates the two domains to produce a larger distance through the substrate region. The RF sector is separated on the lower sector of the chip floorplan. ESD network power clamps are placed in the digital, analog, and RF domains between their power and ground rails. In addition, V_{SS}-to-V_{SS} ESD networks are placed to interconnect the ground rails. The V_{SS}-to-V_{SS} networks use series diode ESD elements, where the number of elements in series is a function of the allowed capacitive coupling between the digital, analog, and RF sectors.

The RF circuitry is surrounded by layers of metal forming a "faraday cage" to isolate the RF signals. The faraday cage is formed by stacking the metal layers, and passing the signals through the breaks in the faraday cage. An example of the RF faraday cage is shown in Figure 7.19. Interconnect layers and via groups are stacked from the silicon substrate to the upper-level metal. In the substrate, the faraday cage is electrically connected to the substrate

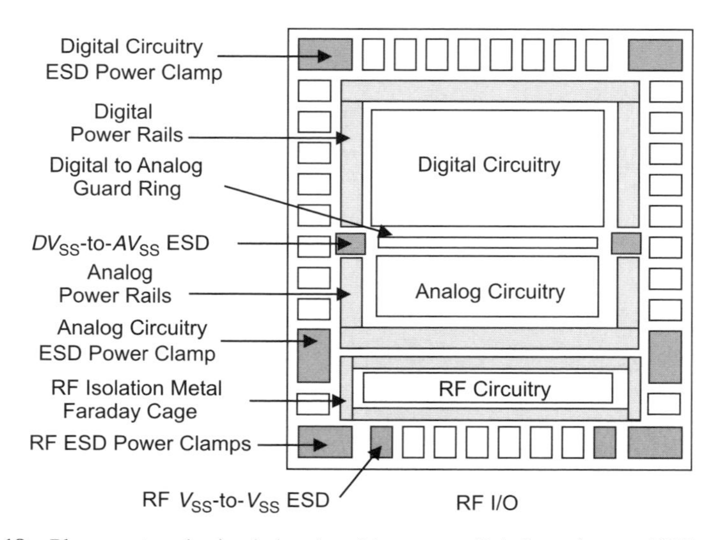

Figure 7.18 Placement and mixed-signal architecture – digital, analog, and RF architecture

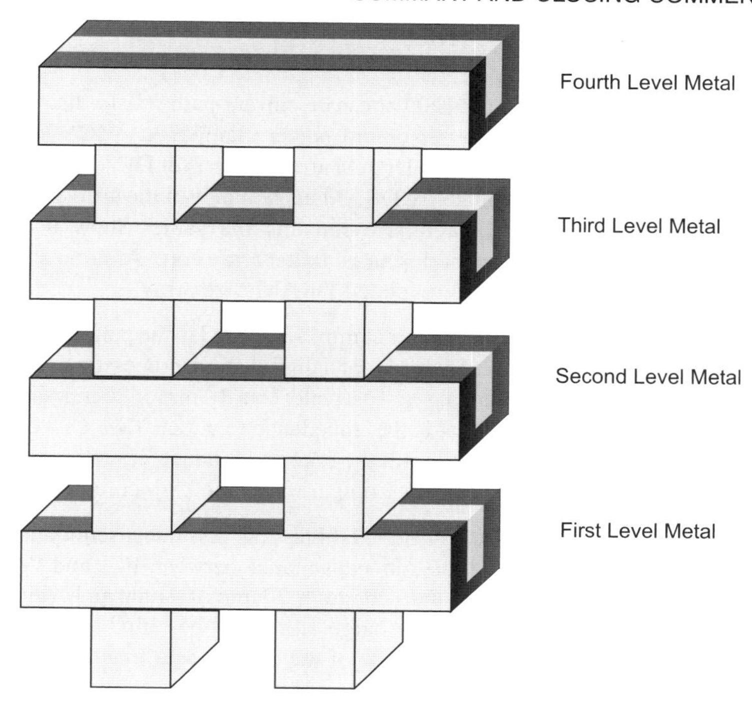

Figure 7.19 RF faraday cage in a mixed-signal architecture – digital, analog, and RF design

region with heavily doped layers. Signal lines are passed from the analog and digital domains through the faraday cage to reach the RF circuitry. This prevents electromagnetic interference (EMI) from the digital and analog domains affecting the RF circuitry.

The RF sector has a V_{CC} and V_{SS} power rail dedicated to RF circuitry. ESD power clamps are placed between the V_{CC} and V_{SS} of the RF domain. ESD power clamps consist of RF components to allow for RF modeling and tracking. ESD networks can be constructed from the RF circuit elements. The circuits are dependent if the technology is RF CMOS, or RF BiCMOS. In an RF CMOS application, the ESD power clamps can be RC-triggered MOSFET ESD power clamps [56–58]. In a BiCMOS technology, the ESD power clamps can be constructed of bipolar transistors [56–58].

7.9 SUMMARY AND CLOSING COMMENTS

In this chapter, the focus has been on examples of design synthesis in full-chip implementations. Examples of DRAM, SRAM, microprocessors, mixed-voltage, mixed-signal, and RF applications were shown. As part of the ESD design synthesis, the layout is key to a successful design implementation for both ESD and CMOS latchup. These examples will provide some understanding of the challenges in the ESD full-chip integration issues. By combining the knowledge of Chapters 1–6 with this chapter, the whole chip design strategy should be better understood for any semiconductor chip architecture.

PROBLEMS

7.1. In early DRAM applications, the ESD alternate current path was to discharge the ESD current into the array, with no exterior rail power clamps. A CMOS DRAM chip is formed with trench DRAM cells. The DRAM array consists of DRAM capacitors of 100 fF per trench DRAM cell. Estimate the DRAM capacitance of the array for 1 Mb to 4 Gb DRAM, assuming the DRAM cell size remains the same. Show a DRAM ESD architecture that allows for current discharge to the array core. Assume an independent peripheral power and ground and regulated DRAM core array.

7.2. In early microprocessors, no ESD power clamps were used in the peripheral area or in the core of the CPU. Estimate the capacitance of a digital microprocessor, assuming 50% of the area was an n-well region on a p-/p++ wafer in a 10 mm × 20 mm chip. Given the frequency of the HBM pulse, what is the impedance (e.g., $Z = |1/\omega C|$)? Given an ESD power clamp was added in parallel with the core capacitance, how low must the series resistance be in order to be effective? Compare with the chip core capacitance.

7.3. A semiconductor chip is charged during CDM testing. Assume a semiconductor chip is 100% digital logic circuits. Plot the chip capacitance between V_{DD} and V_{SS} as the chip size is increased from 1 mm × 1 mm to 20 mm × 20 mm. Assume an n-well to substrate capacitance value. What is the total charge stored in the chip? Assume the chip is discharged in 1 ns, what is the magnitude of the current as a function of chip size?

7.4. In semiconductor chip design, empty areas are "filled" with "fill shapes" for photo, etch, and chemical mechanical polishing (CMP) uniformity. What is the effect of fill shapes in the metal levels when an ESD current pulse is sent through a signal line interconnect surrounded by fill shapes? What is the electrical and thermal influence? If the fill shapes are placed in an array, how does this affect the electrical and thermal profile in the wire interconnect?

7.5. In semiconductor chip design, empty areas are "filled" with "fill shapes" for photo, etch, and CMP uniformity. In some applications, instead of fill shapes, de-coupling capacitors are added near the peripheral I/O areas. The decoupling capacitors can consist of an n-well lower plate that is grounded, and a MOSFET gate structure on the n-well region. Decoupling capacitors are formed by placement of an n-channel MOSFET in an n-well. What is the impact of the de-coupling capacitors when placed between I/O networks? What is the potential concern of placement near p-channel MOSFET driver circuits? What is the impact of the de-coupling capacitors if placed near ESD input circuits? What are the advantages and disadvantages?

7.6. Show the ESD architecture of a semiconductor chip with a digital and analog core. Place back-to-back diodes between the two core sectors on the grounds. Assume a pin-to-pin ESD test is performed between an analog and digital pin. Show the bi-directional current path.

7.7. A mixed-signal chip has a digital core and an analog core. A signal line exists between the two core regions, where a digital driver sends the signal to a receiver in the analog sector. Given an aluminum wire width of 1 μm, and wire film thickness of 1 μm, calculate the resistance from the digital core driver to the receiver assuming the signal is sent from across half of the semiconductor chip size, where the chip size increases from 2 mm × 2 mm to 20 mm × 20 mm. Assume a current of 1 mA in the interconnect, what is the voltage drop across the wire?

7.8. Show the ESD architecture of a semiconductor chip with a digital and analog core. A signal line exists between the two core regions, where a digital driver sends the signal to a receiver in the analog sector. Place back-to-back diodes between the two core sectors on the grounds. Assume an ESD current flows through the signal line. Show the possible current path. Show the maximum number of diodes and voltage across the signal line.

7.9. Show the ESD architecture of a semiconductor chip with a digital and analog core. A signal line exists between the two core regions, where a digital driver sends the signal to a receiver in the analog sector. Place cross-domain ESD networks from the digital V_{DD} to the analog V_{SS}. Also show back-to-back diodes between the two core sectors on the grounds. Assume an ESD current flows through the signal line. Show the possible current paths.

7.10. In an image processing chip, the total I_{DD} current specification is low to avoid impact on the power application. Assume an I_{DD} specification of 100 μA. In the chip, the leakage sources on the V_{DD} power rail are the array, the logic, and the ESD power clamps. What ESD networks provide low leakage? How many are needed to protect an image processing chip? What type of ESD networks are not suitable?

REFERENCES

1. A. Amerasekera and C. Duvvury. *ESD in Silicon Integrated Circuits*. Chichester, UK: John Wiley and Sons, Ltd, 1995.
2. S. Voldman. *ESD: Physics and Devices*. Chichester, UK: John Wiley and Sons, Ltd, 2004.
3. S. Voldman. *ESD: Circuits and Devices*. Chichester, UK: John Wiley and Sons, Ltd, 2005.
4. S. Dabral and T.J. Maloney. *Basic I/O and ESD*. New York: John Wiley and Sons, Inc., 1998.
5. S. Voldman. *ESD: Failure Mechanisms and Models*. Chichester, UK: John Wiley and Sons, Ltd., 2009.
6. E. Adler, J.K DeBrosse, S. F. Geissler, S. J. Holmes, M.D. Jaffe, J.B. Johnson, C.W. Koburger, J.B. Lasky, B. Lloyd, G. L. Miles, J.S. Nakos, W. P. Noble Jr., S. H. Voldman, M. Armacost, and R. Ferguson. The evolution of IBM CMOS DRAM technology. *IBM Journal of Research and Development*, **39** (1–2), Jan/March 1995; 167–188.
7. S. Voldman, V. Gross, M. Hargrove, J. Never, J. Slinkman, M. O'Boyle, T. Scott, and J. Delecki. Shallow trench isolation (STI) double-diode electrostatic discharge (ESD) circuit and interaction with DRAM circuitry. *Proceedings of the Electrical Overstress/Electrostatic Discharge (EOS/ESD) Symposium*, 1992; 277–288.
8. S. Voldman, V. Gross, M. Hargrove, J. Never, J. Slinkman, M. O'Boyle, T. Scott, and J. Delecki. Shallow trench isolation (STI) double-diode electrostatic discharge (ESD) circuit and interaction with DRAM circuitry. *Journal of Electrostatics*, **31** (2–3), 1993; 237–265.
9. S. Voldman and V. Gross. Scaling, optimization, and design considerations of electrostatic discharge protection circuits in CMOS technology. *Proceedings of the Electrical Overstress/Electrostatic Discharge (EOS/ESD) Symposium*, 1993; 251–260.
10. S. Voldman and V. Gross. Scaling, optimization, and design considerations of electrostatic discharge protection circuits in CMOS technology. *Journal of Electrostatics*, **33** (3), October 1994; 327–357.

11. S. Voldman. ESD protection in a mixed voltage interface and multi-rail disconnected power grid environment in 0.5- and 0.25-μm channel length CMOS technologies, *Proceedings of the Electrical Overstress/Electrostatic Discharge (EOS/ESD) Symposium*, 1994; 125–134.
12. S. Voldman. Electrostatic discharge protection circuits for mixed voltage interface and multi-rail disconnected power grid applications. U.S. Patent No. 5,945,713, August 13, 1999.
13. S. Voldman. Voltage regulator bypass circuit. U.S. Patent No. 5,625,280. April 29, 1997.
14. S. L. Jang, M.S. Gau, and C.K. Lin. Novel diode-chain triggering SCR circuits for ESD protection. *Solid State Electronics*, **44**, 2000; 1297–1303.
15. M. Mergens, C. Russ, K. Verhaege, J. Armer, P. Jozwiak, R. Mohn, B. Keppens, and S. Trinh. Diode triggered SCR (DTSCR) for RF ESD protection of BiCMOS SiGe HBT's and CMOS ultra-thin gates oxides. *International Electron Device Meeting (IEDM) Technical Digest*, 2003; 515–518.
16. S. Voldman. The impact of MOSFET technology evolution and scaling on electrostatic discharge protection. Review Paper, *Microelectronics Reliability*, **38**, 1998; 1649–1668.
17. R. D. Adams, R. C. Flaker, K. S. Gray, H.L. Kalter. An 11ns 8K x 18 CMOS static RAM. *Proceedings of the International Solid State Circuits Conference (ISSCC)*, 1988; 242–243.
18. H. Pilo and S. Lamphier. A 300 MHz, 3.3 V 1 Mb SRAM fabricated in a 0.5 μm. CMOS process, *Proceedings of the International Solid State Circuits Conference (ISSCC)*, February 1996; 148–149.
19. H.S. Lee, B. El Kareh, R.C. Flaker, G.G. Gravenities, R.A. Lipa, J.P. Maslack, J.R. Pessetto, W.F. Pokorny, M.A. Roberge, T. Williams, H.A. Zeller, and K.E. Beilstein. An experimental 1Mb CMOS SRAM with configurable organization and operation. *Proceedings of the International Solid State Circuits Conference (ISSCC)*, 1988; 180–181.
20. H. Pilo, A. Allen, J. Covino, P.R. Hansen, S. Lamphier, C. Murphy, T. Traver, P. Yee. An 833-MHz 1.5 W 18-Mb CMOS SRAM with 1.67 Gb/s/pin. *IEEE Journal of Solid State Circuits*, Vol. 35, Issue 11, November 2000; 1641–1647.
21. R.E. Rose, R.C. Szafranski, and S. Voldman. Dual thin oxide ESD network for nonvolatile memory applications. U.S. Patent No. 5,872378, February 16, 1999.
22. C.W. Koburger, W.F. Clark, J.W. Adkisson, E. Adler, P.E. Bakeman, A.S. Bergendahl, A. B. Botula, W. Chang, B. Davari, J.H. Givens, H.H. Hansen, S.J. Holmes, D. V. Horak, C. H. Lam, J.B. Lasky, S.E. Luce, R. W. Mann, G. L. Miles, J. S. Nakos, E.J. Nowak, G. Shahidi, Y. Taur, S. H. Voldman, F.R. White, and M.R. Wordeman. A half-micron CMOS logic generation. *IBM Journal of Research and Development*, **39** (1/2), Jan/March 1995; 215–228.
23. G.G. Shahidi, J.D. Warnock, J. Comfort, S. Fischer, P.A. McFarland, A. Acovic, T.I. Chappell, B. A. Chappell, T.H. Ning, C.J. Anderson, R.H. Dennard, J.Y.-C. Sun, M. R. Polcari, and B. Davari. CMOS scaling in the 0.1 1.X-volt regime for high performance applications. *IBM Journal of Research and Development*, **39** (1/2), Jan/March 1995; 229–244.
24. Y. Taur, Y.J. Mii, D. J. Frank, H.S. Wong, D.A. Buchanan, S. J. Wind, S. A. Rishton, G. A. Sai-Halasz, and E.J. Nowak. CMOS scaling into the 21st century: 0.1 μm and beyond. *IBM Journal of Research and Development*, **39** (1/2), Jan/March 1995; 245–260.
25. S. Voldman. The impact of technology evolution and scaling on electrostatic discharge (ESD) protection in high pin count high performance microprocessors. Invited Talk, *Proceedings of the International Solid State Circuits Conference (ISSCC)*, Session 21, WA 21.4, 1999; 366–367.
26. S. Voldman, W. Anderson, R. Ashton, M. Chaine, C. Duvvury, T. Maloney, and E. Worley. Test structures for benchmarking the electrostatic discharge (ESD) robustness of CMOS technologies. *SEMATECH Technology Transfer Document, SEMATECH TT 98013452A-TR*, May 1998.
27. S. Voldman, W. Anderson, R. Ashton, M. Chaine, C. Duvvury, T. Maloney, and E. Worley. ESD technology benchmarking strategy for evaluation of the ESD robustness of CMOS semiconductor technologies. *Proceedings of the International Reliability Workshop (IRW)*, 1998; October 12–16.
28. S. Voldman, W. Anderson, R. Ashton, M. Chaine, C. Duvvury, T. Maloney, and E. Worley. A strategy for characterization and evaluation of the ESD robustness of CMOS semiconductor technologies.

Proceedings of the Electrical Overstress/Electrostatic Discharge (EOS/ESD) Symposium, 1999; 212–224.

29. S. Voldman. Electrostatic discharge protection, scaling, and ion implantation in advanced semiconductor technologies. Invited Talk, *Proceedings of the Ion Implantation Conference (I^2CON)*, Napa, California, 1999.
30. S. Voldman and G. Gerosa. Mixed voltage interface ESD protection circuits for advanced microprocessors in shallow trench and LOCOS isolation CMOS technology, *International Electron Device Meeting (IEDM) Technical Digest*, December 1994, 811–815.
31. S. Voldman, G. Gerosa, V. Gross, N. Dickson, S. Furkay, and J. Slinkman. Analysis of snubber-clamped diode string mixed voltage interface ESD protection networks for advanced microprocessors, *Proceedings of the Electrical Overstress/Electrostatic Discharge (EOS/ESD) Symposium*, 1995; 43–61.
32. S. Voldman, G. Gerosa, V. Gross, N. Dickson, S. Furkay, and J. Slinkman. Analysis of snubber-clamped diode string mixed voltage interface ESD protection networks for advanced microprocessors, *Journal of Electrostatics*, **38** (1–2), October 1996; 3–32.
33. R. Countryman, G. Gerosa, and H. Mendez. Electrostatic discharge protection device. U.S. Patent No. 5,514,892, May 7, 1996.
34. S. Voldman. Optimization of MeV retrograde wells for advanced logic and microprocessor/PowerPC and electrostatic discharge (ESD). Invited Talk, *Smart and Economic Device and Process Designs for ULSI Using MeV Implant Technology Seminar: SEMICON West, SEMICON West GENUS Seminar*, San Francisco, 1994.
35. G. Gerosa, M. Alexander, J. Alvarez, C. Croxton, M. D'Addeo, A.R. Kennedy, C. Nicoletta, J.P. Nissen, R. Philip, P. Reed, H. Sanchez, S.A. Taylor, and B. Burgess. A 250-MHz 5-W PowerPC microprocessor with on-chip L2 cache controller. *IEEE Journal of Solid-State Circuits*, **32** (11), Nov. 1997; 1635–1649.
36. S. Dabral, R. Aslett, and T. Maloney. Designing on-chip power supply coupling diodes for ESD protection and noise immunity. *Proceedings of the Electrical Overstress/Electrostatic Discharge (EOS/ESD) Symposium*, 1993; 239–249.
37. S. Dabral, R. Aslett, and T. Maloney. Core clamps for low voltage technologies. *Proceedings of the Electrical Overstress/Electrostatic Discharge (EOS/ESD) Symposium*, 1994; 141–149.
38. S. Voldman, R. Schulz, J. Howard, V. Gross, S. Wu, A. Yapsir, D. Sadana, H. Hovel, J. Walker, F. Assaderaghi, B. Chen, J.Y.C. Sun, and G. Shahidi. CMOS-on-SOI ESD protection networks. *Proceedings of the Electrical Overstress/Electrostatic Discharge (EOS/ESD) Symposium*, 1996; 291–302.
39. S. Voldman, F. Assaderaghi, J. Mandelman, L Hsu, and G. Shahidi. Dynamic threshold body- and gate-coupled SOI ESD protection networks. *Proceedings of the Electrical Overstress/Electrostatic Discharge (EOS/ESD) Symposium*, 1997; 210–220.
40. S. Voldman, D. Hui, L. Warriner, D. Young, R. Williams, J. Howard, V. Gross, W. Rausch, E. Leobangdung, M. Sherony, N. Rohrer, C. Akrout, F. Assaderaghi, and G. Shahidi. Electrostatic discharge protection in silicon on insulator (SOI) technology. Invited Talk, *Proceedings of the IEEE International SOI Conference Symposium*, Session 5.1, 1999; 68–72.
41. S. Voldman, D. Hui, D. Young, R. Williams, D. Dreps, M. Sherony, F. Assaderaghi, and G. Shahidi. Electrostatic discharge (ESD) protection in silicon-on-insulator (SOI) CMOS technology with aluminum and copper interconnects in advanced microprocessor semiconductor chips. *Proceedings of the Electrical Overstress/Electrostatic Discharge (EOS/ESD) Symposium*, 1999; 105–115.
42. P. A. Juliano and W.R. Anderson. ESD protection design challenges for a high pin-count Alpha microprocessor in a 0.13 μm CMOS SOI technology *Proceedings of the Electrical Overstress/Electrostatic Discharge (EOS/ESD) Symposium*, 2003; 59–69.
43. S. Voldman. Power sequence independent electrostatic discharge protection circuits U.S. Patent No. 5,610,791, March 11, 1997.

44. J. H. Panner, T.R. Bednar, P.H. Buffet, D.W. Kemerer, D.W. Stout, P.S. Zuchowski. The first copper ASICs: A 12M-gate technology. *Proceedings of the IEEE Custom Integrated Circuits Conference (CICC)*, 1999; 347–350.
45. J. Pequignot, T. Rahman, J.H. Sloan, D.W. Stout, and S. Voldman. Method and apparatus for providing ESD protection. U.S. Patent No. 6,157,530, December 5, 2000.
46. J. Pequignot, T. Rahman, J.H. Sloan, D.W. Stout, and S. Voldman. Method for providing ESD protection for an integrated circuit. U.S. Patent No. 6,262,873, July 17, 2001.
47. J. Pequignot, T. Rahman, J.H. Sloan, D.W. Stout, and S. Voldman. ASIC book to provide ESD protection on an integrated circuit. U.S. Patent No. 6,292,343, September 18, 2001.
48. C. J. Brennan, J. Sloan, and D. Picozzi. CDM failure modes in 130 nm ASIC technology. *Proceedings of the Electrical Overstress/Electrostatic Discharge (EOS/ESD) Symposium*, 2004; 182–186.
49. J. Pequignot, J. Sloan, D. Stout, and S. Voldman. Electrostatic discharge protection networks for triple well semiconductor devices. U.S. Patent Application 7,348,657, July 15, 2004.
50. P. Gray, Hurst, Lewis, and Meyer. *Analysis and Design of Analog Integrated Circuits. 5th Edition.* New York: John Wiley and Sons, Inc., 2009.
51. W.M.C. Sansen. *Analog Design Essentials.* Springer, Netherlands, 2006.
52. A. Hastings. *The Art of Analog Layout.* 2nd Edition. New Jersey: Pearson Prentice Hall, 2006.
53. V. A. Vashchenko and A. Shibkov. *ESD Design in Analog Design*, New York: Springer, 2010.
54. B. J. Baliga. *High Voltage Integrated Circuits.* IEEE Press, New York, 1988.
55. P. Antognetti. *Power Integrated Circuits: Physics, Design, and Applications.* McGraw-Hill, New York, McGraw-Hill, 1986.
56. S. Voldman. The state of the art of electrostatic discharge protection: physics, technology, circuits, designs, simulation and scaling. *Invited Talk. Bipolar/BiCMOS Circuits and Technology Meeting (BCTM) Symposium*, 1998; 19–31.
57. S. Voldman. A review of latchup and electrostatic discharge (ESD) protection in BiCMOS RF silicon germanium technologies: Part I – ESD. *Microelectronics and Reliability* 2005, **45**; 323–343.
58. S. Voldman. *ESD: RF Technology and Circuits.* Chichester, UK: John Wiley and Sons, Ltd, 2006.
59. P. Leroux and M. Steyaert. *LNA-ESD Co-design for Fully Integrated CMOS Wireless Receivers.* New York: Springer: 2005.

Index

ESD: Design and Synthesis, First Edition. Steven H. Voldman.